AF843447

Applications of Nanocomposite Materials in Orthopedics

Related title

Gao, *Advances in Polymer Nanocomposites Types and Applications*
(ISBN: 9780081016305)

Biomaterials and Regenerative Medicine in Ophthalmology
(ISBN: 9780081001479)

Multifunctional Polymeric Nanocomposites Based on Cellulosic Reinforcements
(ISBN: 9780323442480)

Woodhead Publishing Series in Biomaterials

Applications of Nanocomposite Materials in Orthopedics

Edited by

Inamuddin, Abdullah M. Asiri and Ali Mohammad

ELSEVIER

WP
WOODHEAD PUBLISHING
An imprint of Elsevier

Woodhead Publishing is an imprint of Elsevier
The Officers' Mess Business Centre, Royston Road, Duxford, CB22 4QH, United Kingdom
50 Hampshire Street, 5th Floor, Cambridge, MA 02139, United States
The Boulevard, Langford Lane, Kidlington, OX5 1GB, United Kingdom

Library of Congress Cataloging-in-Publication Data
A catalog record for this book is available from the Library of Congress

British Library Cataloguing-in-Publication Data
A catalogue record for this book is available from the British Library

ISBN: 978-0-12-813740-6 (print)
ISBN: 978-0-12-813757-4 (online)

For information on all Woodhead publications
visit our website at https://www.elsevier.com/books-and-journals

Publisher: Matthew Deans
Acquisition Editor: Laura Overend
Editorial Project Manager: Michelle Kubilis
Production Project Manager: Swapna Srinivasan
Cover Designer: Matthew Limbert

Typeset by SPi Global, India

Contents

List of contributors

Syed Anees Ahmad Department of Pathology, King George's Medical University, Lucknow, India

Mudasir Ahmad Department of Chemistry, Faculty of Natural Science, Jamia Millia Islamia, New Delhi, India

Suhail Ahmad Department of Chemistry, Faculty of Natural Science, Jamia Millia Islamia, New Delhi, India

Nadia Akram Department of Chemistry, Government College University, Faisalabad, Pakistan

Tahseen Jahan Ara Department of Chemistry, L.N.M University, Dharbhanga, India

Nahid Chaudhary Guru Gobind Singh Indraprastha University, New Delhi, India

Bor Shin Chee Materials Research Institute, Athlone Institute of Technology (AIT), Athlone, Ireland

Dharmesh R. Chejara Uka Tarsadia University, Bardoli, India

Aline Rossetto da Luz Graduate Program in Engineering and Science of Materials—PIPE, Federal University of Paraná (UFPR), Curitiba, Brazil

Gabriel Goetten de Lima Materials Research Institute, Athlone Institute of Technology (AIT), Athlone, Ireland

Gelson Biscaia de Souza Departament of Physics, State University of Ponta Grossa (UEPG), Ponta Grossa, Brazil

Declan Devine Materials Research Institute, Athlone Institute of Technology (AIT), Athlone, Ireland

Imad A. Disher University of Babylon, Al-Hilla, Iraq

Mehdi Ebrahimi Oral Rehabilitation, Prince Philip Dental Hospital, The University of Hong Kong,Sai Ying Pun, Hong Kong

Vasanth Gopal Department of Physics, School of Advanced Sciences; Centre for Biomaterials, Cellular and Molecular Theranostics (CBCMT), Vellore Institute of Technology, Vellore, India

Manoj Gupta Department of Mechanical Engineering, National University of Singapore, Singapore, Singapore

Mohammad S. Hasnain Shri Venkateshwara University, Amroha, India

Mohammad N. Hoda Hamdard University, New Delhi, India

Saiqa Ikram Department of Chemistry, Faculty of Natural Science, Jamia Millia Islamia, New Delhi, India

Inamuddin Advanced Functional Materials Laboratory, Department of Applied Chemistry, Faculty of Engineering and Technology, Aligarh Muslim University, Aligarh, India

Parth Joshi Uka Tarsadia University, Bardoli, India

Neide Kazue Kuromoto Department of Physics, Federal University of Paraná (UFPR), Curitiba, Brazil

Carlos Maurício Lepienski Graduate Program in Engineering and Science of Materials—PIPE, Federal University of Paraná (UFPR), Curitiba, Brazil

Geetha Manivasagam School of Mechanical Engineering; Centre for Biomaterials Cellular and Molecular Theranostics, Vellore Institute of Technology, Vellore, India

Kaiser Manzoor Department of Chemistry, Faculty of Natural Science, Jamia Millia Islamia, New Delhi, India

Mohammad Akram Minhaj Department of Pharmacology, Maulana Azad Medical College and Hospital, New Delhi, India

Mohsin T. Mohammed Nanotechnology and Advanced Materials Research Unit, Faculty of Engineering, University of Kufa, Najaf, Iraq

Abu Nasar Advanced Functional Materials Laboratory, Department of Applied Chemistry, Faculty of Engineering and Technology, Aligarh Muslim University, Aligarh, India

Amit Kumar Nayak Department of Pharmaceutics, Seemanta Institute of Pharmaceutical Sciences, Mayurbhanj, India

Michael J.D. Nugent Materials Research Institute, Athlone Institute of Technology (AIT), Athlone, Ireland

Bruno Leandro Pereira Graduate Program in Engineering and Science of Materials—PIPE, Federal University of Paraná (UFPR), Curitiba, Brazil

Ruma Perveen Advanced Functional Materials Laboratory, Department of Applied Chemistry, Faculty of Engineering and Technology, Aligarh Muslim University, Aligarh, India

Jignesh P. Raval Uka Tarsadia University, Bardoli, India

Magesh Sankar School of Mechanical Engineering, Vellore Institute of Technology, Vellore, India

M. Saquib Hasnain Department of Pharmacy, Shri Venkateshwara University, Gajraula, India

Eduardo Mioduski Szesz Graduate Program in Engineering and Science of Materials—PIPE, Federal University of Paraná (UFPR), Curitiba, Brazil

Jithin Vishnu Centre for Biomaterials Cellular and Molecular Theranostics, Vellore Institute of Technology, Vellore, India

Preface

These days, there is an increasing requirement for orthopedic implants and bone tissue regeneration worldwide because of the huge number of patients experiencing bone tumor and traffic accidents and other bone fractures and imperfections. The design of new materials that impersonate the structure and properties of the human bone is a key challenge for material scientists. The field of orthopedic tissue engineering is rapidly growing leading to the design of novel materials and methodologies which are intended for quick bone recovery, regeneration, and implants. Nanocomposite materials are recognized to play a significant part as orthopedic replacement and implant materials since the bone composed of collagen matrix and hydroxyapatite nanocrystals itself is a representative example of a nanocomposite. An assortment of nanocomposites with improved properties has been designed to enhance the usefulness and unwavering quality of therapeutic implants. The technological and clinical requirement for orthopedic materials has prompted critical advances in the field of nanomedicine, which grasps the extent of nanotechnology from pharmacology to toxicology of bone tissue regeneration and bone disease treatment. Fundamental science and translational research have uncovered the critical potential applications of nanotechnology in orthopedic surgery, especially with respect to enhancing the interaction between the implant and the host bone. Nanocomposite materials more nearly coordinate to the design of the trabecular bone, in this way enormously enhancing the osseointegration of orthopedic implants. However, in spite of the tremendous advantages of nanocomposites that have been developed, it is unimaginable for them to supplant the naturally developed tissues and organs without any loss of biological function. On the other hand, nanocomposites play a promising role in curing some defects, injuries, and diseases by tissue regeneration and implants. **Applications of Nanocomposite Materials in Orthopedics** provide a solid understanding of the recent developments in the field of nanocomposites used in orthopedics. Related topics on joint replacement, load-bearing capability of fractured bones, bone soft tissue regeneration and hard tissue replacement, artificial bone grafting, bone repair, and bone tissue transplantations are covered to resolve the problems associated with bone fracture and orthopedic surgery in an easy and convenient way. A variety of nanocomposite materials are discussed and their properties and preparation methods are given.

Biodegradable polymer matrix nanocomposites for bone tissue engineering

1

Mohammad S. Hasnain, Syed Anees Ahmad†, Nahid Chaudhary‡, Mohammad N. Hoda§, Amit Kumar Nayak¶*
*Shri Venkateshwara University, Amroha, India, †Department of Pathology, King George's Medical University, Lucknow, India, ‡Guru Gobind Singh Indraprastha University, New Delhi, India, §Hamdard University, New Delhi, India, ¶Department of Pharmaceutics, Seemanta Institute of Pharmaceutical Sciences, Mayurbhanj, India

1.1 Introduction

Nanoparticles are unique and have novel properties compared to large particles having size of 100nm. The word "Nano" is Greek derived means "dwarf" [1,2]. It is a spatial unit of measurement (1.0×10^{-9} m). On reducing the size of the particle to micrometer or nanometer scale, all the physical and chemical properties are changed to those of same large-sized particle [3]. Currently, various kinds of nanoparticular materials (i.e., nanomaterials) are being extensively researched and developed in almost all technological disciplines [4–9]. The recent advancements in the nanomaterial research and development include nanoceramics, nanocomposites, nanofibers, nanofilms, nanotubes, nanorods, nanogels, nanovesicles, etc. [4,5,10,11]. Nanocomposites are the polyphasic materials in which one of the phases has one, two, or three dimensions having nanoscopic size (Fig. 1.1). These are mainly multiple nanomaterials or nanomaterials processed/incorporated within other bulk materials [12,13].

Nanocomposites have better properties that are standard on the small-scale composites and may be produced via astonishingly easy and cheap techniques [5,12]. During the past few years, different compositions of nanocomposites such as organic-organic inorganic-inorganic, and organic-inorganic, nanocomposites are being developed, characterized, and evaluated for the use in a variety of biomedical applications comprising tissue engineering, wound dressings, drug delivery, antimicrobial properties, cardiac prosthesis, stem cell therapy, cancer therapy, artificial blood vessels, biosensors, enzyme immobilization, etc. [5,14,15]. During past few decades, an extensive research progress has been involved in the development of nanocomposites for the use in tissue engineering applications. At the nanoscale, the basic functional cells subunits and tissues are well defined and therefore, the understanding of nanotechnology, nanobiology, and nanomaterials characterizes a new frontline in the tissue engineering research empowering the improvement of new frameworks that copy the perplexing, progressive structure of tissue [16]. In recent years, a variety of nanocomposites made of biodegradable polymers are being explored and exploited for the

Applications of Nanocomposite Materials in Orthopedics. https://doi.org/10.1016/B978-0-12-813740-6.00001-6

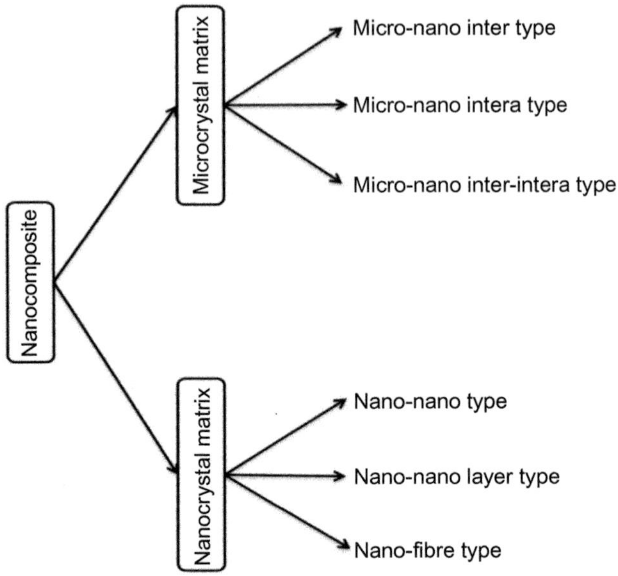

Fig. 1.1 Classification of nanocomposites.

use in tissue engineering applications [17]. Even, these biodegradable polymer matrix nanocomposites have been found effective for tissue generation in the field of bone tissue engineering [18]. This chapter presents a comprehensive review on the use of biodegradable polymer matrix nanocomposites for bone tissue engineering applications.

1.2 Tissue engineering

The term "tissue engineering" was officially defined in a National Science Foundation (NSF) workshop (United States) in 1988 as "the application of principles and methods of engineering and life sciences toward fundamental understanding of structure-function relationships in normal and mammalian tissues and the development of biological substitutes to restore, maintain or improve tissue function" [19]. Tissue engineering is a multidisciplinary field, which mainly focuses on the advancement and use of resources in physics, chemistry, life and clinical sciences, and engineering to overcome the problems of basic therapeutic issues, such as loss of tissue or organ failure [20]. Tissue engineering is one of the newly developed bioengineering area using various biomaterials (including biopolymers, bioceramics, other bioinorganics, etc.), bioactive molecules, cells individually or in combination to induce and/or stimulate the differentiation signals into different surgically transplanted configures and the proliferation enhancements toward the regeneration of tissues in the preferred site of the diseased or damaged areas and organs of the body [21]. It includes the crucial knowledge of structure-function connections in typical and pathological tissues and the advancement of organic substitutes that reestablish, sustain, or enhance the function of tissues [20–22]. For in vitro production of living tissues, cell culture are developed on bioactive degradable substrates that give physical and synthetic signs to manage their separation and get together into three-dimensional structures. One of the major issues in tissue engineering is the acknowledgment of scaffoldings with particular physical, mechanical, and natural properties [21,22]. Platforms go about as substrate for cell development, expansion, and support for new tissue arrangement. Biomaterials and creation advances assume a key part in tissue engineering [19,23]. There may be various causes of existing tissue defects in our human body and there can be various methods outlined in order to rectify the problem with some early motion therapy. The healing of tissues can be achieved, in principle, by the following five possible ways [23]: (i) self-(spontaneous) healing, (ii) autologous tissue transplantation, (iii) cell-free biomaterial implantation, (iv) cell therapy, and (v) tissue engineering approach. In order to select the right method or approach to solve the existing problem of tissue defects, there should be the proper examination of tissues involved, the site of the defect, and healing capacity of the body that varies with age. The generalized combined methods from the material science and the life sciences used to regenerate the artificially developed constructs consisting of matrix (scaffold) along with living cells is called as "tissue engineering" [24]. In order to meet the above requirements, an interdisciplinary field has emerged in the past few decades that include the methods and concepts of engineering, medicine, and biology. Tissue engineering can improve the healthcare quality and has gained a special attention in various developing as well as developed countries. Cell therapies are quite different from tissue engineering, one need to focus on tissue differentiation processes. Basically, the tissue engineering follows two approaches that actually differentiate it with the cell therapies [21]. First, in the in vitro condition, the cells start to communicate and interact with each other in order to synthesize an extracellular matrix (ECM). Second, the in vivo approach includes the seeding onto a scaffold material directly before implantation or when the defects site acts as the center at which the suspended cells are implanted directly before implantation [21,22,24]. The one of the most important factor that is always taken into consideration is biodegradability, which includes the cell restoration and physiological degradation of biomaterial used as scaffold in tissue engineering, so that the newly formed tissue is healthy and completely fulfill the needs of the defect sites [25]. Naturally existing tissues are very well adapted to the local situation, however, the artificially constructed biomaterial should be easily adaptable to the body. Therefore, the synthetic part should be totally eliminated so that smooth biological tissue formation and remodeling take place. Due to the complexity and sensitivity of the host system along with differences between tissues, the selection of biomaterial is a challenge [26]. Irrespective of the host tissue and implantation sites, the basic requirements are as follows [25,26]:

- biodegradability
- porosity
- biocompatibility
- bio-integration
- mechanical properties
- easy manufacture and handling
- cost-effective production

Polymers are the essential materials for platform manufacture in tissue designing applications and different types of biodegradable polymeric materials have been utilized in this field such as [19,27–31]: (1) naturally occurring materials, including polysaccharides [starch, alginate, chitin/chitosan, and hyaluronic acid (HA)] and proteins (soy protein, gelatin, collagen, fibrin gels, silk); (2) synthetic or engineered polymers, for example, poly(lactic acid) (PLA), poly(glycolic acid) (PGA), poly(ε-caprolactone) (PCL), poly(hydroxyl butyrate) (PHB). The PLA, PGA, and their copolymers containing two or more monomers like poly(lactic-*co*-glycolic acid) (PLGA) belong to family of linear aliphatic polyesters, which are most frequently used in tissue engineering [32,33].

1.3 Bone tissue engineering

The component materials of natural bone comprise a nanocomposite structure of three-dimensional matrix. In fact, the natural bone comprises a complex inorganic-organic nanocomposite structure, in which nanocrystalline hydroxyapatite [$Ca_{10}(PO_4)_6(OH)_2$; HAp] and collagen fibrils are organized in a hierarchical architecture [25,34]. The area of bone tissue engineering applications has begun just about three decades ago. Bone tissue engineering applications has spotlighted on the exploration and exploitation of three-dimensional structures (scaffolds), which are capable of supporting, reinforcing, and, in some cases, organizing the bone tissue regeneration and replacements in a natural way [34]. The importance and advancement in the field of bone tissue engineering has noticed a remarkable expansion over the years with an exponentially escalating number of research investigation by various research groups [25]. The applications of bone tissue engineering focus on the substitute treatment alternatives that will preferably get rid of some important issues like inadequate availability, donor-site morbidity, transfer of pathogens, immune rejection, etc. [35]. Currently, the United States, and different nations around the world, is encountering an exceedingly popularity for useful bone grafting. Every year in the United States, about million patients get bone imperfection repairs, at a cost of more than $2.5 billion. It is expected to increase twofold by 2020 in the United States, comprehensively because of an assortment of components, including the developing needs of the population and increased life expectancy [36]. In recent years, bone tissue engineering has been positioned as an impending substitute to the conventional utilization of bone grafts because of their unlimited supply and avoidance of disease transmissions [26]. Bone tissue engineering strategies intend to encourage new functional bone tissue regeneration by means of synergistic combinations of biomaterials, drugs, biomolecules, cells, growth factors, etc. [25]. Even though much advancement has been made, several important impediments come in the way of the bone tissue engineering applications becoming a proper clinical practice in orthopedics. In order to attain the needs for bone tissue engineering, numerous biomimetic composite matrices capable of providing appropriate microarray environment have been researched and developed for promoting osteoblast cell proliferation, thereby stimulating osteogenesis [34].

1.4 Biodegradable polymers used in the design of nanocomposites for bone tissue engineering

Besides bioceramics and metallic materials, several biodegradable polymers are being used for the use in tissue engineering applications. Biodegradable polymers, in general, are classified into two major classes: (i) natural biodegradable polymers and (ii) synthetic biodegradable polymers. Natural biodegradable polymers are derived from natural resources such as plant and animal origins [29,30,37–39]. Therefore, natural biodegradable polymers comprise two major categories: protein originated polymers (e.g., collagen, gelatin, etc.) and polysaccharides (e.g., alginates, chitosan, starch, etc.).

1.4.1 Natural biodegradable polymers

In general, natural polymers comprise highly organized structural features, which may entail ligands (an extracellular substance and necessary to bind with cell receptors). Moreover, natural polymers are more biocompatible in nature and are also able to stimulate bone mineralization compared with the conventionally employed synthetic polymers for the use in bone tissue regeneration. Naturally existing polymers are trending in the field of bone tissue engineering as they are implanted into the structure and they have the ability to easily adapt to the site and do not require second surgical intervention for removal. Some animal and plant polymers have been demonstrated to be utilized as scaffold materials for the tissue engineering applications [19]. For tissue engineering applications, it is expected that the biopolymers degrade (due to their biodegradability) as the new tissues are being formed without causing inflammation, toxic reactions, etc. An essential feature of natural polymers is the stimulation of disagreeable immune reactions owing to the occurrence of impurities and/or endotoxins on the basis of the sources. During the past few decades, natural biodegradable polymer-based composites have been developed with essential benefits such as biodegradability and biocompatibility for the use in the tissue engineering applications.

1.4.1.1 Chitosan

Chitosan is a naturally derived biodegradable polymer obtained from chitin (a major component of crustacean exoskeleton and fungi cell wall) [40]. It is cationic in nature and is composed of α-1,4-linked 2-amino 2-deoxy α-D-glucose (*N*-acetyl glucosamine) [40,41]. It also possesses intrinsic antibacterial, biodegradable, and biocompatible characteristics. Chitosan is notified as a "generally recognized as safe" (GRAS) material by Food and Drug Administration (FDA) and has also been exploited in the formulations of numerous drug delivery dosage forms [6–8,40–42]. In recent years, numerous chitosan-based systems are investigated and developed for the uses in tissue engineering of bone, cartilage, skin, etc., and also in wound healing applications [43]. Over the past few decades, bone tissue engineering has been enhanced by the major role played by chitosan [44]. Chitosan-based composite have been researched and developed for the applications in bone tissue engineering as chitosan do not possess toxic reactions. Moreover, it can be molded into a variety of porous structures, which stimulates osteoconduction [45]. In recent years, numerous chitosan-based nanocomposites are being researched for the use in bone tissue regeneration at the defective and/or diseased bone sites [46–49].

Nazeer et al. [47] prepared and characterized intercalated structured nanocomposites of chitosan and HAp. For the preparation of intercalated chitosan-HAp nanocomposites, nanosized HAp was chemically synthesized via sol-gel method. The self-assembling of chemically synthesized HAp nanoparticles during the drying of solvent-casting films led to the development of various homogeneous chitosan-HAp nanocomposites containing HAp amounts of 5, 10, and 20 wt%. Scanning electron microscopy (SEM) examination of chitosan-HAp nanocomposite film-surface demonstrated the occurrence of disc-shaped HAp nanoparticles with increased mean particle sizes (44 ± 7, 101 ± 16, and 229 ± 33 nm for chitosan-HAp nanocomposite films containing HAp nanoparticles amounts of 5, 10 and 20 wt%, correspondingly) (Fig. 1.2). The cross-sectional SEM photographs and X-ray diffraction (XRD) profiles demonstrated the configuration of layered nanocomposites made of intercalated chitosan-HAp. In addition, thermogravimetric analysis (TGA) suggested intercalated morphologies of these chitosan-based nanocomposites. In the subsequent thermal degradation of intercalated chitosan-HAp nanocomposites, the layered three-dimensional

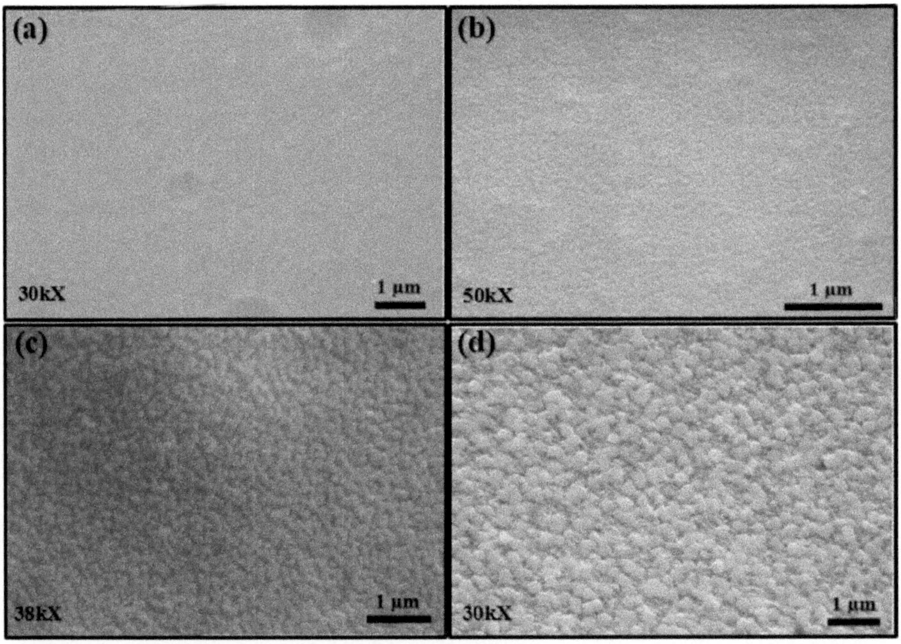

Fig. 1.2 SEM photographs of (A) chitosan film, (B) chitosan-HAp nanocomposite film (containing 5 wt% HAp), (C) chitosan-HAp nanocomposite film (containing 10 wt% HAp), and (D) chitosan-HAp nanocomposite film (containing 20 wt% HAp).
From M.A. Nazeer, E. Yilgör, I. Yilgör, Intercalated chitosan/hydroxyapatite nanocomposites: promising materials for bone tissue engineering applications, Carbohydr. Polym. 175 (2017) 38–46. Copyright © 2017 Elsevier Ltd.

nano-porous scaffold configurations (containing phases of HAp, calcium pyrophosphate and tri calcium phosphate) can be employed as a potential scaffold material for bone tissue engineering because HAp is well known for promoting cell adhesions, cell proliferations, and also osteogenic differentiations of osteoblasts. Nikpour et al. [48] also prepared and characterized chitosan-nanohydroxyapatite (nHAp) composite through in situ hybridization method. These nanocomposites were characterized by SEM, atomic force microscopy (AFM), Fourier transformed infrared (FTIR), and XRD analyses. The results of instrumental characterizations authenticated the homogeneity, interactions as well as integration between chitosan and nHAp within the chitosan-nHAp nanocomposite matrix. In addition, the results of mechanical compressive test suggested satisfactory mechanical performances for the bone tissue replacement by these chitosan-nHAp nanocomposites prepared by in situ hybridization method. Thein-Han and Misra [49] reported preparation and characterization of chitosan-nHAp nanocomposite scaffolds, where high and medium molecular weight chitosan were employed to prepare the chitosan-based nanocomposite scaffold. But, the mechanical behavior of these nanocomposites was found insufficient for the bone repair applications.

In an investigation, Keller et al. [46] have developed chitosan-based nanocomposite scaffolds for the repair of bone defects by bone tissue regeneration. These nanocomposite scaffolds were prepared by using chitosan of different molecular weights, concentrations, and degrees of deacetylation, which are reinforced with silica (SiO_2) nanoparticles. The developed chitosan-SiO_2 nanocomposite scaffolds demonstrated similar pore sizes of not less than 300 μm irrespective of the concentration and deacetylation degrees of chitosan used in the chitosan-SiO_2 nanocomposite formula. The SiO_2 nanoparticles were reinforced as nanofiller materials to enhance the mechanical compression resistance of these chitosan-SiO_2 nanocomposites by 30%. The in vitro biocompatibility of these three-dimensional chitosan-SiO_2 nanocomposite scaffolds was tested by Alamar Blue assay. The results of Alamar Blue assay in human primary osteoblasts suggested the in vitro biocompatibility of chitosan-SiO_2 nanocomposites and the development of cell spheroids signifying the great prospective for the regeneration of bone tissue. The in vivo implantation in the mice calvaria defect model indicated the appropriateness of chitosan-SiO_2 nanocomposite scaffolds for the bone tissue regeneration capability. The in vivo implantation results demonstrated mature as well as denser collagenous tissues with vascularized areas, smaller foci of mineralization, and the infiltration of osteoblast and osteoclast cells (Fig. 1.3). However, the mature bone tissue formation was not observed after 8 weeks of implantation.

1.4.1.2 Alginates

Alginates are one of the marine biopolysaccharide group. Alginates, salts of alginic acid, are anionic linear natural polysaccharidic group extracted from brown algae (including *Laminaria digitata*, *Ascophyllum nodosum*, *Laminaria hyperborea*, and *Macrocystis pyrifera*) and bacteria [50,51]. Alginates are block copolymers made up of 1,4-linked β-D-mannuronic acid (M) with $^4C_1$ ring configuration and α-L-guluronic

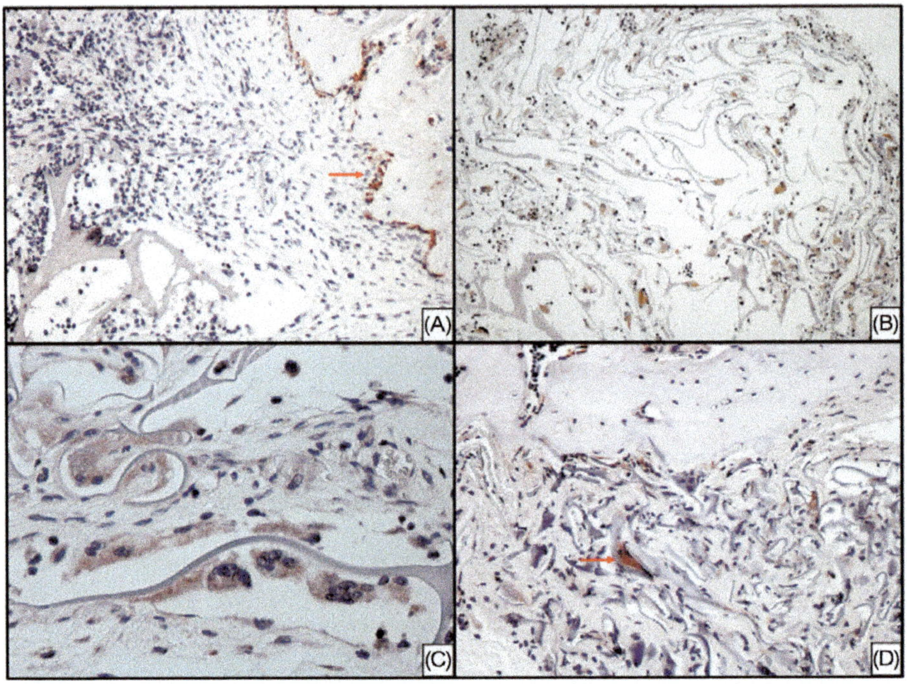

Fig. 1.3 Immuno-histochemical staining of OCN of chitosan-SiO_2 nanocomposite scaffolds over time: (A) 2 weeks: OCN expression in osteoblast cells limited to the bone/scaffold interface *(arrow)* (20×); (B) 4 weeks: numerous osteoblast cells with strong positivity to OCN near the chitosan threads (10×); (C) 4 weeks: positive staining for OCN in multinucleated cells compatible with the osteoclast cells (40×); and (D) 8 weeks: positive staining for OCN with more multifocal and sparse positive cells *(arrow)* (40×).
From L. Keller, A. Regiel-Futyra, M. Gimeno, S. Eap, G. Mendoza, V. Andreu, Q. Wagner, A. Kyzioł, V. Sebastian, G. Stochel, M. Arruebo, N. Benkirane-Jessel, Chitosan-based nanocomposites for the repair of bone defects, Nanomed. Nanotechnol. Biol. Med. 13 (2017) 2231–2240. Copyright © 2017 Elsevier Inc.

acid (G) with $^4C_1$ conformation [52]. It is anionic in chemical nature [53,54]. It is biocompatible, biodegradable, and nonantigenic in nature [50]. Alginates also have the propensity to structure gels with the aqueous medium. Since past few decades, alginates have been expansively investigated as natural biopolymeric excipients or raw materials in various biomedical fields including drug delivery [55–70] and tissue engineering applications [71]. Sodium alginate, sodium salt of alginic acid, is able to produce considerable viscous solutions in the aqueous milieu [72]. The most interesting feature of sodium alginate is to undergo ionic gelation by the influence of various divalent and trivalent metal cations (such as, Ca^{2+}, Zn^{2+}, Ba^{2+}, Al^{3+}, etc.) in the aqueous milieu [73–75]. In recent years, various alginate-based nanocomposite scaffolds are being researched in different tissue engineering applications. Even, majority of these alginate-based nanocomposites have already shown improved biocompatibility, cell proliferation, mechanical strength, porosity, cell adhesion, osteogenic differentiation, and fabulous mineralization, which suggested its effectiveness in bone tissue engineering [71].

Liu et al. [76] developed some composite tissue engineering scaffolds made of alginate/halloysite nanotubes through solution mixing and freeze-drying techniques. In these tissue engineering scaffolds, halloysite nanotubes were reinforced into the alginate-matrix to enhance the mechanical behavior of the composite scaffolds. These alginate/halloysite nanotube composite scaffolds also demonstrated considerable mechanical improvements in the compressive modulus as well as compressive strength compared to that of the scaffolds made of pure alginate in both states, wet and dry. An interconnected porous structural feature with size range of 100–200 μm and over 96% of porosity was noticed in the case of alginate/halloysite nanotube composite scaffolds. The reinforcement of halloysite nanotubes in these nanocomposites led to amplification of the scaffold density and also led to decrease in the swelling rate of alginate in the aqueous milieu. It was also noticed that the incorporation of halloysite nanotubes enhanced the stability of alginate/halloysite nanotube composite scaffolds against the enzymatic degradation in phosphate buffer solution. The cell adhesion characteristics and cell growth behavior of alginate/halloysite nanotube composite scaffolds were characterized by in vitro cytocompatibility test using mouse fibroblast 3T3 cells exhibiting improved cell attachment characteristics by the composite scaffolds compared to that of the scaffolds made of pure alginate. In addition, an alginate/halloysite nanotube composite scaffold was evaluated via MTT test method using mouse fibroblast cells representing active mitochondrial activities of the living cells (Fig. 1.4). Thus, the excellent in vitro cytocompatibility and biocompatibility of the alginate/halloysite nanotube composite scaffolds suggested its use in the tissue engineering applications.

Chae et al. [77] developed alginate/HAp nanocomposite fibrous scaffolds through a novel methodology of electrospinning and in situ synthesis of HAp mimicking bone mineralized collagen fibrils. This novel methodology produced homogeneous deposition of nanocrystalline HAp on the collagen nanofibers, surmounting the rigorous agglomeration of nHAp formed by the usual electrospinning or mechanical blending methodologies. The in vitro cell culture study demonstrated that the rat calvarial osteoblast cells were found appended more steadily on the fibrous scaffold surface as compared to the scaffold made of pure alginate. The osteoblast cells were found to be lengthened into spindle-shaped morphology on the alginate/HAp nanocomposite fibrous scaffolds, whereas in the case of the scaffold made of pure alginate, the osteoblast cells showed morphology of round shape. The distinctive nanofibrous topography of the developed alginate/HAp nanocomposite fibrous scaffolds combined with the hybridization of alginate and nHAp can be beneficial for regenerating bone tissue. Sangeetha et al. [78] also investigated in situ formation of alginate/HAp nanocomposites using gelatin and without gelatin. They found gelatin to improve the crystallinity of HAp in the in situ prepared alginate/HAp nanocomposites. It also affected the microstructural features of the in situ nanocomposites.

Kawaguchi et al. [79] prepared and characterized carbon nanotube-alginate nanocomposite hydrogel for tissue engineering applications. The mechanical elastic deformation characteristics, saline sorption profile, histological study, and in vitro cell

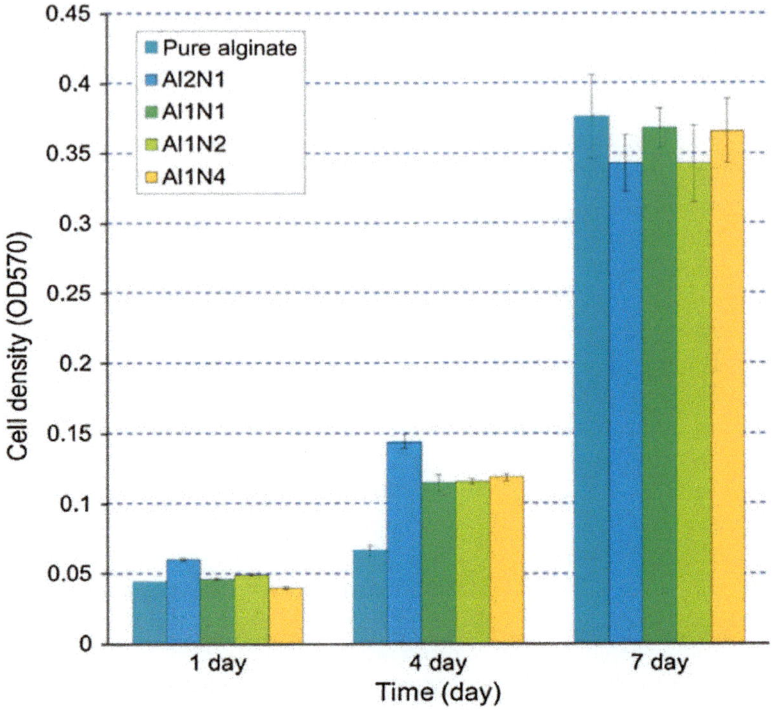

Fig. 1.4 Viability of mouse fibroblast cells on pure alginate scaffold and alginate/halloysite nanotube composite scaffolds (AI1N2) measured by MTT assay representing active mitochondrial activities of the living cells.
From M. Liu, L. Dai, H. Shi, S. Xiong, C. Zhou, In vitro evaluation of alginate/halloysite nanotube composite scaffolds for tissue engineering, Mater. Sci. Eng. C 49 (2015) 700–712. Copyright © 2015 Elsevier B.V.

viability of the carbon nanotube-alginate nanocomposite hydrogel were evaluated. The nanocomposite hydrogel demonstrated improved mechanical behavior and faster gelling characteristics compared to that of the conventional pure alginate gel. The quantity of saline sorption of the freeze-dried carbon nanotube-alginate nanocomposite hydrogel was found similar to that of the conventional pure alginate gel. The results of histological study and in vitro cell viability assay of the nanocomposite hydrogel suggested a mild inflammatory response and noncytotoxicity, respectively. The overall results of the investigation indicated that the carbon nanotube-alginate nanocomposite hydrogel could be beneficial as a bone tissue engineering scaffold material.

1.4.1.3 Starches

Starches are plant-derived biodegradable polymer with excellent biocompatibility [80]. They are the macromolecular biopolymeric group of high molecular weight polymers containing two well-known structural copolymers: amylose and amylopectin [81]. Starches have been already employed for different applications in biomedical areas (e.g., tissue engineering, wound dressing applications, drug delivery, etc.) [80,82–85]. Different potential constraints of starches like deprived processability, lesser moisture resistance, short stability in the acidic milieu, etc., are being managed by modifications of starches [80,81,86]. In recent years, various starch-based composites have been developed for biomedical uses including tissue engineering and drug delivery [86–89]. Even, various starch-based composites have already been investigated for the fabrication of tissue engineering scaffolding of bone tissue facilitating good porous structure for the cell migration [89]. Recently, several starch-based nanocomposites have been fabricated via reinforcing other biopolymers and bioinorganics to improve scaffolding properties [90].

Meskinfam et al. [90] synthesized and characterized nHAp-starch biocomposites via a biomimetic process. The SEM and transmission electron microscopy (TEM) studies suggested that the shape as well as morphological features of nHAp were influenced by the occurrence of starch as template material in these nanocomposites. This demonstrated the rod-like nHAp formation within the starch matrix. The in vitro biocompatibility of nHAp-starch biocomposites was evaluated by in vitro cell culture and MTT assay. The results of the in vitro cell culture and MTT assay demonstrated that nHAp influenced the cell proliferations and the nHAp-starch biocomposites did not show any adverse result on the cellular morphology of the cultured cell, cell viability, and cell proliferations. Sadjadi et al. [91] also synthesized similar starch-based nanocomposite for the use in bone tissue engineering through in situ biomimetic procedure of bone-like nHAp formation in the presence of wheat starch. Wheat starch as template material influenced the shape and morphology of the nHAp. It also led to rod-like nHAp formation (similar to HAp present in the inorganic component in natural bone).

In another study, Raafat et al. [92] synthesized starch/*N*-vinylpyrrolidone-HAp nanocomposite-based hydrogels for the use in bone tissue engineering applications through γ-radiation-induced graft-copolymerization and cross-linking methodology. In situ precipitation of HAp was attained through alternate soaking method. To develop an optimal formula of starch-based nanocomposite-based hydrogels, different preparation conditions were optimized to make sure the highest degree of gelation. The synthesized starch/*N*-vinyl pyrrolidone-HAp nanocomposite-based hydrogels were characterized using FTIR spectroscopy, energy dispersive X-ray (EDX) spectroscopy, TGA, XRD, SEM, and by investigating in vitro swelling performance. The results of the mechanical testing demonstrated that the compressive strength of the starch/*N*-vinyl pyrrolidone-HAp nanocomposite was measured to be augmented primarily after the first cycle of HAp deposition. Afterward, it was found reduced progressively through increasing the HAp deposition cycle number. In vitro bioactivity as well as haemato-compatibility of the developed starch-based nanocomposite-based hydrogels was evaluated, which designated that these starch-based nanocomposites were bioactive as well as biocompatible in nature.

1.4.1.4 Cellulose

Cellulose is a straight long-chain polysaccharide comprising a few hundred to over 10,000 β-(1–4)-connected D-glucose units [93]. It is a naturally occurring raw material occurs in abundance as well as renewable resource crude material [94]. It is widely present in most of living species, commonly obtained from cotton, hemp, wood, and linen. The mechanical strength, chemical stability, biodegradability, and biocompatibility of cellulose are due to its high molecular weight, intra- and intermolecular hydrogen bonding. Its use in the tissue engineering is complicated as the dissolution of cellulose is difficult in commonly used solvents due to its chemical properties [95,96]. This problem is overcome by the use of cellulose derivatives such as carboxymethyl cellulose (CMC) and bacterial cellulose for the preparation of tissue engineering scaffolds [97,98]. The bacterial cellulose is obtained from the biosynthesis of *Acetobacter xylinum* [99].

Saber-Samandari et al. [100] synthesized nanocomposite scaffold of semiinterpenetrated networking cellulose-*graft*-polyacrylamide/nHAp via freeze-drying method. The cellulose-graft-polyacrylamide was synthesized via free radical polymerization. In addition, some important mechanical properties such as elastic modulus and compressive strength profiles of the nanocomposite scaffolds were measured. Under the optimal conditions, the newly synthesized semiinterpenetrated networking cellulose-graft-polyacrylamide/nHAp nanocomposite scaffolds showed 47.37% porosity, 0.29 GPa elastic modulus, and 4.80 MPa compressive strength. The SEM analysis demonstrated the interconnected pores (120–190 μm) of the nanocomposite scaffolds (Fig. 1.5). To determine the apatite-forming capability, these cellulose-based nanocomposite scaffolds were soaked in the simulated body fluid (SBF) for 7, 14, and 28 days. The result suggested that the new apatite growth occurred on the scaffold surface after 14 days of soaking in SBF. The overall results of the study recommended that the synthesized semiinterpenetrated networking cellulose-graft-polyacrylamide/nHAp nanocomposite scaffolds can be combined with the living natural bone tissue through the generation of apatite layers on the scaffold-surface and can be employed in the bone tissue regeneration purpose.

He and colleagues [101] manufactured uniaxially adjusted cellulose nanocomposite nanofibers to all around and oriented to cellulose nanocrystals (CNCs) by means of electrospinning. The integration of CNCs into the gyrating dope brought about more uniform morphology of the electrospun cellulose/CNC nanocomposite nanofibers (ECCNN) and outstanding development of their physical properties. Cell culture tests showed that cells could multiply quickly at first glance as well as somewhere inside the composite material and the adjusted nanofibers displayed a solid impact on coordinating cell association.

1.4.1.5 Collagen

Collagen is a natural polymer and is also recognized as one of the major abundantly available natural occurring structural protein of the living body, being a principal element of the extracellular matrix (ECM) [102,103]. It usually occurs in the tissues of cartilage, tendon, and bone [102]. Collagen is already reported as biodegradable proteinous biomaterial [104,105]. Because of its intrinsic biocompatibility concerning lower antigenic property, cytotoxic responses and noninflammatory action, collagen is widely used for various biomedical applications including wound dressing, soft and hard tissue engineering, drug delivery, etc. [104]. However, it possesses some

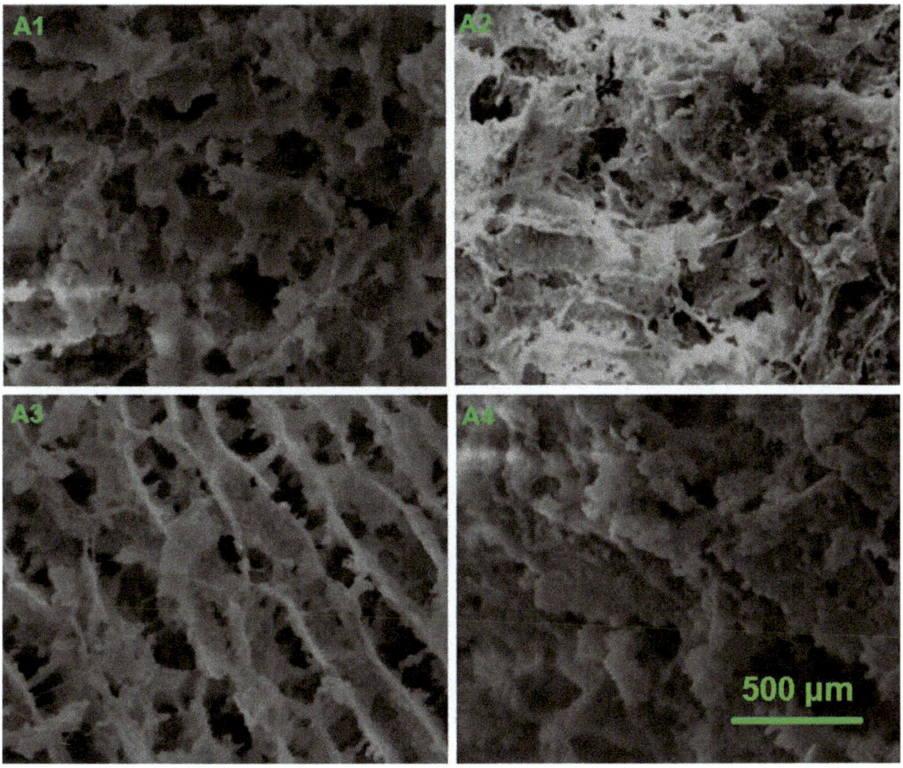

Fig. 1.5 SEM morphologies of the prepared semiinterpenetrated networking cellulose-*graft*-polyacrylamide/nHAp nanocomposite scaffolds.
From S. Saber-Samandari, S. Saber-Samandari, S. Kiyazar, J. Aghazadeh, A. Sadeghi, In vitro evaluation for apatite-forming ability of cellulose-based nanocomposite scaffolds for bone tissue engineering, Int. J. Biol. Macromol. 86 (2016) 434–442. Copyright © 2016 Elsevier B.V.

limitations like lower elastic behavior and reduced mechanical strength. These limitations of collagen as biomaterial are being managed by different modification approaches and one of these is the preparation/synthesis of collagen-based composites [105]. A recent technology is trying to develop different collagen-based scaffolds for bone tissue engineering that can provide biological response in the body, for example, interaction between the bone cells and the intension is to make collagen-based scaffolds for the artificial biomimicking of ECM helping the bone tissue regeneration [97,106]. Recently, numerous collagen-based nanocomposites have been developed to improve the scaffolding properties for the use in bone tissue regeneration [105,107].

Pek et al. [108] developed porous collagen-apatite nanocomposite scaffold foams for the use as bone tissue engineering scaffold, which were reported to be as structurally, chemically, and mechanically comparable with the natural bone. The nanocomposite scaffold foams contained collagen fibers and apatite nanocrystals. These were shown to have higher mechanical strength and therefore, can be beneficial to provide support for the adjacent hard tissue. The porous collagen-apatite nanocomposite scaffolds displayed comparable structural features of trabecular bone as they possessed nanosized structure. In addition, the apatitic phase of these collagen-based nanocomposites showed chemical composition comparable with that of the trabecular bone. Also, the crystalline phase as well as the grain size of the synthesized apatite were comparable with that of the apatite of trabecular bone. The porous collagen-apatite nanocomposite scaffold showed good bioactivity, which can promote excellent cell adhesion and cell proliferation. The nanocomposite scaffolds were also found osteoconductive in nature as they effectively repaired a nonunion bone fracture in the rat femur and a critical-sized bone defect in the pig tibia.

Kikuchi et al. [109] synthesized bone-like HAp/collagen nanocomposite scaffolds and in vitro-in vivo evaluation of the synthesized nanocomposites were performed. To synthesize the HAp/collagen nanocomposite, simultaneous titration coprecipitation technique was carried out using calcium hydroxide, orthophosphoric acid, and porcine atelocollagen as raw materials. These nanocomposites exhibited a self-organized nanostructure feature comparable to the bone brought together via the chemical interaction between collagen and HAp. The synthesized bone-like HAp-collagen nanocomposite material was resorbed through phagocytosis of the osteoclast-like cells and also accomplished osteoblast cells to generate new bone tissue in the adjacent region. This HAp-collagen nanocomposite showed similar nanostructure morphology and the composition can substitute the autologous bone graft materials. In another work, Zou et al. [110] prepared and characterized porous β-tricalcium phosphate-collagen composites for the use as bone tissue engineering scaffold. These collagen-based composites revealed an integrated structure.

1.4.1.6 Gelatin

Gelatin is a commonly natural occurring protein, which is also recognized as biodegradable polymer [111]. It is biocompatible, nonantigenic and also possesses good plasticity. Gelatin is one of the important biopolymer capable of mimicking certain useful characteristics of the natural ECM for the cell adhesion and cell proliferation [112]. It is also being investigated as raw biomaterial in the fabrication of tissue engineering scaffolds for the growth of damaged bone tissues. It is being employed as substitute in the bone grafting [113]. In recent years, nanofibrous gelatin-based composite scaffolds have been fabricated in order to mimic the bone composition as well as the architectural behavior of living bone. Nanofibrous gelatin matrix can be fabricated by thermally induced phase separation (TIPS) method. Nanofibrous gelatin-based composite scaffolds being investigated for bone tissue engineering applications possess higher surface area as well as enhanced mechanical power as compared to the commercially available gelatins of various grades with identical features of porosity and pore size. These nanofibrous gelatin-based composite scaffolds have already displayed prospective results for the use in bone tissue engineering applications [114].

Recently, Johari et al. [115] have investigated a three-dimensional bioglass (bioactive SiO_2-CaO-P_2O_5 glasses)/gelatin nanocomposite (which was also osteoblast seeded) as a prospective bone substitute scaffold material. The SEM analysis of these nanocomposite scaffold materials demonstrated homogeneity as well as enhanced cell adhesion features. These gelatin-based nanocomposites were also studied for the in vitro biocompatibility as well as in vivo bone regeneration (new bone formation) ability through implantation in the critical-sized calvarial defect repair model in rats. The in vitro biocompatibility results suggested the bioglass/gelatin nanocomposite as good cytocompatible material as it enhanced osteoblast cell adhesion and cell proliferation. At the various time point postmortem, the in vivo implantation of the nanocomposite was analyzed by means of histological, histo-morphometric, and immune-histochemical techniques. The in vivo implantation results indicated the higher level of biodegradability and biocompatibility of these gelatin-based nanocomposites. These osteoblast seeded nanocomposite scaffolds displayed improved repairing of bone defects through osteogenesis in the critical-sized calvarial defect repair model in rats.

In another recent research, Samadikuchaksaraei et al. [116] developed osteoblast cell conditioned nHAp/gelatin composite scaffold for the use in bone tissue regeneration applications. The gelatin-based composite scaffold was prepared by means of layer solvent casting/freeze-drying/lamination methods. The composites were conditioned by the osteoblast culture on the composite surface and were removed via the repeated freeze-thaw method. The outcomes of the mechanical and in vitro biological analyses demonstrated that the mechanical profiles as well as in vitro biological characteristics were not affected by the procedure of osteoblast cell conditioning. The prospective of osteoblast cell conditioned nHAp/gelatin composite scaffold was to maintain the cell adhesion as well as cell growth. In addition, the in vitro cytotoxicity of these composite scaffolds was evaluated by using rat mesenchymal stem cells. These developed osteoblast cell conditioned nHAp/gelatin nanocomposite was evaluated for in vivo implantation studies in the critical-sized calvarial defect repair model in rats' bone, which suggested osteoblast cell conditioning has improved the biocompatibility as well as osteoinductivity of the developed nHAp/gelatin nanocomposite scaffold. The osteoblast cell conditioning with these nanocomposite scaffolds speeded up the amounts of collagen throughout the bone healing procedure. In addition, the in vivo results also showed that the newly generated bone tissue occupied approximately the total bone defect (i.e., $93.40 \pm 3.30\%$) within the implantation of 3 months.

1.4.1.7 Hyaluronic acid (HA)

The HA is an anionic, nonsulfated glycosaminoglycan, comprising repeated units of D-glucuronic acid-β-1,3-*N*-acetyl-D-glucosamine-β-1,4 units. It is found in extracellular matrix of every connective tissue in the body and have several properties like excellent viscoelasticity, water solubility, biocompatibility, and nonimmunogenicity [105]. However, it is degradable by the enzymatic activity. Schanté et al. [117] reported that on grafting amino acids with HA, its enzymatic stability was enhanced. For the use in bone tissue engineering applications, material with few structures as hydrogels, fibers, meshes, and foams has been made [118,119]. In addition, modifications in HA-based scaffolds have been extensively investigated for bone tissue engineering to improve the structural integrity, toughness, and mechanical strength [120].

In a recent research by Subramaniam et al. [121], HAp-calcium sulfate-HA composite encapsulated with collagenase was developed and evaluated for its efficacy for the use as bone substitute material in bone regeneration applications. The nanocomposite was evaluated for its biocompatibility by carrying out WST-1 assay. The in vivo bone tissue regeneration was evaluated in alveolar bone defective rat model and the bone tissue regeneration and growth was authenticated by analyzing various micro-CT photographs as well as histological assessments. The HA-based nanocomposite possessing 6.69 MPa of mechanical force exhibited good biocompatibility. The collagenase releasing pattern had provided the speeding up of bone tissue remodeling course, which was authenticated by the obtained results of micro-computed tomography (micro-CT) analyses as well as hematoxylin/eosin staining studies. In the alveolar bone defects, the HAp-calcium sulfate-HA nanocomposite encapsulated with 2 mg/mL collagenase type I was implanted in rat model and the in vivo implantation results demonstrated newly bone tissue development with the matured bone tissue morphological features as compared with that of the nanocomposite without encapsulation of collagenase and the porous HAp particles.

Huang et al. [122] synthesized an injectable nHAp/glycol chitosan/HA nanocomposite hydrogel for the use in bone tissue engineering. The synthesized HA-based nanocomposite hydrogel displayed a porous structural features (with pore size 100–350 μm) connected with numerous nHAp particles and it was evidenced by the SEM analysis. In addition, the SEM analysis also suggested the aggregation of these HA-based nanocomposites within the developed hydrogel. It was noticed that with an increment in the concentration of HA, the porosity as well as swelling ratio of the nanocomposite hydrogel was reduced consequently. The synthesized injectable nHAp/glycol chitosan/HA nanocomposite hydrogels were found to be affected by enzymatic hydrolysis. This also demonstrated a more rapid rate of degradation in phosphate buffer solution containing lysozyme of 2.5 mg/mL compared to that of in phosphate buffer solution without lysozyme. In vitro cytocompatibility efficiency of these HA-based injectable nanocomposite hydrogel was estimated using MC-3T3 E1 cell lines and it was found cytocompatible. Moreover, good cell adhesion on the nanocomposite hydrogel even after 7 days of incubation was also noticed. All the results of this study clearly suggested the bone tissue engineering prospective features of the developed injectable nHAp/glycol chitosan/HA nanocomposite hydrogel.

1.4.1.8 Dextran

Dextran is a natural polysaccharide having 1,6 linked glucose molecules, with two –OH groups per glucose units. As it is biocompatible as well as biodegradable, it is suitable biomaterial for the delivery of numerous drug molecules [123]. Dextran has lower unspecific binding and also has applications in biosensors and in receptor binding studies. It is stable a biomaterial and is being used in implants, tissue engineering applications, etc. Dextran is commonly synthesized using sucrose by lactic acid bacteria, like *Streptococcus mutans*, *Leuconostoc mesenteroides*, etc. It is employed as antithrombotic material in various medical practices. In combination with pullulan (a natural hydrophilic polysaccharide), dextran is used in the tissue engineering applications as it helps in cell differentiation, cell proliferation, and in viability of human endothelial progenitor cells. In bone tissue engineering, it promotes in vitro osteogenic cell differentiation and in vivo mineralized bone tissue regeneration [123].

In a research by Fricain et al. [123], a new type of dextran-based nanocomposite scaffolds were developed. This dextran-based nanocomposite scaffolds also contained pullulan and also supplemented with or without nHAp particles. The nanocomposites were of microporous in nature. In vitro analysis of these nHAp-pullulan/dextran polysaccharide nanocomposite-based macroporous material demonstrated that they were capable of encouraging the development of multicellular aggregations. These nanocomposite scaffolds also displayed the late as well as early expressions of bone-specific markers with the human bone-marrow stromal cells in the medium of deprived osteoinductive factors. In the deficiency of any cultured cells, in vivo heterotopic implantations in the experimental animals like mice as well as goats were evaluated. The outcomes of in vivo evaluations in mice and goats demonstrated that only the dextran-based nanocomposite macroporous scaffold (containing polysaccharide matrix and nHAp) maintained subcutaneous (s.c.) local growth factor agents such as bone morphogenetic protein-2 (BMP-2), VEGF 165, etc. The results also encouraged the deposition of biologically active apatite layer. In addition, it also supported the denser bone mineralized tissue developments of in the mice s.c. and osteoid tissue regenerations even after carrying out intramuscular (i.m.) implantation in goats. The dextran-based nanocomposite scaffolds were then implanted in the orthotopic animal models of critical-size defects for the preclinical experimentations, in smaller (mice) and larger animals (goats), and in three various bone defective sites. The polysaccharide matrix and nHAp within these nanocomposites encouraged a well bone mineralized tissues in the three various bone defective models with direct contact between the implantation sites and osteoid tissues. Thus, nHAp-pullulan/dextran polysaccharide nanocomposite-based macroporous material can be employed as bone tissue engineering scaffolds in the orthopedic applications through encouraging the differentiations of host mesenchymal stem cells.

Some other examples of natural biodegradable polymer-based nanocomposites for bone tissue engineering applications are presented in Table 1.1.

1.4.2 Synthetic biodegradable polymers

The synthetic polymers are among the major group of biodegradable polymers that are manufactured under controlled conditions, having conventionally reproducible physical and mechanical properties like rigidity, versatile modulus, and certain rate of degradation [154]. They are mostly less expensive than biological scaffolds and promote angiogenesis by favorable interactions with the endothelial cells. The commonly used synthetic biodegradable polymers, which are being used in the development of tissue engineering scaffolds, are PLA, PCL, and PLGA [154,155]. To improve the mechanical characteristics, cell adhesion potential, and cell proliferation property, various kinds of nanomaterials (e.g., metallic nanoparticles, ceramic nanoparticles, nanophase apatite materials, carbon nanotubes, etc.) were being reinforced within these synthetic biodegradable polymer-based matrices. And, these nanocomposites have been extensively prepared and evaluated for the use in bone tissue engineering applications [155].

Table 1.1 Some examples of natural biodegradable polymer-based nanocomposites for bone tissue engineering applications

Natural biodegradable polymer-based nanocomposites for bone tissue engineering applications	References
nHAp/chitosan composite scaffolds	Kong et al. [124]
Chitosan-natural nHAp-fucoidan nanocomposites	Lowe et al. [125]
Chitosan/montmorillonite/HAp nanocomposite	Katti et al. [126]
Chitosan/HAp nanocomposite rods	Hu et al. [127]
β-Type chitosan and nHAp nanocomposites	El-Sherbiny et al. [128]
Hybrid nanostructured HAp-chitosan composite scaffold	Guo et al. [129]
Chitosan-nanocrystalline calcium phosphate composite scaffold	Chesnutt et al. [45]
Carbon nanotube-grafted-chitosan—natural HAp composite	Venkatesan et al. [130]
Chitosan-PEG-HAp-ZnO nanocomposites	Bhowmick et al. [131]
nHAp/β-CD/chitosan nanocomposite	Shakir et al. [132]
Chitosan/graphene oxide 3D scaffold composites	Dinescu et al. [133]
Organically modified clay supported chitosan/HAp-ZnO nanocomposites	Bhowmick et al. [134]
nHAp/chitosan/CMC composite scaffold	Jiang et al. [135]
Nanobiocomposite scaffold containing chitosan/nHAp/nano-silver	Saravanan et al. [136]
Nanobiocomposite scaffolds containing chitosan/nHAp/nano-copper-zinc	Tripathi et al. [137]
nHAp/chitosan-starch nanocomposite	Shakir et al. [138]
Biocomposite scaffolds containing chitosan/alginate/nano-SiO_2	Sowjanya et al. [139]
Bioinspired nanostructured HAp/collagen three-dimensional porous scaffolds	Guan et al. [140]
Bioinspired collagen-apatite nanocomposite	Liu et al. [141]
Calcium alginate-HAp nanocomposite hydrogels	Bouropoulos et al. [142]
Alginate/HAp composite scaffolds	Brun et al. [143]
Single-walled carbon nanotube-reinforced alginate composite scaffolds	Yildirim et al. [144]
Porous alginate/HAp composite scaffolds	Lin et al. [145]
Injectable 3D-formed β-tricalcium phosphate bead/alginate composite	Matsuno et al. [146]
Nanobiocomposite scaffold of chitosan-gelatin-alginate-HAp	Sharma et al. [147]
Starch/multiwalled carbon nanotubes composites	Famá et al. [89]
Nanocomposite scaffolds based on HAp entrapped in cellulose network	Beladi et al. [148]
Biomimetic nanocomposites of carboxymethyl cellulose-HAp	Garai and Sinha [149]
Bacterial cellulose-HAp nanocomposites	Saska et al. [150]
Porous HAp/gelatin nanocomposite scaffold	Azami et al. [112]
Biphasic calcium phosphate/gelatin nanocomposite scaffolds	Bakhtiari et al. [151]
Collagen-HA/bioactive glass nanocomposite scaffold	Wu et al. [152]
Pullulan/dextran/nHAp macroporous composite beads	Schlaubitz et al. [153]
Porous β-TCP/collagen composites	Zou et al. [110]

1.4.2.1 Polylactic acid (PLA)

The PLA was first synthesized in 1932 by Carothers. The PLA chemistry involves the processing and polymerization of monomer of lactic acid. The PLA is synthesized by various polymerization processes like enzymatic polymerization and azeotopic dehydration [156]. The most common methods used for the synthesis of PLA are ring opening polymerization and direct polymerization. As lactic acid has chiral carbon, PLA possesses stereoisomers like poly(L-lactide) (PLLA), poly(D,L-lactide) (PDLLA), poly(D-lactide) (PDLA), etc. [154–156]. Since it can cause inflammation owing to its degradation along with formation of crystalline fragments, PLLA is used along with D,L-lactic and L-lactic acid monomers to overcome this problem [157]. The PLA is highly versatile, low molecular weight, biodegradable, and biocompatible biopolymer and owing to these properties, PLA is extensively studied for the use in medical applications [158]. It is widely used in various biomedical fields like tissue engineering, suture, drug delivery, bone fixation, etc. [155,158]. The PLA is being extensively used as tissue engineering scaffold material due to its biocompatibility and biodegradability potentials [159,160]. The PLLA is a favorable material used as stent for heart surgery, for repairing tendon and ligament, and in urological surgery [161,162].

In a research, Nejati et al. [161] synthesized nHAp rods/PLLA composite scaffolds for bone tissue regeneration applications. The rod-shaped nHAp with a mean length of about 100–400 nm and width 37–65 nm was almost analogous to the natural bone apatite in accordance with the chemical composition as well as structural morphological features. To prepare nHAp rods/PLLA composite scaffolds, nHAP were employed via thermally induced phase separation technique. The porosity of nHAp rods/PLLA composite scaffolds was measured to be 85.06%; whereas the mean macropore diameter was measured to be 64–175 μm. The XRD and FTIR studies demonstrated various chemical interactions between PLLA matrix and nHAp particles. The mechanical features of these nanocomposite scaffolds were evaluated and the compressive strength could high up to 14.90 MPa, which was higher than that for pure PLLA (1.79 MPa) and microcomposite scaffolds (13.68 MPa), respectively. These results of the mechanical behavior analysis demonstrated the encouraging influence of nHAp particles as fillers for the enhancement of the mechanical profile of the PLLA composite matrix. In vitro biocompatibility of nHAp rods/PLLA composite scaffolds was evaluated by using mesenchymal stem cells, which demonstrated that the biocompatibility as well as cell affinity was observed to be greater than that for pure PLLA and microcomposite scaffolds. It was noticed that the round-shaped cells connected as well as proliferated to the surface of nanocomposite scaffolds, and happen to spindle-like structure and then migrated via the pores The round-shaped cell number was evident on the pure PLLA scaffold surface as the proliferated cells on these scaffolds displayed a spindle-shaped morphological view (Fig. 1.6). These nanocomposite scaffolds were showed to be in vitro biocompatible to the cells.

Eftekhari et al. [163] fabricated and characterized a novel biomimetic and porous nanocomposite biomaterial composed of PLLA, nHAp, and microcrystalline cotton cellulose. The PLLA/cellulose/nHAp nanocomposites were prepared through pretreatment of reinforcing materials by a coupling agent and fabrication of nanocomposites. The influence of different weight ratios of reinforcing materials was also

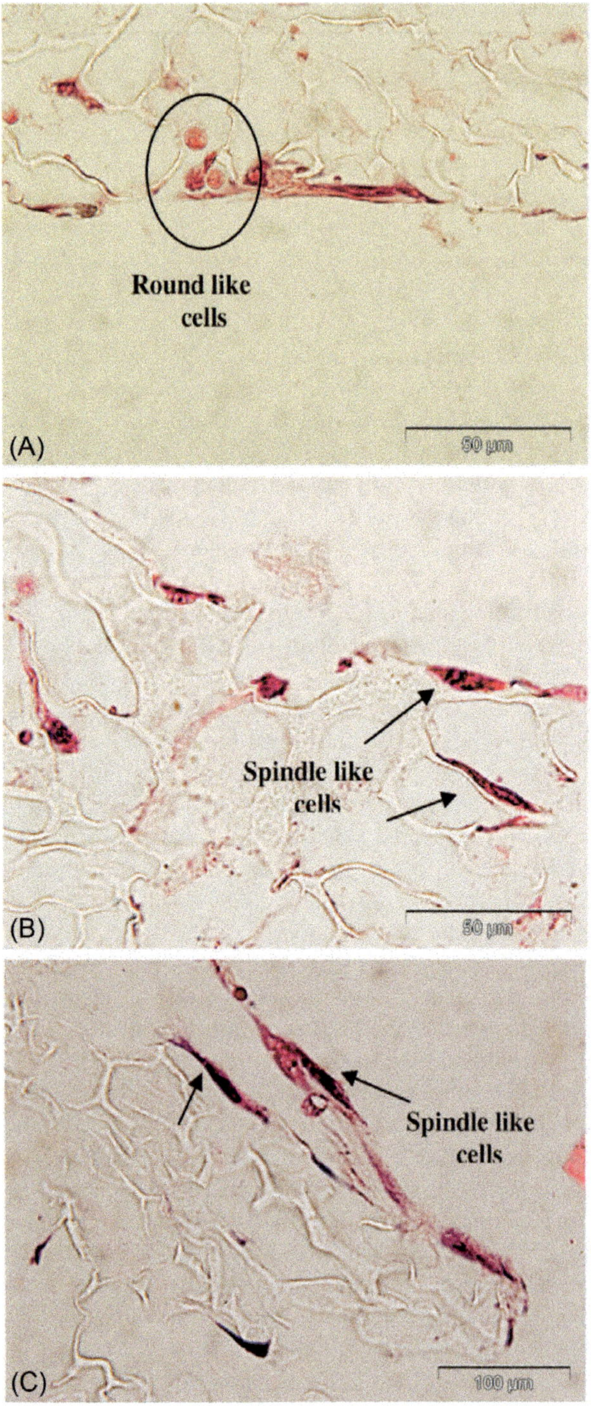

Fig. 1.6 Optical microscopy images of the colored mesenchymal stem cells (H&E staining) attached to the (A) pure PLLA, (B) mHAP/PLLA, and (C) nHAP/PLLA.
From E. Nejati, H. Mirzadeh, M. Zandi, Synthesis and characterization of nano-hydroxyapatite rods/poly(L-lactide acid) composite scaffolds for bone tissue engineering, Compos. Part A. 39 (2008) 1589–1596. Copyright © 2008 Elsevier Ltd.

estimated. The prepared PLLA/cellulose/nHAp nanocomposites were characterized by FTIR spectroscopy, SEM, DSC analyses, and compression test. The obtained results designated the occurrence of molecular interactions between all the constituents guiding to increment of the crystalline nature of the constituent polymers from 50% to 80%. The obtained results of compression testing designated that the reinforcement of microcrystalline cotton cellulose and nHAp increased the mechanical behavior of these PLLA/cellulose/nHAp nanocomposites. The increment of mechanical behavior might be because of the homogenous dispersion of the reinforcing materials in the PLLA template and the effectual interface between the constituent polymers and the reinforcement materials due to the pattern of the chemical bonding(s) between the functional moieties at the interface by using the coupling agent (sodium dodecyl sulfate). The overall results of the study indicated that these novel biomimetic and porous PLLA/cellulose/nHAp nanocomposites was comparable with the trabecular bone in terms of the composition, structural features, and mechanical behavior. Therefore, it can be an appropriate contender for bone repairing applications.

In a recent research by Mao et al. [164], novel porous and stable PLA/ethyl cellulose/HAp composite scaffolds were synthesized by a combination methodology of particulate leaching, higher concentration solvent casting, and room temperature compression molding approaches. These composite scaffolds were characterized for functional, structural, and mechanical characteristics. The characterization suggested that the porous and stable PLA/ethyl cellulose/HAp composite scaffolds containing 20 wt% of HAp demonstrated the optimal mechanical behavior as well as preferred porous structural features. The contact angle and the porosity of these composite scaffolds containing 20 wt% of HAp after 56 days were measured 45.13 ± 2.40 % and 84.28 ± 7.04%, respectively. The loss of weight and the compressive strength after 56 days were measured to be 4.77 ± 0.32% and 1.57 ± 0.09 MPa, respectively. These results convinced that the physiological requirements promote the bone tissue regeneration in the defective bone sites and can utilized as bone substitute materials for the use in bone tissue engineering.

1.4.2.2 Poly(lactic-co-glycolic acid) (PLGA)

The PLGA is a synthetic biodegradable polymer possessing a linear polymeric structure [154,155]. It is a copolymer of glycolic acid and lactic acid, which is synthesized by copolymerizing with glycolide monomers (i.e., glycolic acid) using various ratios of glycolic acid and lactic acid [165]. The degradation rate of PLGA depends on different issues like molecular weight, lactic acid/glycolic acid ratio, structure, and shape of the matrix. The PLGA copolymer accomplishes the recognition due to the approval from US FDA for human use and its good processability [154,165]. The solubility of pure PLA and polyglycolic acid are poor, whereas PLGA can be made more soluble in wide ranges of solvents that are commonly used, like tetrahydrofuran (THF), chlorinated solvents, ethyl acetate, acetone, etc. [165]. The physical properties on which PLGA depends are storage temperature, monomers, molecular weight, time of exposure to water, and lactic acid/glycolic acid ratio. The PLGA has been used in biomedical applications in different forms like porous scaffolds, films, microspheres, and hydrogels [154,155,165]. The current trend of the biomedical applications of PLGA has extensively been used in bone tissue regeneration and bone drug delivery because of its higher rate of biodegradability and physicochemical properties [165,166].

Jose et al. [167] developed aligned nanofibrous nanocomposite scaffolds made of PLGA and nHAp via electrospinning technique for bone regeneration. The morphology of these nanofibrous nanocomposite scaffolds was studied by SEM and the results suggested that the incorporation of various quantities of nHAp (1–20 wt%) enlarged the mean diameter of nanofibers from 300 nm (scaffolds of neat PLGA) to 700 nm (scaffolds of PLGA +10% nHAp). The agglomeration of nHAp was noted at the elevated concentrations, and also, a noticeable influence at the 20% nHAp within these PLGA/nHAp nanofibrous scaffolds whereby the occurrence of nHAp caused the breaking of fiber (Fig. 1.7). The thermal studies of these PLGA-based nanofibrous nanocomposites demonstrated faster processing of electrospinning locked in the amorphous characteristics of PLGA, which caused reduction in glass-transition temperature of the nanocomposite scaffold matrices. In addition, a rise in the glass-transition temperature was noticed with the rising nHAp quantity within the nanofibrous nanocomposite formula. The dynamic mechanical performance of these nanofibrous nanocomposite scaffolds showed that nHAp operated as the reinforcement material at the low concentrations (1%–5% nHAp). However, these nanocomposite scaffolds operated as shortcomings at the high concentrations (10%–20% nHAp). The storage modulus of these scaffolds was measured and was found to be increased at the low concentrations (1%–5% nHAp); on the other hand, further enhancement of the nHAp concentration led to reduction in the storage modulus of these scaffolds. And, for the PLGA-based nanofibrous nanocomposite scaffolds made using 20% nHAp, the storage modulus of 371 MPa was measured. The degradation properties of these PLGA-based nanofibrous nanocomposite scaffolds showed that the incorporated nHAp controlled the uptake of phosphate buffer saline as well as the reduction of mass. In addition, the mechanical performance of these nanocomposite scaffolds demonstrated a sinusoidal tendency with a small reduction in the storage modulus by the first week and a succeeding decline in the storage modulus by the sixth week because of the degradation.

Kim et al. [168] prepared a type of bone tissue regeneration nanocomposite scaffolds made of β-tricalcium phosphate and PLGA. These PLGA-based nanocomposite scaffold system with two specific architectural features were manufactured through a fused deposition modeling technique (i.e., a kind of extrusion freeform fabrication methodology). The depositions of microfilaments (at 0–90 degrees) were characterized as the "simple" type scaffolding architectural features of these PLGA-based nanocomposites, while those placed due to deposition at the alternating angles (between −45 degrees, 0 degrees, 45 degrees, and 90 degrees) were characterized as the "complex" type scaffolding architectural feature. In addition, a coat of HAp was placed onto both the simple- and complex-type scaffolding architectural scaffolds. The morphological structures of these rapid prototyped PLGA/β-tricalcium phosphate/HAp nanocomposite scaffold surfaces were evaluated before and after HAp coating by SEM analyses, which suggested uniform sharing of HAp coating onto nanocomposite scaffold surfaces. The in vivo implantations of PLGA/β-tricalcium phosphate/HAp nanocomposite scaffolds into the femoral bone defective model (unicortical) in rabbits were carried out. The implanted nanocomposite scaffolds and the host bone of the treated rabbits were collected after 6 and 12 weeks of in vivo implantations to perform histological observation studies. The obtained results of this investigation suggested that these

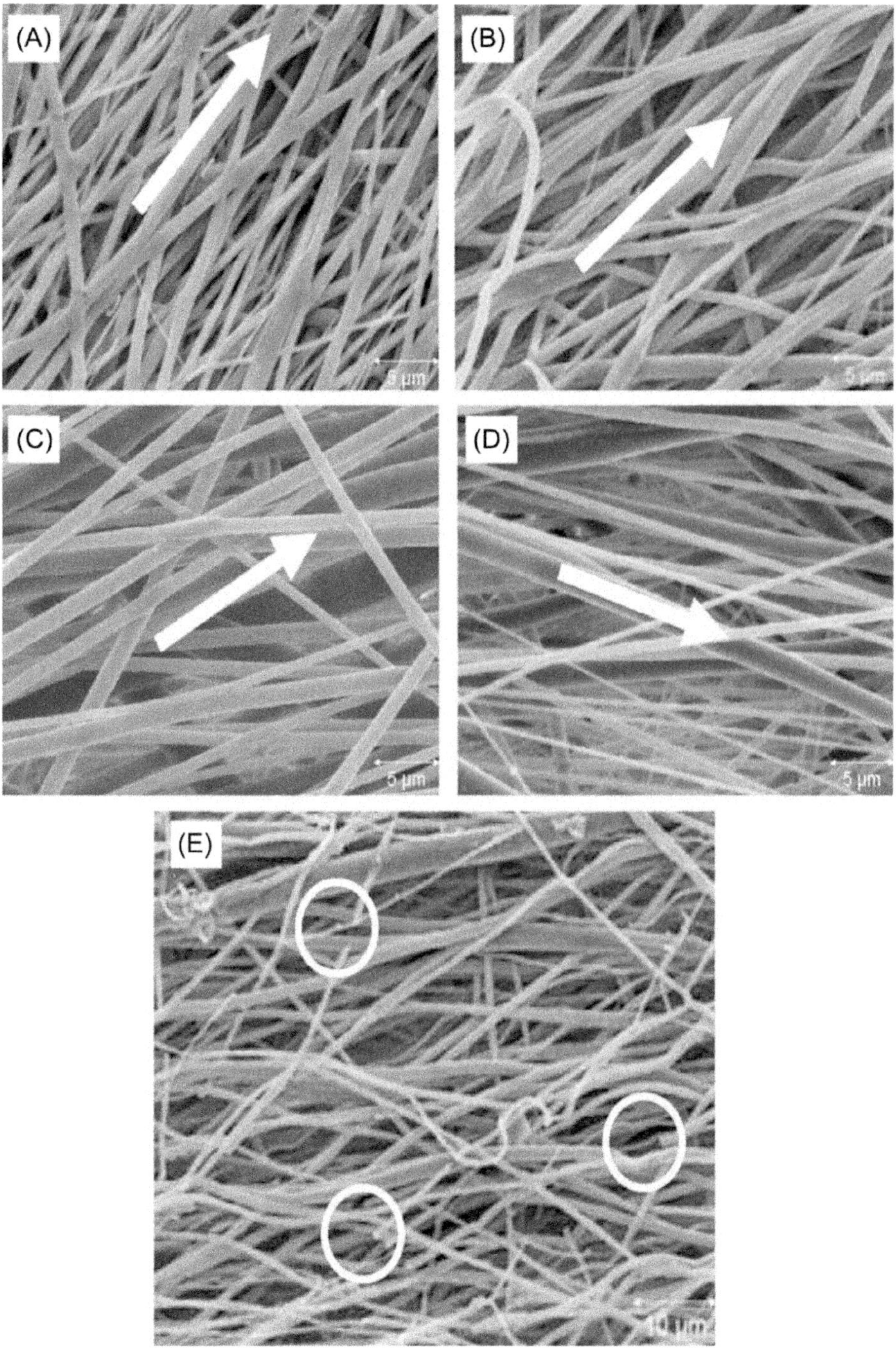

Fig. 1.7 SEM images of nanocomposite scaffolds: (A) neat PLGA, (B) PLGA +1% nHAp, (C) PLGA +5% nHAp, (D) PLGA +10% nHAp, and (E) PLGA +10% nHAp. The *arrows* designate the orientation direction and the *circles* signify the broken fibers.
From M.V. Jose, V. Thomas, K.T. Johnson, D.R. Dean, E. Nyairo, Aligned PLGA/HA nanofibrous nanocomposite scaffolds for bone tissue engineering, Acta Biomater. 5 (2009) 305–315. Copyright © 2008 Acta Materialia Inc.; Published by Elsevier Ltd.

rapid prototyped PLGA/β-tricalcium phosphate/HAp nanocomposite scaffolds were of biocompatible in nature and therefore, these PLGA-based nanocomposite scaffold system can be employed for bone tissue regeneration and also, for the delivery of different kinds of biological materials for bone tissue engineering applications.

Krucińska et al. [169] fabricated PLGA/HAp and PLGA/poly(hydroxyl butyrate) (PHB)/HAp nanofibrous nanocomposite scaffold as osteoconductive as well as osteoinductive biomaterials for the use in bone tissue regeneration applications. The nanofibrous nanocomposite structures were attained through PLGA electrospinning with HAp incorporation and a blend of PLGA with PHB with HAp incorporation. The biopolymers used in the experiment were synthesized by means of a novel technique using zirconium catalyst. To develop the nanofibrous nanocomposite scaffold, variable factors of the optimized electrospinning process were chosen. For the characterization of bone tissue regeneration nanofibrous nanocomposite scaffolds, physical, mechanical, functional, and thermal characteristics were estimated and analyzed.

1.4.2.3 Poly(propylene fumarate) (PPF)

The PPF is a synthetic biodegradable polyester biomaterial [170]. It possesses an unsaturated linear polymeric structure having double bonds internally, which permits chemical, thermal, and photo-initiated cross-linking. The average molecular weight of PPF ranges between 500 and 5000 g/mol. The PPF is degraded hydrolytically into nontoxic and biocompatible propylene glycol and fumaric acid [170]. The degradation rate and time of PPF depends on the polymer properties like molecular weight, cross-linker types, and densities [170,171]. The mechanical properties of PPF are enhanced significantly on the cross-linking and introduction of pores is done by inclusion of leachable salts. It is potentially used in biopotential grafting of bones [170]. The cross-linked PPF has been found appropriate for tissue engineering applications. Recently, various PPF-based scaffolds have found their prospective applications in bone tissue engineering applications [171,172].

Lee et al. [172] synthesized a series of cross-linkable PPF/HAp nanocomposites. These PPF/HAp nanocomposites with four different weight fractions of nHAp were evaluated for thermal characteristics as well as mechanical behaviors. To evaluate the surface chemistry of these PPF-based nanocomposites, the hydrophilicity was evaluated by measuring static contact angles. In addition, the ability of adsorbed proteins was assessed using MicroBCA protein assay kit after the incubation with fetal bovine serum (10%). The hydrophilicity as well as protein adsorption pattern on the nanocomposite surface was raised, significantly. The mechanical characteristics of these PPF/HAp nanocomposites were not found amplified primarily because of the higher modulus of the cross-linked PPF by the reinforcement of HAp nanoparticles to the PPF. In vitro cell culture experiment were carried out using mouse preosteoblast (MC3T3-E1) cell lines to examine the capability of PPF/HAp nanocomposites to support the cell adhesion and proliferations of cells after 1, 4, and 7 days of in vitro cell culture. After 4 days of cell culture, an improved cell adhesion and proliferations was noted. These PPF/HAp nanocomposites can be used for bone tissue replacement therapy considering the outstanding mechanical behavior and osteoconductive character.

Alge et al. [173] developed and evaluated PPF-reinforced dicalcium phosphate dihydrate (DCPD) cement-based three-dimensional macroporous composite scaffolds via the rapid prototype technology for the use in bone tissue engineering application. The characterization of PPF-reinforced DCPD cement composites suggested a considerable improvement in the mechanical behavior of cement composites with powder/liquid ratio of 1. Flexural modulus and flexural strength values of these cement composites were found to be enhanced in comparison with that of the nonreinforced control samples. In addition, as compared to the nonreinforced control samples, the values of maximum displacement during the mechanical behavior testing of these cement composites were found to be enhanced; whereas the work of fracture was found to be raised. The result of the compressive strength assessment indicated that the reinforcement of PPF augmented the scaffold strength. In vivo implantation of these PPF-reinforced DCPD cement-based three-dimensional macroporous composite scaffolds was tested by the calvarial defective model in rabbits. The in vivo implantation period of this experiment was 6 weeks. Even though the addition of mesenchymal stem cells to these PPF-based macroporous composite scaffolds did not considerably enhance the degree of tissue regeneration as many bone nodules with active osteoblast cells were detected within the pores of these macroporous composite scaffold, particularly in the marginal areas. All these results of this investigation propose that PPF-reinforced DCPD cement-based three-dimensional macroporous composite scaffolds can be used as promising bone tissue engineering composite scaffold biomaterial.

1.4.2.4 Poly(ε-caprolactone) (PCL)

The PCL is an aliphatic polyester and possesses a regular structure [174]. It has recently gained a lot of interest for the use as biomedical polymer owing to its higher degrees of toughness, good processability, cost-efficient prospective, etc. [174,175]. It is semicrystalline in nature. Its melting point is above the body temperature. The PCL takes very long time to degrade compared to various natural biopolymers. Since past few years, PCL is being used as scaffold in tissue engineering applications [155,174]. However, the use of pure PCL is not sufficient to meet the requirement of bone tissue engineering due to poor mechanical potential and less biodegradability [175]. These shortcomings of pure PCL force the researchers to modify the PCL for the use in tissue engineering. The modification of PCL leads to better improvement in its applications and could have a better broad spectrum in tissue engineering. Modified PCL have shown many applications and suitable scaffold cartilage, skin, bone etc. [175,176]. Recently, various modified PCL-based scaffold composites and nanocomposites have been investigated for the use in bone tissue regeneration applications [174–176].

In a research by Johari et al. [177], PCL/nano-fluoridated HAp nanocomposite scaffolds were developed via solvent casting-particulate leaching methodology. The nano-fluoridated HAp, which was used in the preparation of these PCL-based nanocomposite scaffolds, had chemical composition: $Ca_{10}(PO_4)_6OH_{2-x}F_x$ (where $x=0.5$, 1, 1.5, and 2.0). To prepare these nanocomposite scaffolds, different weight percentages (10%–40%) of the nano-fluoridated HAp were reinforced to the PCL matrix and NaCl

Table 1.2 Some examples of synthetic biodegradable polymer-based nanocomposites for bone tissue engineering applications

Synthetic biodegradable polymer-based nanocomposites for bone tissue engineering applications	References
Nanocomposite of PLA and surface grafted HAp	Hong et al. [178]
Dense nHAp/PLLA composites	Gay et al. [179]
nHAp/collagen/PLA composite as hierarchically biomimetic bone scaffold	Liao et al. [180]
Porous PLLA/apatite composites	Zhang and Ma [162]
PLA/calcium metaphosphate composite	Ung et al. [181]
PDLLA/TiO_2-bioglass foam composites	Boccaccini et al. [182]
PLGA/HAp composite scaffolds for delivery of BMP-2 plasmid DNA	Nie and Wang [183]
PLGA/HAp composite scaffold	Kim et al. [184]
PCL/TiO_2 nanocomposites	Tamjid et al. [185]
HAp/PCL nanocomposite	Rezaei and Mohammadi [186]
PCL-PDIPF-HAp composite scaffold	Fernández et al. [187]
3D PCL/HAp nanocomposite tissue scaffolds	Lauren et al. [188]
3D fiber-deposited PCL/iron-doped HAp nanocomposite magnetic scaffolds	De Santis et al. [189]
Magnetic PCL/iron-doped HAp nanocomposite substrates	Gloria et al. [190]
Poly(3-hydroxybutyrate)/nHAp composite scaffolds	Hayati et al. [191]

particles (300–500 mm diameter) was employed as porogen materials. The XRD and FTIR spectroscopy analyses were performed to characterize the phase configuration and interactions between excipient materials (if any) of the PCL-based nanocomposite scaffolds. The influence of fluorine contents on the mechanical behavior of these nanocomposites was studied. The mechanical behavior of these PCL-based nanocomposite scaffolds demonstrated that the measured compressive strength values increased with reducing the fluorine contents in nano-fluoridated HAp. In another research by Baykan et al. [176], biomimetic PCL/β-tricalcium phosphate multispiral scaffold was developed and evaluated. These multispiral scaffold was evaluated by in vitro and in vivo experiments. The results of these experiments suggested its prospective use in bone tissue engineering applications.

Some other examples of synthetic biodegradable polymer-based nanocomposites for bone tissue engineering applications are presented in Table 1.2.

1.5 Conclusion

The overall aim of the development and research of nanocomposite scaffolds was to generate the potential for the use in bone tissue engineering applications. For the successful applications of bone tissue engineering, the area of research, which is still clinically critical, require further research. There should be deep knowledge and understanding of molecular biology of bone and cartilage and their interaction with the extracellular matrix. In recent years, a variety of nanocomposites made of biodegradable polymers are being explored and exploited for the use in tissue engineering applications. Even, these biodegradable polymer matrix nanocomposites have been found effective for tissue generation in the field of bone tissue engineering. The technology of bone tissue engineering using various nanocomposite scaffolds has recently gained a greater importance as it benefits the people who have lost their organs or tissues accidently or due to some bone tissue affected diseases like osteomyelitis, osteoporosis, etc.

References

[1] A.K. Nayak, A.K. Dhara, Nanotechnology in drug delivery applications, Arch. Appl. Sci. Res. 2 (2010) 284–293.
[2] D. Pal, A.K. Nayak, Nanotechnology for targeted delivery in cancer therapeutics, Int. J. Pharm. Sci. Rev. Res. 1 (2010) 1–7.
[3] A.C. Balazs, T. Emrick, T.P. Russell, Nanoparticle polymer composites: where two small worlds meet, Science 314 (2006) 1107–1110.
[4] J. Malakar, A. Ghosh, A. Basu, A.K. Nayak, Nanotechnology: a promising carrier for intracellular drug delivery system, Int. Res. J. Pharm. 3 (2012) 36–40.
[5] Y. Zheng, J. Monty, R.J. Linhardt, Polysaccharide-based nanocomposites and their applications, Carbohydr. Res. 405 (2015) 23–32.
[6] S. Jana, A. Gangopadhaya, B.B. Bhowmik, A.K. Nayak, A. Mukhrjee, Pharmacokinetic evaluation of testosterone-loaded nanocapsules in rats, Int. J. Biol. Macromol. 72 (2015) 28–30.
[7] S. Jana, S. Manna, A.K. Nayak, K.K. Sen, S.K. Basu, Carbopol gel containing chitosan-egg albumin nanoparticles for transdermal aceclofenac delivery, Colloids Surf. B: Biointerfaces 114 (2014) 36–44.
[8] S. Jana, N. Maji, A.K. Nayak, K.K. Sen, S.K. Basu, Development of chitosan-based nanoparticles through inter-polymeric complexation for oral drug delivery, Carbohydr. Polym. 98 (2013) 870–876.
[9] J. Malakar, S.O. Sen, A.K. Nayak, K.K. Sen, Formulation, optimization and evaluation of transferosomal gel for transdermal insulin delivery, Saudi Pharm. J. 20 (2012) 355–363.
[10] B. Das, S.O. Sen, R. Maji, A.K. Nayak, K.K. Sen, Transferosomal gel for transdermal delivery of risperidone, J. Drug Deliv. Sci. Technol. 38 (2017) 59–71.
[11] P.H.C. Camargo, K.G. Satyanarayana, F. Wypych, Nanocomposites: synthesis, structure, properties and new application opportunities, Mater. Res. 12 (2009) 1–39.
[12] E.T. Thostenson, C. Li, T.W. Chou, Nanocomposites in context, Compos. Sci. Technol. 65 (2005) 491–516.
[13] P.M. Ajayan, Nanocomposite. Science and Technology, Wiley-VCH Verlag, Weinheim, 2000.
[14] C.J. Grande, F.G. Torres, C.M. Gomez, M.C. Bañó, Nanocomposites of bacterial cellulose/hydroxyapatite for biomedical applications, Acta Biomater. 5 (2009) 1605–1615.
[15] H. Ullah, F. Wahid, H.A. Santos, T. Khan, Advances in biomedical and pharmaceutical applications of functional bacterial cellulose-based nanocomposites, Carbohydr. Polym. 150 (2016) 330–352.

[16] H. Gleiter, Nanostructured materials: basic concepts and microstructure, Acta Mater. 48 (2000) 1–29.
[17] S.K. Swain, I. Gotman, R. Unger, E.Y. Gutmanas, Bioresorbable β-TCP-FeAg nanocomposites for load bearing bone implants: high pressure processing, properties and cell compatibility, Mater. Sci. Eng. C 78 (2017) 88–95.
[18] Y. Zhang, Y. Hu, Z. Luo, X. Shen, C. Mu, K. Cai, Simultaneous delivery of BMP-2 factor and anti-osteoporotic drugs using hyaluronan-assembled nanocomposite for synergistic regulation on the behaviors of osteoblasts and osteoclasts *in vitro*, J. Biomater. Sci. Polym. Ed. 26 (2015) 290–310.
[19] M.S. Hasnain, A.K. Nayak, R. Singh, F. Ahmad, Emerging trends of natural-based polymeric systems for drug delivery in tissue engineering applications, Sci. J. UBU 1 (2010) 1–13.
[20] R. Langer, J. Vacanti, Tissue engineering, Science 260 (1993) 920–926.
[21] A. Samadikuchaksaraei, Scientific and industrial status of tissue engineering, Afr. J. Biotechnol. 6 (2007) 2897–2909.
[22] B.P. Chan, K.W. Leong, Scaffolding in tissue engineering: general approaches and tissue-specific considerations, Eur. Spine J. 2008 (17) (2008) 467–479.
[23] M.K. Sah, K. Pramanik, Surface modification and characterisation of natural polymers for orthopaedic tissue engineering: a review, Int. J. Biomed. Eng. Technol. 9 (2012) 101–121.
[24] P. Pàmies, Tissue engineering: paper scaffolds, Nat. Mater. 15 (2016) 3.
[25] A.R. Amini, C.T. Laurencin, S.P. Nukavarapu, Bone tissue engineering: Recent advances and challenges, Crit. Rev. Biomed. Eng. 40 (2012) 363–408.
[26] C.T. Laurencin, A.M. Ambrosio, M.D. Borden, J.A. Cooper, Tissue engineering: orthopedic applications, Annu. Rev. Biomed. Eng. 1 (1999) 19–46.
[27] A.K. Nayak, D. Pal, Natural polysaccharides for drug delivery in tissue engineering, Everyman's Sci. XLVI (2012) 347–352.
[28] M. Slokolsky-Papkov, K. Agashi, A. Olaye, K. Shakesheff, A.J. Domb, Polymer carriers for drug delivery in tissue engineering, Adv. Drug Deliv. Rev. 59 (2007) 187–206.
[29] P.B. Malafaya, G.A. Silva, R.L. Reis, Natural-origin polymers as carriers and scaffolds for biomolecules and cell delivery in tissue engineering applications, Adv. Drug Deliv. Rev. 59 (2007) 207–233.
[30] J.F. Mano, G.A. Silva, H.S. Azevedo, P.B. Malafaya, R.A. Sousa, S.S. Silva, L.F. Boesel, J.M. Oliveira, T.C. Santos, A.P. Marques, N.M. Neves, R.L. Reis, Natural origin biodegradable systems in tissue engineering and regenerative medicine: present status and some moving trends, J. R. Soc. Interface 4 (2007) 999–1030.
[31] H. Shin, S. Jo, A.G. Mikos, Biomimetic materials for tissue engineering, Biomaterials 24 (2003) 4353–4364.
[32] B.J. Bolland, J.M. Kanczler, P.J. Ginty, S.M. Howdle, K.M. Shakesheff, D.G. Dunlop, R.O. Oreffo, The application of human bone marrow stromal cells and poly (dl-lactic acid) as a biological bone graft extender in impaction bone grafting, Biomaterials 29 (2008) 3221–3227.
[33] L. Fambri, C. Migliaresi, K. Kesenci, E. Piskin, Biodegradable polymers, in: Integrated Biomaterials Science, Kluwer Academic, Plenum, Berlin, 2002, pp. 119–187.
[34] M. Swetha, K. Sahithi, A. Moorthi, N. Srinivasan, K. Ramasamy, N. Selvamurugan, Biocomposites containing natural polymers and hydroxyapatite for bone tissue engineering, Int. J. Biol. Macromol. 47 (2010) 1–4.
[35] R.J. O'Keefe, J. Mao, Bone tissue engineering and regeneration: from discovery to the clinic-an overview, Tissue Eng. Part B Rev. 17 (2011) 389–392.

[36] B. Baroli, From natural bone grafts to tissue engineering therapeutics: brainstorming on pharmaceutical formulative requirements and challenges, J. Pharm. Sci. 98 (2009) 1317–1375.

[37] A.K. Nayak, D. Pal, Plant-derived polymers: ionically gelled sustained drug release systems, in: M. Mishra (Ed.), Encyclopedia of Biomedical Polymers and Polymeric Biomaterials, vol. VIII, Taylor & Francis Group, New York, NY, 2016, pp. 6002–6017.

[38] A.K. Nayak, H. Bera, M.S. Hasnain, D. Pal, Graft-copolymerization of plant polysaccharides, in: V.K. Thakur (Ed.), Biopolymer Grafting, Elsevier, Netherlands, 2018, pp. 1–62.

[39] D. Pal, A.K. Nayak, Plant polysaccharides-blended ionotropically-gelled alginate multiple-unit systems for sustained drug release, in: V.K. Thakur, M.K. Thakur, M.R. Kessler (Eds.), Handbook of Composites From Renewable Materials, Polymeric Composites, vol. 6, WILEY-Scrivener, USA, 2017, pp. 399–400.

[40] A.K. Nayak, D. Pal, Chitosan-based interpenetrating polymeric network systems for sustained drug release, in: A. Tiwari, H.K. Patra, J.-W. Choi (Eds.), Advanced Theranostics Materials, WILEY-Scrivener, USA, 2015, pp. 183–208.

[41] S. Jana, A. Saha, A.K. Nayak, K.K. Sen, S.K. Basu, Aceclofenac-loaded chitosan-tamarind seed polysaccharide interpenetrating polymeric network microparticles, Colloids Surf. B: Biointerfaces 105 (2013) 303–309.

[42] A. Verma, J. Dubey, N. Verma, A.K. Nayak, Chitosan-hydroxypropyl methylcellulose matrices as carriers for hydrodynamically balanced capsules of moxifloxacin HCl, Curr. Drug Deliv. 14 (2017) 83–90.

[43] M. Dash, F. Chiellini, R.M. Ottenbrite, E. Chiellini, Chitosan—a versatile semi-synthetic polymer in biomedical applications, Prog. Polym. Sci. 36 (2011) 981–1014.

[44] S.K.L. Levengood, M. Zhang, Chitosan-based scaffolds for bone tissue engineering, J. Mater. Chem. B 2 (2014) 3161.

[45] B.M. Chesnutt, A.M. Viano, Y. Yuan, Y. Yang, T. Guda, M.R. Appleford, J.L. Ong, W.O. Haggard, J.D. Bumgardner, Design and characterization of a novel chitosan/nanocrystalline calcium phosphate composite scaffold for bone regeneration, J. Biomed. Mater. Res. A 88A (2009) 491–502.

[46] L. Keller, A. Regiel-Futyra, M. Gimeno, S. Eap, G. Mendoza, V. Andreu, Q. Wagner, A. Kyzioł, V. Sebastian, G. Stochel, M. Arruebo, N. Benkirane-Jessel, Chitosan-based nanocomposites for the repair of bone defects, Nanomed. Nanotechnol. Biol. Med. 13 (2017) 2231–2240.

[47] M.A. Nazeer, E. Yilgör, I. Yilgör, Intercalated chitosan/hydroxyapatite nanocomposites: promising materials for bone tissue engineering applications, Carbohydr. Polym. 175 (2017) 38–46.

[48] M.R. Nikpour, S.M. Rabiee, M. Jahanshahi, Synthesis and characterization of hydroxyapatite/chitosan nanocomposite materials for medical engineering applications, Compos. Part B 43 (2012) 1881–1886.

[49] W.W. Thein-Han, R.D.K. Misra, Biomimetic chitosan–nanohydroxyapatite composite scaffolds for bone tissue engineering, Acta Biomater. 5 (2009) 1182–1197.

[50] C.H. Goh, P.W.S. Heng, L.W. Chan, Alginates as a useful natural polymer for microencapsulation and therapeutic applications, Carbohydr. Polym. 88 (2012) 1–12.

[51] D. Pal, A.K. Nayak, Alginates, blends and microspheres: controlled drug delivery, in: M. Mishra (Ed.), Encyclopedia of Biomedical Polymers and Polymeric Biomaterials, vol. I, Taylor & Francis Group, New York, NY, 2015, pp. 89–98.

[52] A. Jejurikar, X.T. Seow, G. Lawrie, D. Martin, A. Jayakrishnan, L. Grøndahl, Degradable alginate hydrogels crosslinked by the macromolecular crosslinker alginate dialdehyde, J. Mater. Chem. 22 (2012) 9751–9758.

[53] J. Malakar, A.K. Nayak, Theophylline release behavior for hard gelatin capsules containing various hydrophilic polymers, J. Pharm. Educ. Res. 3 (2012) 10–16.

[54] J. Malakar, K. Das, A.K. Nayak, Statistical optimization in formulation development of *in situ* cross-linked matrix tablets for sustained salbutamol sulfate release, Polym. Med. 44 (2014) 221–230.

[55] P.R. Guru, A.K. Nayak, R.K. Sahu, Oil-entrapped sterculia gum-alginate buoyant systems of aceclofenac: development and *in vitro* evaluation, Colloids Surf. B: Biointerfaces 104 (2013) 268–275.

[56] M.S. Hasnain, A.K. Nayak, M. Singh, M. Tabish, M.T. Ansari, T.J. Ara, Alginate-based bipolymeric-nanobioceramic composite matrices for sustained drug release, Int. J. Biol. Macromol. 83 (2016) 71–77.

[57] A.K. Nayak, D. Pal, K. Santra, *Plantago ovata* F. mucilage-alginate mucoadhesive beads for controlled release of glibenclamide: development, optimization, and *in vitro-in vivo* evaluation, J. Pharm. 2013 (2013) 151035.

[58] A.K. Nayak, B. Das, R. Maji, Calcium alginate/gum Arabic beads containing glibenclamide: development and *in vitro* characterization, Int. J. Biol. Macromol. 51 (2012) 1070–1078.

[59] J. Malakar, A.K. Nayak, A. Das, Modified starch (cationized)-alginate beads containing aceclofenac: formulation optimization using central composite design, Starch/Stärke 65 (2013) 603–612.

[60] J. Malakar, A.K. Nayak, P. Jana, D. Pal, Potato starch-blended alginate beads for prolonged release of tolbutamide: development by statistical optimization and *in vitro* characterization, Asian J. Pharm. 7 (2013) 43–51.

[61] J. Malakar, P. Dutta, S.D. Purokayastha, S. Dey, A.K. Nayak, Floating capsules containing alginate-based beads of salbutamol sulfate: *in vitro-in vivo* evaluations, Int. J. Biol. Macromol. 64 (2014) 181–189.

[62] A.K. Nayak, D. Pal, M.S. Hasnain, Development, optimization and *in vitro-in vivo* evaluation of pioglitazone-loaded jackfruit seed starch-alginate beads, Curr. Drug Deliv. 10 (2013) 608–619.

[63] J. Malakar, A.K. Nayak, D. Pal, Development of cloxacillin loaded multiple-unit alginate-based floating system by emulsion–gelation method, Int. J. Biol. Macromol. 50 (2012) 138–147.

[64] H. Bera, S.G. Kandukuri, A.K. Nayak, S. Boddupalli, Alginate-sterculia gum gel-coated oil-entrapped alginate beads for gastroretentive risperidone delivery, Carbohydr. Polym. 120 (2015) 74–84.

[65] H. Bera, S. Boddupalli, S. Nandikonda, S. Kumar, A.K. Nayak, Alginate gel-coated oil-entrapped alginate–tamarind gum–magnesium stearate buoyant beads of risperidone, Int. J. Biol. Macromol. 78 (2015) 102–111.

[66] A.K. Nayak, D. Pal, K. Santra, Swelling and drug release behavior of metformin HCl-loaded tamarind seed polysaccharide-alginate beads, Int. J. Biol. Macromol. 82 (2016) 1023–1027.

[67] A.K. Nayak, D. Pal, Ionotropically-gelled mucoadhesive beads for oral metformin HCl delivery: formulation, optimization and antidiabetic evaluation, J. Sci. Ind. Res. 72 (2013) 15–22.

[68] A.K. Nayak, D. Pal, Formulation optimization of jackfruit seed starch-alginate mucoadhesive beads of metformin HCl, Int. J. Biol. Macromol. 59 (2013) 264–272.

[69] A.K. Nayak, D. Pal, J. Malakar, Development, optimization and evaluation of emulsion-gelled floating beads using natural polysaccharide-blend for controlled drug release, Polym. Eng. Sci. 53 (2013) 338–350.

[70] A.K. Nayak, D. Pal, J. Pradhan, M.S. Hasnain, Fenugreek seed mucilage-alginate mucoadhesive beads of metformin HCl: design, optimization and evaluation, Int. J. Biol. Macromol. 54 (2013) 144–154.

[71] J. Venkatesan, I. Bhatnagar, P. Manivasagan, K.-H. Kang, S.-K. Kim, Alginate composites for bone tissue engineering: a review, Int. J. Biol. Macromol. 72 (2015) 269–281.

[72] J.S. Yang, S.B. Chen, Y. Fang, Viscosity study of interactions between sodium alginate and CTAB in dilute solutions at different pH values, Carbohydr. Polym. 75 (2009) 333–337.

[73] B. Das, S. Dutta, A.K. Nayak, U. Nanda, Zinc alginate-carboxymethyl cashew gum microbeads for prolonged drug release: development and optimization, Int. J. Biol. Macromol. 70 (2014) 505–515.

[74] P. Sinha, U. Ubaidulla, A.K. Nayak, Okra (*Hibiscus esculentus*) gum-alginate blend mucoadhesive beads for controlled glibenclamide release, Int. J. Biol. Macromol. 72 (2015) 1069–1075.

[75] P. Sinha, U. Ubaidulla, M.S. Hasnain, A.K. Nayak, B. Rama, Alginate-okra gum blend beads of diclofenac sodium from aqueous template using $ZnSO_4$ as a cross-linker, Int. J. Biol. Macromol. 79 (2015) 555–563.

[76] M. Liu, L. Dai, H. Shi, S. Xiong, C. Zhou, *In vitro* evaluation of alginate/halloysite nanotube composite scaffolds for tissue engineering, Mater. Sci. Eng. C 49 (2015) 700–712.

[77] T. Chae, H. Yang, V. Leung, F. Ko, T. Troczynski, Novel biomimetic hydroxyapatite/alginate nanocomposite fibrous scaffolds for bone tissue regeneration, J. Mater. Sci. Mater. Med. 24 (2013) 1885–1894.

[78] K. Sangeetha, A. Thamizhavel, E.K. Girija, Effect of gelatin on the *in situ* formation of alginate/hydroxyapatite nanocomposite, Mater. Lett. 91 (2013) 27–30.

[79] M. Kawaguchi, T. Fukushima, T. Hayakawa, N. Nakashima, Y. Inoue, S. Takeda, K. Okamura, K. Taniguchi, Preparation of carbon nanotube-alginate nanocomposite gel for tissue engineering, Dental Mater. J. 25 (2006) 719–725.

[80] A.K. Nayak, D. Pal, Natural starches-blended ionotropically-gelled microparticles/beads for sustained drug release, in: V.K. Thakur, M.K. Thakur, M.R. Kessler (Eds.), Handbook of Composites From Renewable Materials, Nanocomposites: Advanced Applications, vol. 8, WILEY-Scrivener, USA, 2017, pp. 527–560.

[81] H.F. Zobel, Molecules to granules: a comprehensive starch review, Starch/Stärke 40 (1988) 44–50.

[82] P.F. Builders, M.I. Arhewoh, Pharmaceutical applications of native starch in conventional drug delivery, Starch/Stärke 68 (2016) 864–873.

[83] A.K. Nayak, D. Pal, Blends of jackfruit seed starch-pectin in the development of mucoadhesive beads containing metformin HCl, Int. J. Biol. Macromol. 62 (2013) 137–145.

[84] N. Biswas, R.K. Sahoo, Tapioca starch blended alginate mucoadhesive-floating beads for intragastric delivery of metoprolol tartrate, Int. J. Biol. Macromol. 83 (2016) 61–70.

[85] A.K. Nayak, D. Pal, K. Santra, *Artocarpus heterophyllus* L. seed starch-blended gellan gum mucoadhesive beads of metformin HCl, Int. J. Biol. Macromol. 65 (2014) 329–339.

[86] Y. Chung, S. Ansari, L. Estevez, S. Hayrapetyan, E.P. Giannelis, H. Lai, Preparation and properties of biodegradable starch–clay nanocomposites, Carbohydr. Polym. 79 (2010) 391–396.

[87] S. Bala Subramanian, A.P. Francis, T. Devasena, Chitosan-starch nanocomposite particles as a drug carrier for the delivery of bis-desmethoxy curcumin analog, Carbohydr. Polym. 114 (2014) 170–178.

[88] F.G. Torres, S. Commeaux, O.P. Troncoso, Starch-based biomaterials for wound-dressing applications, Starch/Starke 65 (2013) 543–551.

[89] L.M. Famá, V. Pettarin, S.N. Goyanes, C.R. Bernal, Starch/multi-walled carbon nanotubes composites with improved mechanical properties, Carbohydr. Polym. 83 (2011) 1226–1231.
[90] M. Meskinfam, M.A.S. Sadjadi, H. Jazdarreh, K. Zare, Biocompatibility evaluation of nano hydroxyapatite-starch biocomposites, J. Biomed. Nanotechnol. 7 (2011) 455–459.
[91] M. Sadjadi, M. Meskinfam, H. Jazdarreh, Hydroxyapatite-starch nano biocomposites synthesis and characterization, Int. J. Nano Dimens. 1 (2010) 57–63.
[92] A.I. Raafat, A.A. Saad Eldin, A.A. Salama, N.S. Ali, Characterization and bioactivity evaluation of (starch/*N*-vinylpyrrolidone)—hydroxyapatite nanocomposite hydrogels for bone tissue regeneration, J. Appl. Polym. Sci. 128 (2013) 1697–1705.
[93] Y. Raymond, Cellulose Structure Modification and Hydrolysis, Wiley, New York, 1986.
[94] K. Dieter, H. Brigitte, F. Hans-Peter, B. Andreas, Cellulose: Fascinating biopolymer and sustainable raw material, Angew. Chem. Int. Ed. 44 (B) (2005) 3358–3393.
[95] S.J. Eichhorn, A. Dufresne, M. Aranguren, et al., Review: current international research into cellulose nanofibres and nanocomposites, J. Mater. Sci. 45 (2010) 1–33.
[96] S. Saber-Samandari, S. Saber-Samandari, M. Gazi, F.C. Cebeci, E. Talasaz, Synthesis, characterization and application of cellulose based nano-biocomposite hydrogels, J. Macromol. Sci. Pure Appl. Chem. 50 (2013) 1133–1141.
[97] A. Aravamudhan, D.M. Ramos, J. Nip, M.D. Harmon, R. James, M. Deng, C.T. Laurencin, X. Yu, S.G. Kumbar, Cellulose and collagen derived micro-nano structured scaffolds for bone tissue engineering, J. Biomed. Nanotechnol. 9 (2013) 719–731.
[98] K. Novotna, P. Havelka, T. Sopuch, K. Kolarova, V. Vosmanska, V. Lisa, V. Svorcik, L. Bacakova, Cellulose-based materials as scaffolds for tissue engineering, Cellulose 20 (2013) 2263–2278.
[99] H.G. de Oliveira Barud, R.R. da Silva, H. da Silva Barud, A. Tercjak, J. Gutierrez, W.R. Lustri, O.B. de Oliveira, S.J. Ribeiro, A multipurpose natural and renewable polymer in medical applications: bacterial cellulose, Carbohydr. Polym. 153 (2016) 406–420.
[100] S. Saber-Samandari, S. Saber-Samandari, S. Kiyazar, J. Aghazadeh, A. Sadeghi, *In vitro* evaluation for apatite-forming ability of cellulose-based nanocomposite scaffolds for bone tissue engineering, Int. J. Biol. Macromol. 86 (2016) 434–442.
[101] X. He, Q. Xiao, C. Lu, Y. Wang, X. Zhang, J. Zhao, W. Zhang, X. Zhang, Y. Deng, Uniaxially aligned electrospun all-cellulose nanocomposite nanofibers reinforced with cellulose nanocrystals: scaffold for tissue engineering, Biomacromolecules 15 (2014) 618–627.
[102] P. Fratzl, Collagen: Structure and Mechanics, Springer, New York, 2008.
[103] M.J. Mienaltowski, D. Birk, Structure, physiology, and biochemistry of collagens, Adv. Exp. Med. Biol. 802 (2014) 5–29.
[104] S. Meghezi, D.G. Seifu, N. Bono, L. Unsworth, K. Mequanint, D. Mantovani, Engineering 3D cellularized collagen gels for vascular tissue regeneration, J. Vis. Exp. 100 (2015) 1–12.
[105] S. Pina, J.M. Oliveira, R.L. Reis, Natural-based nanocomposites for bone tissue engineering and regenerative medicine: a review, Adv. Mater. 27 (2015) 1143–1169.
[106] G. Cunniffe, F. O'Brien, Collagen scaffolds for orthopedic regenerative medicine, J. Miner. Met. Mater. Soc. 63 (2011) 66–73.
[107] B. Sarker, J. Hum, S.N. Nazhat, A.R. Boccaccini, Combining collagen and bioactive glasses for bone tissue engineering: a review, Adv. Health Mater. 4 (2015) 176–194.
[108] Y.S. Pek, S. Gao, M.S. Mohamed Arshad, K.-J. Leck, J.Y. Ying, Porous collagen-apatite nanocomposite foams as bone regeneration scaffolds, Biomaterials 29 (2008) 4300–4305.

[109] M. Kikuchi, S. Itoh, S. Ichinose, K. Shinomiya, J. Tanaka, Self-organization mechanism in a bone-like hydroxyapatite/collagen nanocomposite synthesized *in vitro* and its biological reaction *in vivo*, Biomaterials 22 (2001) 1705–1711.
[110] C. Zou, W. Weng, X. Deng, K. Cheng, X. Liu, P. Du, G. Shen, G. Han, Preparation and characterization of porous b-tricalcium phosphate/collagen composites with an integrated structure, Biomaterials 26 (2005) 5276–5284.
[111] K.B. Djagnya, Z. Wang, S. Xu, Gelatin: a valuable protein for food and pharmaceutical industries: review, Crit. Rev. Food Sci. Nutr. 41 (2010) 481–492.
[112] M. Azami, S. Tavakol, A. Sanadikuchaksarae, M.S. Hashjin, N. Baheiraei, M. Kamali, M.R. Nourani, A porous hydroxyapatite/gelatin nanocomposite scaffold for bone tissue repair: *in vitro* and *in vivo* evaluation, J. Biomater. Sci. Polym. Ed. 23 (2012) 2353–2368.
[113] K.Y. Chen, C.H. Yao, Repair of bone defects with gelatin-based composites: a review, BioMed 1 (2011) 29–32.
[114] X. Liu, L.A. Smith, J. Hu, P.X. Ma, Biomimetic nanofibrous gelatin/apatite composite scaffolds for bone tissue engineering, Biomaterials 30 (2009) 2252–2258.
[115] B. Johari, M. Kadivar, S. Lak, M. Gholipourmalekabadi, A.M. Urbanska, M. Mozafari, M. Ahmadzadehzarajabad, A. Azarnezhad, S. Afshari, J. Zargan, S. Kargozar, Osteoblast-seeded bioglass/gelatin nanocomposite: a promising bone substitute in critical-size calvarial defect repair in rat, Int. J. Artif. Org. 39 (2016) 524–533.
[116] A. Samadikuchaksaraei, M. Gholipourmalekabadi, E.E. Ezadyar, M. Azami, M. Mozafari, B. Johari, S. Kargozar, S.B. Jameie, A. Korourian, A.M. Seifalian, Fabrication and in vivo evaluation of an osteoblast-conditioned nano-hydroxyapatite/gelatin composite scaffold for bone tissue regeneration, J. Biomed. Mater. Res. A 104 (2016) 2001–2010.
[117] C.E. Schanté, G. Zuber, C. Herlin, T.F. Vandamme, Improvement of hyaluronic acid enzymatic stability by the grafting of amino-acids, Carbohydr. Polym. 87 (2012) 2211–2216.
[118] M.S. Bae, J.-Y. Ohe, J.B. Lee, D.N. Heo, W. Byun, H. Bae, Y.-D. Kwon, I.K. Kwon, Photo-cured hyaluronic acid-based hydrogels containing growth and differentiation factor 5 (GDF-5) for bone tissue regeneration, Bone 59 (2014) 189–198.
[119] M.S. Bae, D.H. Yang, J.B. Lee, D.N. Heo, Y.-D. Kwon, I.C. Youn, K. Choi, J.H. Hong, G.T. Kim, Y.S. Choi, Photo-cured hyaluronic acid-based hydrogels containing simvastatin as a bone tissue regeneration scaffold, Biomaterials 32 (2011) 8161–8171.
[120] M.N. Collins, C. Birkinshaw, Hyaluronic acid based scaffolds for tissue engineering—a review, Carbohydr. Polym. 92 (2013) 1262–1279.
[121] S. Subramaniam, Y.-H. Fang, S. Sivasubramanian, F.-H. Lin, C. Lin, Hydroxyapatite-calcium sulfate-hyaluronic acid composite encapsulated with collagenase as bone substitute for alveolar bone regeneration, Biomaterials 74 (2016) 99–108.
[122] Y. Huang, X. Zhang, A. Wua, H. Xu, An injectable nano-hydroxyapatite (n-HA)/glycol chitosan (G-CS)/hyaluronic acid (HyA) composite hydrogel for bone tissue engineering, RSC Adv. 6 (2016) 33529–33536.
[123] J.C. Fricain, S. Schlaubitz, C. Le Visage, I. Arnault, S.M. Derkaoui, R. Siadous, S. Catros, C. Lalande, R. Bareille, M. Renard, A nano-hydroxyapatite–pullulan/dextran polysaccharide composite macroporous material for bone tissue engineering, Biomaterials 34 (2013) 2947–2959.
[124] L. Kong, Y. Gao, W. Cao, Y. Gong, N. Zhao, X. Zhang, Preparation and characterization of nano-hydroxyapatite/chitosan composite scaffolds, J. Biomed. Mater. Res. A 75 (2005) 275–282.
[125] B. Lowe, J. Venkatesan, S. Anil, M.S. Shim, S.-K. Kim, Preparation and characterization of chitosan-natural nanohydroxyapatite-fucoidan nanocomposites for bone tissue engineering, Int. J. Biol. Macromol. 93 ((2016) 1479–1487.

[126] K. Katti, D. Katti, R. Dash, Synthesis and characterization of a novel chitosan/montmorillonite/hydroxyapatite nanocomposite for bone tissue engineering, Biomed. Mater. 3 (2008) 4122.
[127] Q. Hu, B. Li, M. Wang, J. Shen, Preparation and characterization of biodegradable chitosan/hydroxyapatite nanocomposite rods via *in situ* hybridization: a potential material as internal fixation of bone fracture, Biomaterials 25 (2004) 779–785.
[128] I.M. El-Sherbiny, S. Yahia, M.A. Messiery, F.M. Reicha, Preparation and physicochemical characterization of new nanocomposites based on β-type chitosan and nanohydroxyapatite as potential bone substitute materials, Int. J. Polym. Mater. 63 (2014) 213–220.
[129] Y.P. Guo, J.J. Guan, J. Yang, Y. Wang, C.Q. Zhang, Q.F. Ke, Hybrid nanostructured hydroxyapatite-chitosan composite scaffold: bioinspired fabrication, mechanical properties and biological properties, J. Mater. Chem. B 3 (2015) 4679–4689.
[130] J. Venkatesan, Z.J. Qiań, B. Ryu, N. Ashok Kumar, S.K. Kim, Preparation and characterization of carbon nanotube-grafted-chitosan—natural hydroxyapatite composite for bone tissue engineering, Carbohydr. Polym. 83 (2011) 569–577.
[131] A. Bhowmick, N. Pramanik, P.J. Manna, T. Mitra, T.K.R. Selvaraj, A. Gnanamani, M. Das, P.P. Kundu, Development of porous and antimicrobial CTS-PEG-HAP-ZnO nano-composites for bone tissue engineering, RSC Adv. 5 (2015) 99385–99393.
[132] M. Shakir, R. Jolly, M.S. Khan, A. Rauf, S. Kazmi, Nano-hydroxyapatite/beta-CD/chitosan nanocomposite for potential applications in bone tissue engineering, Int. J. Biol. Macromol. 93 (2016) 276–289.
[133] S. Dinescu, M. Ionita, A.M. Pandele, B. Galateanu, H. Iovu, A. Ardelean, M. Costache, A. Hermenean, *In vitro* cytocompatibility evaluation of chitosan/graphene oxide 3D scaffold composites designed for bone tissue engineering, Biomed. Mater. Eng. 24 (2014) 2249–2256.
[134] A. Bhowmick, S.L. Banerjee, N. Pramanik, P. Jana, T. Mitra, A. Gnanamani, M. Das, P.P. Kundu, Organically modified clay supported chitosan/hydroxyapatite-zinc oxide nanocomposites with enhanced mechanical and biological properties for the application in bone tissue engineering, Int. J. Biol. Macromol. 106 (2018) 11–19.
[135] L. Jiang, Y. Li, X. Wang, L. Zhang, J. Wen, M. Gong, Preparation and properties of nano-hydroxyapatite/chitosan/carboxymethyl cellulose composite scaffold, Carbohydr. Polym. 74 (2008) 680–684.
[136] S. Saravanan, S. Nethala, S. Pattnaik, A. Tripathi, A. Moorthi, N. Selvamurugan, Preparation, characterization and antimicrobial activity of a bio-composite scaffold containing chitosan/nano-hydroxyapatite/nano-silver for bone tissue engineering, Int. J. Biol. Macromol. 49 (2011) 188–193.
[137] A. Tripathi, S. Saravanan, S. Pattnaik, A. Moorthi, N.C. Partridge, N. Selvamurugan, Bio-composite scaffolds containing chitosan/nano-hydroxyapatite/nano-copper-zinc for bone tissue engineering, Int. J. Biol. Macromol. 50 (2011) 294–299.
[138] M. Shakir, R. Jolly, M.S. Khan, N. Iram, H.M. Khan, Nano-hydroxyapatite/chitosan-starch nanocomposite as a novel boneconstruct: synthesis and *in vitro* studies, Int. J. Biol. Macromol. 80 (2015) 282–292.
[139] J.A. Sowjanya, J. Singh, T. Mohita, S. Sarvanan, A. Moorthi, N. Srinivasan, N. Selvamurugan, Biocomposite scaffolds containing chitosan/alginate/nano-silica for bone tissue engineering, Colloids Surf. B: Biointerfaces 109 (2013) 294–300.
[140] J. Guan, J. Yang, J. Dai, Y. Qin, Y. Wang, Y. Guo, Q. Ke, C. Zhang, Bioinspired nanostructured hydroxyapatite/collagen three-dimensional porous scaffolds for bone tissue engineering, RSC Adv. 5 (2015) 36175–36184.

[141] S. Liu, Y. Sun, Y. Fu, D. Chang, C. Fu, G. Wang, Y. Liu, F.R. Tay, Y. Zhou, Bioinspired collagen-apatite nanocomposites for bone regeneration, J. Endod. 42 (2016) 1226–1232.
[142] N. Bouropoulos, A. Stampolakis, D.E. Mouzakis, Dynamic mechanical properties of calcium alginate-hydroxyapatite nanocomposite hydrogels, Sci. Adv. Mater. 2 (2009) 239–242.
[143] F. Brun, G. Turco, A. Accardo, S. Paoletti, Automated quantitative characterization of alginate/hydroxyapatite bone tissue engineering scaffolds by means of micro-CT image analysis, J. Mater. Sci. 22 (2011) 2617–2629.
[144] E.D. Yildirim, X. Yin, K. Nair, W. Sun, Fabrication, characterization, and biocompatibility of single-walled carbon nanotube-reinforced alginate composite scaffolds manufactured using free form fabrication technique, J. Biomed. Mater. Res. B Appl. Biomater. 87B (2008) 406–414.
[145] H.-R. Lin, Y.-J. Yeh, Porous alginate/hydroxyapatite composite scaffolds for bone tissue engineering: preparation, characterization, and *in vitro* studies, J. Biomed. Mater. Res. B Appl. Biomater. 71B (2004) 52–65.
[146] T. Matsuno, Y. Hashimoto, S. Adachi, K. Omata, Y. Yoshitaka, Y. Ozeki, Y. Umezu, Y. Tabata, M. Nakamura, T. Satoh, Preparation of injectable 3D-formed β-tricalcium phosphate bead/alginate composite for bone tissue engineering, Dental Mater. J. 27 (2008) 827–834.
[147] C. Sharma, A.K. Dinda, P.D. Potdar, C.F. Chou, N.C. Mishra, Fabrication and characterization of novel nano-biocomposite scaffold of chitosan-gelatin-alginate-hydroxyapatite for bone tissue engineering, Mater. Sci. Eng. C Mater. Biol. Appl. 64 (2016) 416–427.
[148] F. Beladi, S. Saber-Samandari, S. Saber-Samandari, Cellular compatibility of nanocomposite scaffolds based on hydroxyapatite entrapped in cellulose network for bone repair, Mater. Sci. Eng. C 75 (2017) 385–392.
[149] S. Garai, A. Sinha, Biomimetic nanocomposites of carboxymethyl cellulose-hydroxyapatite: novel three dimensional load bearing bone grafts, Colloids Surf. B: Biointerfaces 115 (2014) 182–190.
[150] S. Saska, H.S. Barud, A.M. Gaspar, R. Marchetto, S.J. Ribeiro, Y. Messaddeq, Bacterial cellulose-hydroxyapatite nanocomposites for bone regeneration, Int. J. Biomater. 2011 (2011) 175362.
[151] L. Bakhtiari, H.R. Rezaie, S.M. Hosseinalipour, M.A. Shokrgozar, Investigation of biphasic calcium phosphate/gelatin nanocomposite scaffolds as a bone tissue engineering, Ceram. Int. 36 (2010) 2421–2426.
[152] L. Wu, F. Zhu, G. Tao, *In-vitro* biocompatibility evaluation of collagen-hyaluronic acid/bioactive glass nanocomposite scaffold, J. Macromol. Sci., Part A 50 (2013) 1121–1125.
[153] S. Schlaubitz, S.M. Derkaoui, L. Marosa, et al., Pullulan/dextran/nHA macroporous composite beads for bone repair in a femoral condyle defect in rats, PLoS ONE 9 (2014) e110251.
[154] L.S. Nair, C.T. Laurencin, Biodegradable polymers as biomaterials, Prog. Polym. Sci. 32 (2007) 762–798.
[155] M. Okamoto, B. John, Synthetic biopolymer nanocomposites for tissue engineering scaffolds, Prog. Polym. Sci. 38 (2013) 1487–1503.
[156] D. Garlotta, A literature review of poly (lactic acid), J. Polym. Environ. 9 (2001) 63–84.
[157] J. Goswami, N. Bhatnagar, S. Mohanty, A.K. Ghosh, Processing *and* characterization *of* poly (lactic acid) based bioactive composites for biomedical scaffold application, Express Polym. Lett. 7 (2013) 767–777.
[158] K.M. Nampoothiri, N.R. Nair, R.P. John, An overview of the recent developments in polylactide (PLA) research, Bioresour. Technol. 101 (2010) 8493–8501.

[159] B. Gupta, N. Revagade, J. Hilborn, Poly (lactic acid) fiber: an overview, Prog. Polym. Sci. 32 (2007) 455–482.
[160] M.S. Lopes, A. Jardini, R. Maciel Filho, Poly (lactic acid) production for tissue engineering applications, Proced. Eng. 42 (2012) 1402–1413.
[161] E. Nejati, H. Mirzadeh, M. Zandi, Synthesis and characterization of nano-hydroxyapatite rods/poly(L-lactide acid) composite scaffolds for bone tissue engineering, Compos. Part A 39 (2008) 1589–1596.
[162] R. Zhang, P.X. Ma, Porous poly(l-lactic acid)/apatite composites created by biomimetic process, J. Biomed. Mater. Res. 45 (1999) 285–293.
[163] S. Eftekhari, I.E. Sawi, Z.S. Bagheri, G. Turcotte, H. Bougherara, Fabrication and characterization of novel biomimetic PLLA/cellulose/hydroxyapatite nanocomposite for bone repair applications, Mater. Sci. Eng. C 39 (2014) 120–125.
[164] D. Mao, Q. Li, N. Bai, H. Dong, D. Li, Porous stable poly(lactic acid)/ethyl cellulose/hydroxyapatite composite scaffolds prepared by a combined method for bone regeneration, Carbohydr. Polym. 180 (2018) 104–111.
[165] H.K. Makadia, S.J. Siegel, Poly lactic-co-glycolic acid (PLGA) as biodegradable controlled drug delivery carrier, Polymers 3 (2011) 1377–1397.
[166] Z. Meng, H. Li, Z. Sun, W. Zheng, Y. Zheng, Fabrication of mineralized electrospun PLGA and PLGA/gelatin nanofibers and their potential in bone tissue engineering, Mater. Sci. Eng. C 33 (2013) 699–706.
[167] M.V. Jose, V. Thomas, K.T. Johnson, D.R. Dean, E. Nyairo, Aligned PLGA/HA nanofibrous nanocomposite scaffolds for bone tissue engineering, Acta Biomater. 5 (2009) 305–315.
[168] J. Kim, S. McBride, B. Tellis, P. Alvarez-Urena, Y.H. Song, D.D. Dean, V.L. Sylvia, H. Elgendy, J. Ong, J.O. Hollinger, Rapid-prototyped PLGA/β-TCP/hydroxyapatite nanocomposite scaffolds in a rabbit femoral defect model, Biofabrication 4 (2012) 025003.
[169] I. Krucińska, O. Chrzanowska, M. Boguń, M. Kowalczuk, P. Dobrzyński, Fabrication of PLGA/HAp and PLGA/PHB/HAp fibrous nanocomposite materials for osseous tissue regeneration, AUTEX Res. J. 14 (2014) 95–110.
[170] S. Wang, L. Lu, M.J. Yaszemski, Bone tissue-engineering material poly(propylene fumarate): correlation between molecular weight, chain dimensions, and physical properties, Biomacromolecules 7 (2006) 1976–1982.
[171] R. Mishra, B. Roux, M. Posukonis, E. Bodamer, E.M. Brey, J.P. Fisher, D. Dean, Effect of prevascularization on in vivo vascularization of poly(propylene fumarate)/fibrin scaffolds, Biomaterials 77 (2016) 255–266.
[172] K.-W. Lee, S. Wang, M.J. Yaszemski, L. Lu, Physical properties and cellular responses to crosslinkable poly(propylene fumarate)/hydroxyapatite nanocomposites, Biomaterials 29 (2008) 2839–2848.
[173] D.L. Alge, J. Bennet, T. Treasure, S. Voytik-Harbin, W.S. Goebel, T.G. Chu, Poly(propylene fumarate) reinforced dicalcium phosphate dihydrate cement composites for bone tissue engineering, J. Biomed. Mater. Res. A 100 (2012) 1792–1802.
[174] M. Abedalwafa, F. Wang, L. Wang, C. Li, Biodegradable poly-epsilon-caprolactone (PCL) for tissue engineering applications: a review, Rev. Adv. Mater. Sci. 34 (2013) 123–140.
[175] J.P. Chen, Y.S. Chang, Preparation and characterization of composite nanofibers of polycaprolactone and nanohydroxyapatite for osteogenic differentiation of mesenchymal stem cells, Colloids Surf. B: Biointerfaces 86 (2011) 169–175.
[176] E. Baykan, A. Koc, A. Eser Elcin, Y. Murat Elcin, Evaluation of a biomimetic poly (ε-caprolactone)/β-tricalcium phosphate multispiral scaffold for bone tissue engineering: *in vitro* and *in vivo* studies, Biointerphases 9 (2014) 029011.

[177] N. Johari, M.H. Fathi, M.A. Golozar, The effect of fluorine content on the mechanical properties of poly (e-caprolactone)/nano-fluoridated hydroxyapatite scaffold for bone-tissue engineering, Ceram. Int. 37 (2011) 3247–3251.

[178] Z. Hong, P. Zhang, C. He, X. Qiu, A. Liu, L. Chen, X. Chen, X. Jing, Nano-composite of poly(l-lactide) and surface grafted hydroxyapatite: mechanical properties and biocompatibility, Biomaterials 26 (2005) 6296–7304.

[179] S. Gay, S. Arostegui, J. Lemaitre, Preparation and characterization of dense nanohydroxyapatite/PLLA composites, Mater. Sci. Eng. C 29 (2009) 172–177.

[180] S.S. Liao, F.Z. Cui, W. Zhang, Q.L. Feng, Hierarchically biomimetic bone scaffold materials: nano-HA/collagen/PLA composite, J. Biomed. Mater. Res. Part B.: Appl. Biomater. 89B (2004) 158–165.

[181] Y. Ung, S.S. Kim, H.K. Young, S.H. Kim, B.S. Kim, S. Kim, Y.C. Cha, H.K. Soo, A poly (lactic acid)/calcium metaphosphate composite for bone tissue engineering, Biomaterials 26 (2005) 6314–6322.

[182] A.R. Boccaccini, J.J. Blaker, M. Maquet, W. Chung, R. Jerome, S.N. Nazhat, Poly-(dl-lactide) (PDLLA) foams with TiO_2 nanoparticles and PDLLA/TiO_2-bioglass foam composites for tissue engineering scaffolds, J. Mater. Sci. 41 (2006) 3999–4008.

[183] H. Nie, C.-H. Wang, Fabrication and characterization of PLGA/HAp composite scaffolds for delivery of BMP-2 plasmid DNA, J. Control. Release 120 (2007) 111–121.

[184] S.S. Kim, K.M. Ahn, M.S. Park, J.H. Lee, C.Y. Choi, B.S. Kim, A poly(lactide-co-glycolide)/hydroxyapatite composite scaffold with enhanced osteoconductivity, J. Biomed. Mater. Res. A 80 (2007) 206–215.

[185] E. Tamjid, R. Bagheri, M. Vossoughi, A. Simchi, Effect of TiO_2 morphology on the in vitro bioactivity and mechanical properties of polycaprolactone/TiO_2 nanocomposites for tissue engineering, Mater. Lett. 65 (2011) 2530–2533.

[186] A. Rezaei, M.R. Mohammadi, *In vitro* study of hydroxyapatite/polycaprolactone nanocomposite synthesized by an *in situ* sol-gel process, Mater. Sci. Eng. C 33 (2013) 390–396.

[187] J.M. Fernández, M.S. Molinuevo, M.S. Cortizo, A.M. Cortizo, Development of an osteoconductive PCL–PDIPF–hydroxyapatite composite scaffold for bone tissue engineering, J. Tissue Eng. Regen. Med. 5 (2011) e126–e135.

[188] S. Lauren, G. Selcuk, W. Xuejun, G. Milind, S. Wei, Fabrication of three dimensional polycaprolactone/hydroxyapatite tissue scaffolds and osteoblast-scaffold interactions *in vitro*, Biomaterials 28 (2007) 5291–5297.

[189] R. De Santis, A. Russo, A. Gloria, U. D'Amora, T. Russo, S. Panseri, M. Sandri, A. Tampieri, M. Marcacci, V.A. Dediu, C.J. Wilde, L. Ambrosio, Towards the design of 3D fiber-deposited poly(ε-caprolactone)/iron-doped hydroxyapatite nanocomposite magnetic scaffolds for bone regeneration, J. Biomed. Nanotechnol. 11 (2015) 1236–1246.

[190] A. Gloria, T. Russo, U. D'Amora, S. Zeppetelli, T. D'Alessandro, M. Sandri, M. Bañobre-López, Y. Piñeiro-Redondo, M. Uhlarz, A. Tampieri, J. Rivas, T. Herrmannsdörfer, V.A. Dediu, L. Ambrosio, R. De Santis, Magnetic poly(ε-caprolactone)/iron-doped hydroxyapatite nanocomposite substrates for advanced bone tissue engineering, J. R. Soc. Interface 10 (2013) 1–11.

[191] A.N. Hayati, H.R. Rezaie, S.M. Hosseinalipour, Preparation of poly(3-hydroxybutyrate)/nano-hydroxyapatite composite scaffolds for bone tissue engineering, Mater. Lett. 65 (2011) 736–739.

Electrospun hydrogels composites for bone tissue engineering

2

Bor Shin Chee, Gabriel Goetten de Lima, Declan Devine, Michael J.D. Nugent
Materials Research Institute, Athlone Institute of Technology (AIT), Athlone, Ireland

2.1 Introduction

Nanotechnology has the power to transform society. It is the science of changing properties of materials at the molecular and the atomic level [1]. The drive for materials with specific sizes and geometry has made an enormous impact on biomedical applications in terms of nanotechnology. The integration of nanotechnology into biomedical applications is called "nanomedicine." Electrospinning is one of the foremost nanotechnology applications to fabricate materials with sizes in the nanometer range. It has widespread applications in nanomedicine and the potential to solve unmet medical needs in the future. In addition, the incorporation of the electrospinning technology with the biocompatibility and controlled biodegradable rate of hydrogels have driven the research of electrospun hydrogel composites for successful tissue engineering. These electrospun hydrogels with hydrophilic polymeric network are beneficial because they are capable of trapping a large amount of water or biological fluid without dissolving in the polymer matrix and consequently giving a moist environment for cell seeding.

The core objective of this book chapter is to examine the utilization of different polymer-based high-functional and high-performance electrospun nanomaterials for tissue engineering, particularly in the field of bone tissue regeneration. Bone tissue engineering is a combination of biology and engineering that involves the use of cells (e.g., osteoblasts), bioactive molecules (e.g., bone morphogenetic protein 2), and biomaterials (e.g., electrospun hydrogels composites) under an in vitro culture system. Bone tissue engineering aims to develop substitute bone tissue to either replace or restore the function of impaired bone tissues. Bone tissue is considered hard tissue; it is a type of mineralized tissue or calcified tissue. Traditionally, electrospun hydrogels have merely been used for soft tissue engineering, however, successful efforts have been made to render hydrogels appropriate for hard tissue engineering [2].

Bone tissue is a nanostructure composed of tough and yet flexible collagen fibers (1–10 μm) reinforced with calcium phosphate nanocrystals as well as minor proteins and growth factors. Numerous methods have been used to generate biomaterials in nanoscales. One such method is electrospinning, which is regarded as a versatile, inexpensive, and simple methodology with positive outcomes for mimicking the nano-architecture of bones. It allows the construction of very compact and high-performance nanomaterials. This chapter captures an in-depth overview of the current state of research and knowledge related to electrospun composites in terms of theory, electrospinning processing conditions, and mechanisms for bone tissue

Applications of Nanocomposite Materials in Orthopedics. https://doi.org/10.1016/B978-0-12-813740-6.00003-X

engineering applications. The electrospinning technology outlined in this chapter includes (i) electrospun nanofiber-reinforced hydrogels, (ii) electrospun hydrogels with biological electrospray cells, and (iii) electrospun hydrogels with antimicrobial activity. The effects of the attachment, migration, and proliferation of cells and bioactive molecules for in vivo bone regeneration have also been investigated among these three primary hydrogel nanocomposites.

When discussing how to produce biomaterials via electrospinning, it is not possible to omit the parameters (i.e., polymer solution, processing, and ambient parameters) for tuning fibers morphology and size. Through appropriate adjustment of different variables from the parameters, the variations in properties, morphologies, structures, and diameters of electrospun nanofibers can be generated. This chapter also reviews some of the recent patents issued in the field of electrospinning and bone tissue engineering. As evidenced by the slow growth in the number of patents for the electrospun hydrogels for bone tissue engineering applications in the 21st century, it is expected that there would be a significant increase in granted patents for bone tissue repair and regeneration in the future. Finally, the chapter concludes with the future applications and possible research opportunities of electrospun hydrogels.

2.1.1 General principles of electrospinning

There is an increasing recognition that a nano-sized fiber is most likely to be more bioactive than a micro-sized one [3]. Undeniably, the most applicable and controllable nanofiber production method is electrospinning. This method allows the fabrication of filament forming, super lightweight polymers, and electrospun nanofibers from solutions and melts (polymer or polymer mixed) in the presence of an electric field. It is one of the few techniques used to create composite materials made of two or more different materials [4]. The first development of electrostatic attraction of a fluid was investigated in the 17th century by physician William Gilbert [5]. After a period of dramatic growth, the first patent for electrospun nanofibers was issued in the year 1902 by William James Morton titled "Method of Dispersing Fluid" [6].

An electrospinning apparatus setup (Fig. 2.1) comprises three main parts: a high-voltage generator, a syringe pump, and a collector plate. Initially, a syringe pump containing polymer solution with a metal needle tip is connected to the positive electrode of the power supply generator, while the grounded electrode is connected to a collector plate on the opposite end. Subsequently, a high voltage is applied in the system to create an electric field between the tip of the needle and the collector plate, which helps to cross-link the polymer chains during the electrospinning process. The polymer solution acts as a charged carrier transferring the electricity over to the side of the collector with no charge and this can result in a potential voltage difference between the polymer solution and the collection plate [7]. When the relatively weak surface tension of the charged solution droplet is overwhelmed by a strong electrostatic force, the droplet is distorted and forms a conical shape known as a Taylor cone. The distortion is continued and leads to an electrically charged jet ejection which draws the aligned thin polymer fibers to accelerate toward the collector [8–12]. In the electrospinning process, the solvent evaporates and leaves behind dry nanofibers deposited on the collector [7].

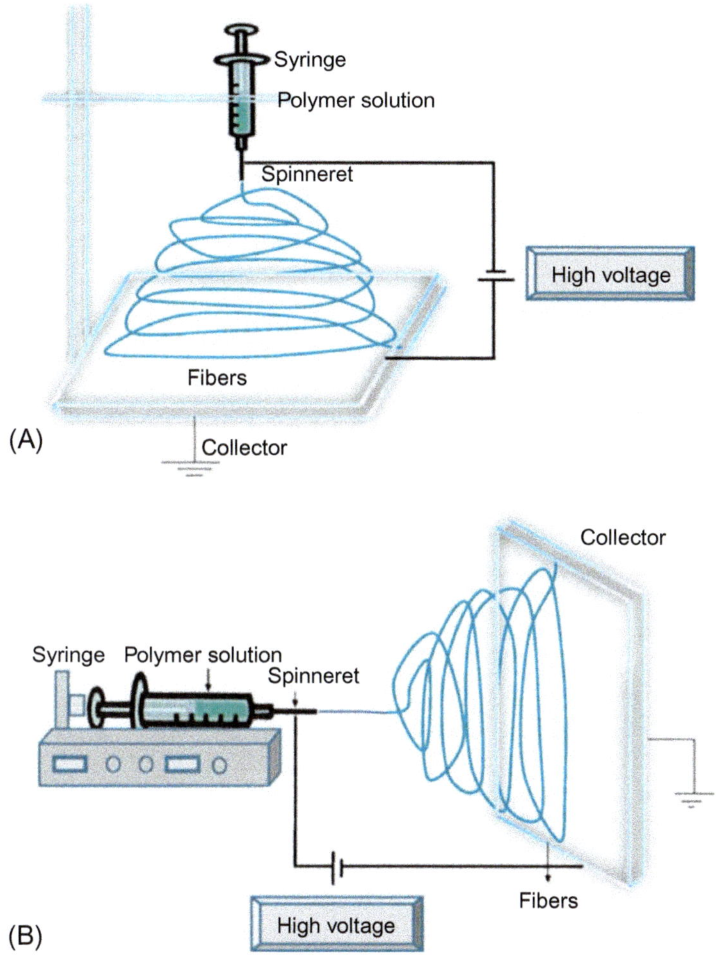

Fig. 2.1 Schematic illustration of setting up of electrospinning apparatus: (A) typical vertical setup and (B) horizontal setup of electrospinning apparatus.
Data from N. Bhardwaj, S.C. Kundu, Electrospinning: a fascinating fiber fabrication technique, Biotechnol. Adv. 28 (3) (2010) 325–347, Copyright 2010, with permission from Elsevier.

There are a number of conventional techniques available for the fabrication of nanofibers for use in bone tissue engineering, including electrospinning [13], rotary jet spinning [14], self-assembly [15], sol-gel methods [16], phase separation [17], melt-blow technology [18], melt spinning [19], and template synthesis [20]. Of all these methods, electrospinning tends to have the most remarkable effects in relation to the size and shape of fibers, which is similar to the extracellular matrix (ECM). Unlike the use of mechanical forces to draw nanofibers via conventional spinning processes (i.e., melt spinning) [8], the electrostatic force acts as a driving mechanism in electrospinning to produce nanofibers which have uniform, nonbeaded, and ultrafine morphologies (Fig. 2.2) [10,22].

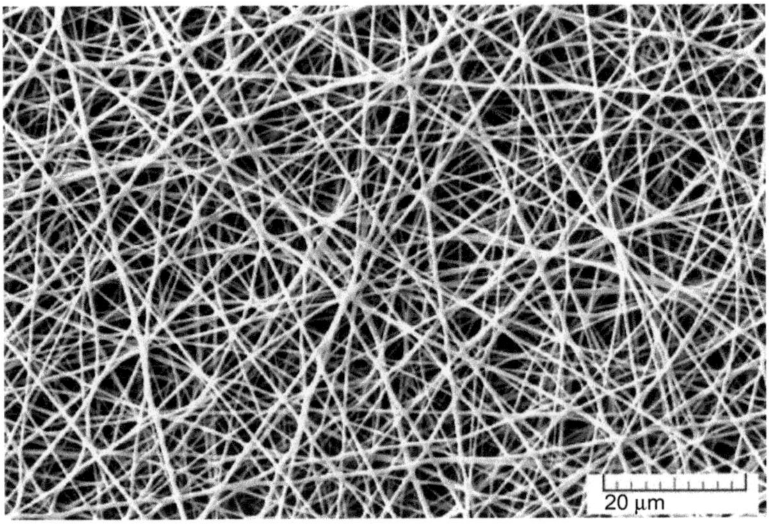

Fig. 2.2 Scanning electron microscopy (SEM) photograph of PVA electrospun nanofibers at 2.00 KX magnification [21].

Electrospun nanofibers tend to contribute excellent control over nanofiber dimensions and alignment with lengths measuring up to kilometers as well as having an average diameter of submicrometers to tens of nanometers [8,23,24]. However, electrospun materials are not very cost effective and economical due to the slow production rate. In order to increase the reproducible nanofibers production rate for scale-up commercialization, the sluggish single jet spinning has been substituted by several methods such as multiple jets (Fig. 2.3) [25] or needleless electrospinning from multiple rings [26].

2.2 Electrospun nanocomposites for medical applications

Electrospun nanomaterials have some outstanding properties, such as high surface area-to-volume ratio compared to film [24,27], light weight [22], high porosity [28], controllable membrane thickness [10], and bioavailability [29]. Due to the unique structure of these biomaterials, these are ideal candidates for use in biomedical structured elements such as wound dressings [22], drug delivery systems, and scaffolding used in tissue engineering [8], and are not limited to use for filtration operation [30], textiles field [31], and electronic component coating [8] in industrial applications.

Research activities in electrospinning became more prevalent when novel polymer/drug nanofibers were successfully fabricated [27]. The initial patent describing this new drug delivery system was granted in 2010. In the invention, Kim and Yun have demonstrated a drug delivery system using electrospinning of biodegradable polymers: poly (ε-caprolactone) (PCL) and polyethylene oxide (PEO) [32]. In addition,

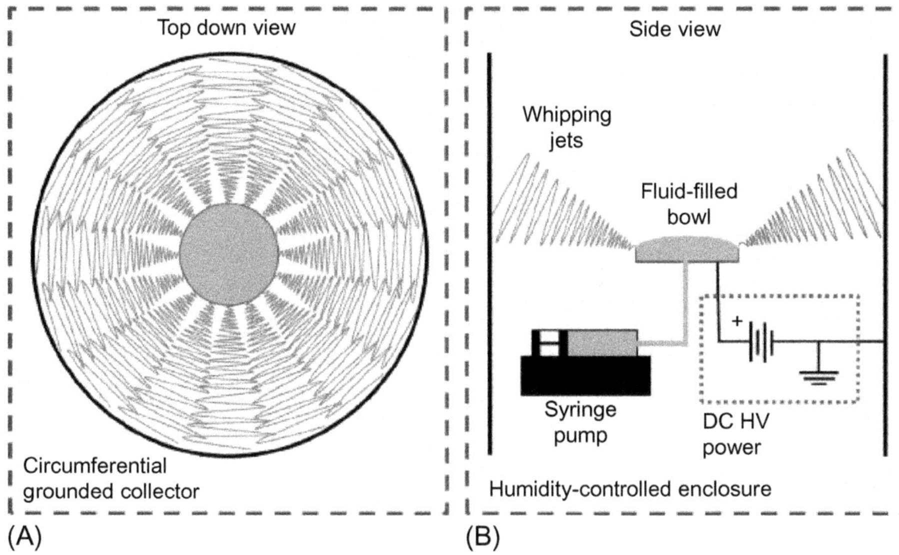

Fig. 2.3 Schematic illustration of the bowl edge electrospinning apparatus. (A) Top-looking down view of jets spinning directly from the bowl lip to the collector. (B) A side view of the refill system and power supply.
Data from N.M. Thoppey, R.E. Gorga, L.I. Clarke, J.R. Bochinski, Control of the electric field-polymer solution interaction by utilizing ultra-conductive fluids, Polymer (United Kingdom) 55 (24) (2014) 6390–6398, Copyright 2014, with permission from Elsevier.

according to research papers, Li et al. and Bahrainian et al. have studied the fast-dissolving drug delivery systems. They have reported that the introduction of drug ingredients in an amorphous or in a nanocrystal state into the electrospinning polymer aqueous solutions can give the medicines a fast-wetting surface property, enhancing both drug solubility and dissolution rates, especially for poorly water soluble components [11,29]. If multiple nano-formulated fibers are processed together to form a mesh, the surface area-to-volume ratio of the polymer increases manyfold, making them ideal for the active ingredients entrapment inside and for the rapid release from the polymer membranes when these are swallowed by the patients.

In addition, drug delivery systems can also be applied via an electrospun nanofiber implant in dentistry, orthopedic, tissue engineering applications, etc. [7]. The effectiveness in drug delivery has the potential for implantation of electrospun biomaterials into the human body as an alternative for the use of injections or tablets every day, providing a long-term drug release systems from the electrospun mesh into the body over a period time in a controlled way [11].

The human body has many functions and almost every part of our body can be replaced. Indeed Professor Seeram predicted "In the future, maybe 20–30 years from now, every human would have 20% of their body replaced with these devices" at the International Conference on Nanofibers, Applications, and Related Technologies (NART) on August 31, 2015.

2.2.1 Electrospun nanocomposite for bone tissues regeneration via osteoconduction, osteoinduction, and osteogenesis

Bone has an innate ability to heal well after mild fractures without the need for surgical intervention. Nevertheless, patients do not have the potential for self-healing with large bone defects because the bone lacks an orchestrated regeneration. This normally can lead to the need for an autologous bone transplant procedure called autografting [33] and there are estimated 2.2 million of such surgeries happening worldwide every year [34]. Despite, its widespread use, complications have been reported due to the short supply of bone stock, especially in patients who suffered fractures from osteoporosis or patients who already underwent a similar procedure that requires a second surgery. This may result in a significant donor site morbidity [35] and increases the risk of infection at the defect site [36]. Therefore, the search for new bone regeneration strategies to substitute bone-grafting procedure is desirable for bone tissue engineering.

The ECM is an essential part of the natural habitat of cells and tissues. An electrospun scaffold is one of the strategies recommended to produce structurally and functionally similar ECMs for bone implant [24]. Nanofibrous scaffolds are found to match both mechanical and biological contexts of real bone tissue matrix because they are able to resemble the architecture of the native ECM at a nanometer scale to assist the migration, organization, and survival of cells in bone regeneration. Moreover, the extraordinarily high porosity of scaffolds provides a substrate for better cell adhesion, proliferation, and differentiation to achieve better cell growth for postsurgical implantations [11,37]. In order to recreate biofunctional nanoscale scaffolds to repair damaged bone tissues, electrospinning technique is emerging as an interesting candidate for biomimetic materials [33,38].

There are three main elements as depicted in Fig. 2.4 that support the promotion of bone formation process in bone tissue engineering: osteoconduction (electrospun scaffolds), osteoinduction (bioactive molecules), and osteogenesis (stem cells) [39,40]. The current challenge for the scientific community is the engineering of biocompatible and bioactive nanomaterials for the creation of osteoconductive three-dimensional (3D) ECM scaffolds that promote osteoinduction and osteogenesis of bone tissue [40]. Osteoconduction is the capability to allow the growth and rearrangement of bone tissue on the surface of implanted nanomaterials. These materials act as an adjunct for the occurrence of both osteoinduction and osteogenesis. In the absence of stem cells and bioactive molecules, the osteoconductive scaffolds cannot form any new bone tissues [41].

Osteoinduction is the capability to induce new bone formation by signaling molecules and growth factors. Appropriate osteoinductive agents, essentially the bone morphogenic proteins (BMPs) involved in electrospun scaffolds, can help to attract the recipients' own mesenchymal stem cells. The stem cells are consequently stimulated to develop into preosteoblasts for ultimate bone formation [33,41]. However, osteogenesis is the capability to produce new bone tissue by cells called osteoblasts. The introduction of osteogenesis into osteoconductive scaffolds mimics the composition and structure of human trabecular bone for early-stage bone colonization by osteoblasts [42]. It simply means electrospun scaffolds act as a substrate for in vitro cell growth. The osteogenetic and osteoinductive potential of electrospun scaffolds have shown strong effects to facilitate bone regeneration.

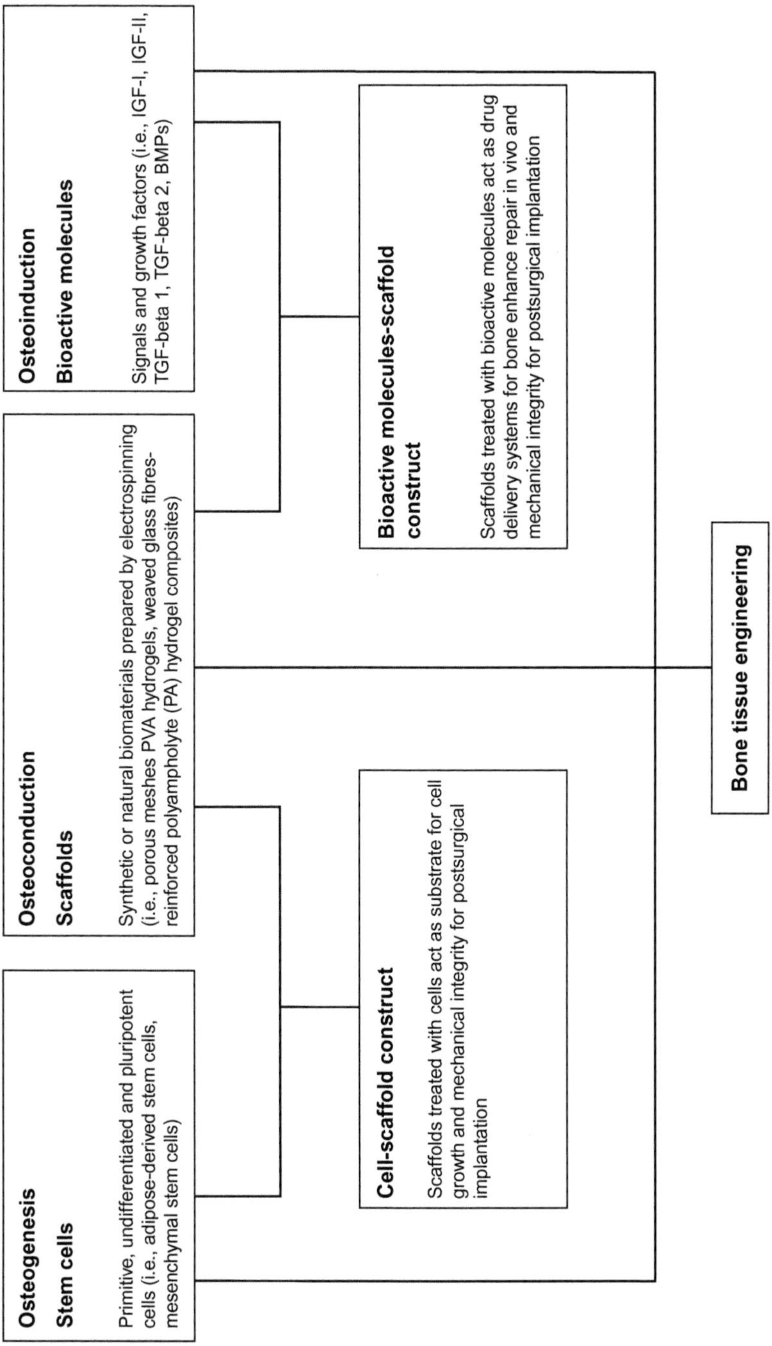

Fig. 2.4 Osteoconductive scaffolds promote osteoinduction and osteogenesis of bone tissue.

2.2.1.1 *The effect of osteogenesis and osteoinduction on osteoconductive electrospun scaffolds*

An electrospun scaffold with osteoconductive, osteoinductive, and osteogenic properties is regularly seen in the bone healing process. The porous nature of bone tissue engineering scaffold offers a place to accommodate cells and to direct the cells to grow in the correct physical form [1]. Initially, the cells that need to be multiplied are predominantly from precultured preosteoblasts/osteoblasts that are stimulated by trauma or generated from cells collected from primitive mesenchymal cells via osteoinduction [43,44]. These cells are pretreated with media containing serum and bioactive molecules (e.g., signaling molecules), then used within the specific region of engineered 3D electrospun biomaterial sector. The cell-treated nanocomposite is then cultured at a condition which normally exists in the human body (in vivo conditions) for 1–2 weeks before reimplantation to allow cells differentiate into functioning tissues [45]. The nanocomposite serves as a mechanical integrity and a shape-determining biomaterial. This whole process is called tissue engineering, which involves cells, bioactive molecules, and biomaterials under an in vitro culture system (Fig. 2.5).

There are a multitude of techniques illustrating the process of delivering bioactive molecules (e.g., growth factors) into bones. For example, (i) chemical immobilization of the growth factor bound to the scaffold via chemical binding or affinity interactions [46], and (ii) physical encapsulation of growth factors into polymeric microspheres via water/oil/water emulsion [47,48], oil/water emulsion or spray drying, and subsequently incorporation of the microspheres within scaffolds [49]. However, some of these synthetic biomaterials implanted into bone defects lack the capability to contribute to load bearing. They are mostly encapsulated by a fibrous tissue and do not interact with bones due to a lack of bioactivity. Consequently, the scaffold remains as a foreign body, resulting in the isolation of the adjacent bone which can possibly lead to severe complication of the fractures [50]. Thus, the load bearing of the scaffold may be improved by using nanocomposite technology that mimics natural bone architecture.

For instance, Li et al. fabricated a nanofibrous scaffold seeded with growth factor, BMP-2 using the electrospinning method, and examined its biological properties with human bone marrow-derived mesenchymal stem cells (hMSCs). The researchers concluded that the co-processed BMP-2 in electrospun scaffold had sustained a high level of calcium deposition and boosted transcript levels of bone-specific markers (i.e., osteocalcin, bone-specific alkaline phosphatase) than in controls [51]. This provides a positive result for further development, the nanofibrous scaffolds that are treated with cell and bioactive molecules can act as an ideal drug delivery systems to enhance bone healing in vivo and thus immediately treating the patient in a single surgery [52,53].

2.3 Electrospun biomaterials for bone tissue engineering

As a biomaterial, the biocompatibility and functionality of hydrogels play an important role in biomedical application because hydrogels closely resemble the ECM, making them the best materials of choice as tissue engineering scaffolds [54,55]. These

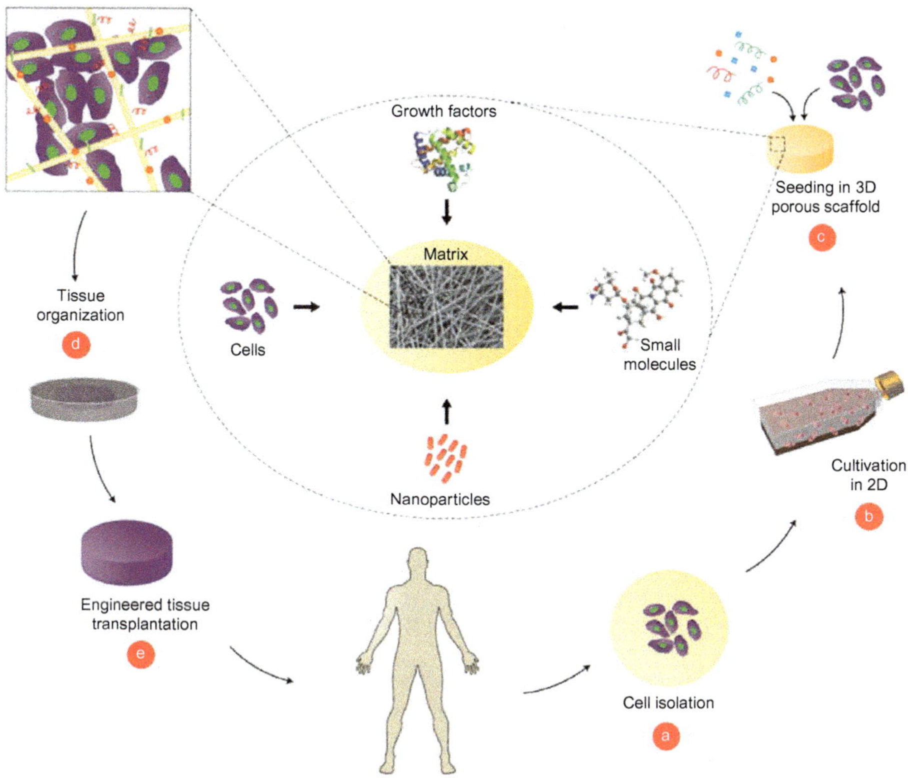

Fig. 2.5 Schematic illustration of a tissue engineering concept that involves seeding cells within electrospun scaffolds (a) Cells are isolated from the patient and may be cultivated. (b) In vitro on two-dimensional surfaces for efficient expansion. (c) Next, the cells are seeded in porous scaffolds together with growth factors, small molecules, and micro- and/or nanoparticles. The scaffolds serve as a mechanical support and a shape-determining material, and their porous nature provides high mass transfer and waste removal. (d) The cell constructs are further cultivated in bioreactors to provide optimal conditions for organization into a functioning tissue. (e) Once a functioning tissue has been successfully engineered, the construct is transplanted on the defect to restore function.
Data from T. Dvir, B.P. Timko, D.S. Kohane, R. Langer, Nanotechnological strategies for engineering complex tissues, Nat. Nanotechnol. 6 (1) (2011) 13–22, Copyright 2010, with permission from Nature Publishing Group.

materials are able to assist cell proliferation when new bone tissues are assembled [56]. A recently published Allied Market Research report predicted the global hydrogel market to be of approximately 27.2 billion USD by 2022, with a compound annual growth rate (CAGR) of 6.3% from 2016 to 2022.

In the case of biomimetic scaffold implants, bone conduction depends on the biomaterial used and its reactions to the site of injury on the bone. A wide range of materials are recommended for bone tissue engineering as listed in Table 2.1, most of these are biocompatible and biodegradable polymers. The major advantage of biodegradable

Table 2.1 Electrospun hydrogel for bone tissue engineering

Type of electrospun hydrogel	Cells or bioactive molecules seeded on hydrogel	Electrospinning conditions				Reference
		Voltage (kV)	Tip-to-collector distance (cm)	Flow rate (μL/min)	Feed rate (mL/h)	
Chitosan/alginate/poly(ethylene oxide) (CTS/Alg/PEO)	MC3T3 cells	13	10	20.0	–	[57]
Medical grade poly(ε-caprolactone)/collagen (mPCL/Col)	Mesenchymal stem cells (MSCs)	12.5	12.5	25	–	[58]
Poly(ε-caprolactone) (PCL)	Human primary osteoblasts (HOB)	15	17	20	–	[59]
Medical grade poly(ε-caprolactone)/collagen/poly(ethylene oxide) (mPCL/Col/PEO) Medical grade poly(ε-caprolactone)/collagen/gelatine (mPCL/Col/Gel)	Human fetal osteoblasts (hFOBs)	mPCL-col = –9 PEO = 7 Gel = 7	6	–	mPCL-col = 0.75 PEO = 1.25 Gel = 1.25	[60]
Polycaprolactone/poly(methyl methacrylate) (PCL/PMMA)	MG-63 osteoblast cells	25	–	25	–	[61]
Polycaprolactone/hydroxyapatite/gelatine (PCL/HAp/Gel)	Human fetal osteoblast cells (hFOB)	13	12	16.7	–	[62]
Poly(L-lactic acid) (PLLA)	Bone morphogenetic protein 2 (BMP-2)	20–30	15	14.0	–	[55]
Poly(L-lactic acid)/collagen/hydroxyapatite (PLLA/Col/HAp)	Human fetal osteoblasts (hFOB)	12	15	–	–	[63]

Poly(L-lactic acid)/collagen I (PLLA/Col I)	Human mesenchymal stem cells (hMSC)	10–18	15	0.2–0.3	–	[64]
Gelatine/polycaprolactone (Gel/PCL) Gelatine/polycaprolactone/nanohydroxyapatite (Gel/PCL/nHAp) Gelatine/polycaprolactone/bone powder (Gel/PCL/bone powder)	Human adipose-derived stem cells (hASCs)	17	13	16.7	–	[65]
Gelatine/siloxane (GS)	Bone marrow-derived mesenchymal stem cells (BMSCs)	15–25	15–30	–	–	[66]
Poly(vinyl alcohol)/gelatine (PVA/Gel)	MG-63 cells	22	10	–	0.5–1.0	[67]
Poly(vinyl alcohol)/chitosan (PVA/CTS)	Umbilical cord blood (UCB)-derived mesenchymal stem cells (MSCs)	50	15	–	–	[68]

scaffolds is the controlled therapeutic release with slow degradation occurring on implantation and gradually substituted by the growth of ECM proteins secreted from the adhered cells [69,70]. The polymeric materials are classified into (i) natural biopolymers, such as chitosan, collagen, gelatin, (ii) synthetic polymers, such as PLGA,[1] PLA,[2] PCL,[3] PVA,[4] (iii) inorganic materials and metals, such as ceramics, bioactive glasses, MgO,[5] and (iv) polymer composites, such as PCL/collagen [71,72]. The synthetic polymers exhibit excellent chemical and physical properties, while biopolymers produced from animal- or plant-derived proteins or carbohydrates are biocompatible and are of great importance in shaping cells [24]. Hence, the blends of synthetic polymers and natural biopolymers are of particular significance for improving or modifying the physicochemical properties of constructed polymer materials [73].

The PVA is an interesting biomaterial, widely used in practical applications such as implantation [39] because of its outstanding chemical properties such as good chemical resistance, nontoxicity, high hydrophilicity, biocompatibility; and physical properties such as good film forming capacity, processability, thermal stability, complete biodegradability, and high crystal modulus. The PVA has been identified as the suitable candidate to be electrospun as hydrogel nanofibers and to implant in the body of the human beings [73–75]. The abundant hydroxyl groups present on the backbone of electrospun PVA ease the attachment of drugs or cell signaling molecules [76] and also helps to release drugs or biological materials in a controlled manner [77]. In an ongoing study by De Lima et al. electrospinning of PVA with a ceramics component plays an important role for successful tissue regeneration [21].

2.3.1 Electrospun nanofiber-reinforced hydrogels

An electrospun nanofiber-reinforced hydrogel is a blend of two or more dissimilar polymer constituents, which provide an environment similar to ECM. Hydrogels allow free cell movement through the matrix, inertness to interact with the body, and resistance to protein adsorption. For instance, PEO is bioinert and highly resistant to nonspecific protein adsorption [78,79]. However, hydrogels have some disadvantages including poor mechanical integrity when the water content is high, to provide for encapsulated cells. Hence, the addition of meshes of nanofibers embedded in hydrogel forms a complex structure that can solve problems that are unattainable by any monolithic material. The nanofibers infiltration with hydrogel matrix can enhance the biological activity of new tissue by fixing the gel to living tissues (Fig. 2.6C) [56].

The electrospun fibers can be cut into shorter strands and physically mixed with a hydrogel. In this mixture, the electrospun fibers networks act in a way similar to fibrous proteins of natural bone tissue to provide stiffness and strength in tension as well as promote directional cell growth, while the hydrogel acts as a gelatinous ground substance to provide a highly hydrated 3D environment with comparable complex

[1] Poly(lactic-*co*-glycolic acid).
[2] Poly(lactic acid).
[3] Polycaprolactone.
[4] Poly(vinyl) alcohol.
[5] Magnesium oxide.

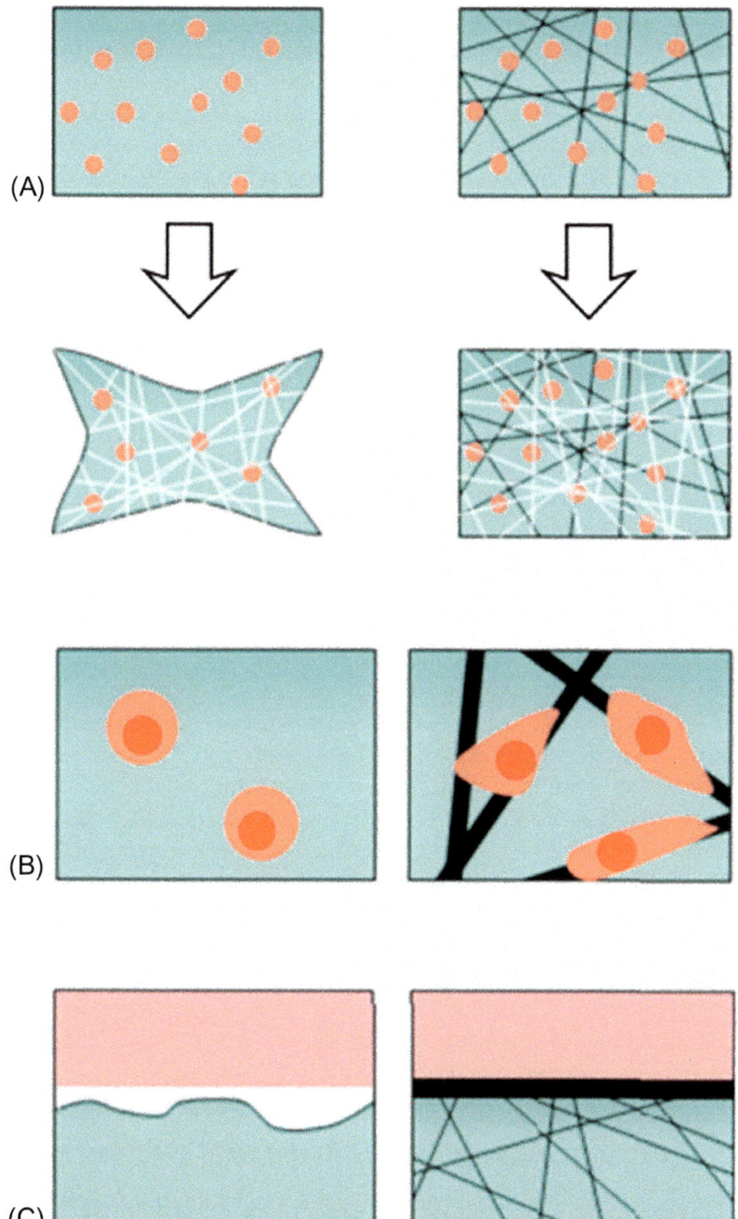

Fig. 2.6 Electrospun nanofibers embedded in hydrogels provide enhanced biological activity by (A) resistance to contraction during the development of new tissue, (B) provision of attachment sites and contact directionality to cells, and (C) improved binding to body tissues. Data from A.L. Butcher, G.S. Offeddu, M.L. Oyen, Nanofibrous hydrogel composites as mechanically robust tissue engineering scaffolds, Trends Biotechnol. 32 (11) (2014) 564–570, Copyright 2014, with permission from Elsevier.

mechanical behavior and nutrient transport [80]. The biocompatibility of electrospun fibers is inversely proportional to the fiber diameter, as the decrease in fiber diameter leads to an increase in biocompatibility due to the larger surface area, which offers more cell adhesion sites (Fig. 2.6B). As evident from Table 2.2, PCL has been widely selected to manipulate into a large range of orthopedic implants because it was approved by the US Food and Drug Administration (FDA) as the biological inert fiber material for direct contact with biological fluid among the synthetic polymers [66]. Moreover, PCL is hydrolyzable in the human body, it is more attractive for long-term implants and drug delivery systems because PCL degrades at a much slower rate than PLA, PGA, and PLGA in vivo [85].

More biomimetic environments can be created as the binding sites provide cells with contact guidance and directionality that are vital for cellular differentiation. According to Sakai's research group, "Electrospun nanofiber-reinforced hydrogels provide a constant volume and surface area to the adhered cells, allowing more extensive cell proliferation compared to the hydrogel alone" [86]. This is partially due to the resistance of contractions among fibrous components during the growth of new tissue by seeded cells (Fig. 2.6A) and the pore size of the mesh nanofibers. The lack of vascular supply in bone defects is likely to cause high cell morbidity immediately after implantation of a cell-seeded scaffold [33]. Hence, the pores formed by electrospun fiber meshes are highly interconnected which prohibit cell penetration along the structure's thickness, diminishing integration between the electrospun scaffolds with cells and ECM, as well as aiding the diffusion of oxygen and nutrients. The increased rate of cell growth then results in greater cellular expression of osteogenic markers and more distinct cell mineralization of osteoblasts [56,87].

In addition, the nanofibers have high strength-to-weight ratios. The electrospun nanofiber-reinforced hydrogels can be much more mechanically robust to failure than pure hydrogels, even at a very low volume fraction of nanofibers [38,56]. In a recent patent, Koh et al. have demonstrated a novel multilayered electrospun fiber incorporated hydrogel that consists of at least two bioactive ingredients for independently controlled release in vivo [88]. For example, Keller et al. had studied the bilayered nanoactive implant with the incorporation of two bioactive molecules: hyaluronic acid and BMP-7 for osteoarticular repair (Table 2.2). The upper and lower layers were composed of electrospun PCL membrane with BMP-7 and PCL electrospun nanofiber-reinforced alginate hydrogel with hyaluronic acid, respectively [82].

2.3.2 Electrospun hydrogels with biological electrospray cells

The continuous techniques of scaffold fabrication and subsequent cell seeding or the fabrication of blend scaffold/cell have been optimized by the microintegration of two devices, electrospinning and electrospraying, to produce one biofunctional scaffold [4]. The microintegration system tends to directly integrate the cells into the hydrogels during their production stages [89]. "Bio-electrospraying" or "cell-electrospinning" is the term specifically describes the process of producing fine micrometer-sized cell-bearing droplets and delivering the cell suspensions on a scaffold via electrospraying. Table 2.3 presents a combination of nanofibers and nanobeads scaffolds.

Table 2.2 Electrospun fibers coupled with hydrogels for bone tissue engineering

Type of hydrogel	Type of electrospun fiber	Cells or bioactive molecules seeded on hydrogel	Electrospinning conditions				Reference
			Voltage (kV)	Tip-to-collector distance (cm)	Flow rate (μL/min)	Feed rate (mL/h)	
Alginate	Polycaprolactone (PCL)	Recombinant bone morphogenetic protein-2 (rhBMP-2)	13–20	20–23	12.5	–	[81]
Alginate/hyaluronic acid	Polycaprolactone (PCL)	Mesenchymal stem cells (hMSCs) and human chondrocytes (hCHs)	15	17	20	–	[82]
Polyethylene glycol (PEG)	Polycaprolactone (PCL)	PC12 cells	30	20	50	–	[83]
Poly (lactide-*co*-ethylene oxide fumarate) (PLEOF)	Poly-L-lactic acid (PLLA)	Bone marrow stromal cells (BMS)	25	7	16.7	–	[84]
Heprasil	Medical grade poly(ε-caprolactone)/collagen (mPCL/Col)	Human fetal osteoblasts (hFOBs)	9	6	–	4	[60]

Table 2.3 Electrospun hydrogel with electrospray nanoparticles or bio-electrospray cells for bone tissue engineering

Type of electrospun hydrogel	Type of bio-electrospray cell	Electrospinning conditions			Electrospraying conditions			Reference
		Voltage (kV)	Flow rate (μL/min)	Tip-to-collector distance (cm)	Voltage (kV)	Flow rate (μL/min)	Tip-to-collector distance (cm)	
Poly(lactic-*co*-glycolic acid) (PLGA)	Mesenchymal stem cells (MSCs)	15	9	7.5	15	44	4	[89]
Poly(etherurethane urea) (PEUU)	Vascular smooth muscle cells (SMCs)	10	25	23	5	250	5	[90]

Braghirolli et al. studied the electrospun PLGA with bioelectrospraying of MSCs elicited appropriate physicochemical characteristics and the scaffold offered a favorable response toward MSC differentiation into osteogenic lineages [58,89]. This has demonstrated the viability of electrospun hydrogel/cell construct in tissue repair and functional restoration.

The combination of bio-electrospraying and electrospinning has been recommended as more suitable for cell growth and proliferation than a single electrospinning technique. This method can increase the osteoconductive and osteoinductive effect, with particular regards to improved cell-scaffold interaction [71]. The incorporation of the bio-electrospray method has promoted cells evenly 3D distributed along the fine-tuned hydrogel nanofibers scaffolds, especially at the beginning of the cultivation stage without causing any momentous deleterious effects on engineered cell construction at a molecular level [13]. In addition, the cells can interact well with the surface and pore inner walls of the scaffolds, resulting in a favorable surface topography and an osteophilic condition for the cells stay under a balanced material and energy exchange environment to promote cell penetration [71,91].

2.3.3 Electrospun hydrogels with antimicrobial activity

Infection is the biggest global challenge in medicine complications. *Escherichia coli*, *Salmonella typhi*, *Vibrio cholera*, *Pseudomonas aeruginosa*, *Rhodococcus rhodochrous*, *Proteus vulgaris*, *Aeromonas hydrophila*, and *Bacillus cereus* are human pathogenic bacteria that can cause bone infections (osteomyelitis) [92]. These bacteria are resistant to various antibiotics which can probably cause the extension of hospitalization associated with high morbidity. Therefore, the goal of finding the right antibiotics to kill the antibiotic-resistant bacteria in vivo has driven the development of biodegradable, infection-resistant bone repair implant scaffolds [93]. These scaffolds should not only be tailored for biodegradable applications to eradicate the need for surgical removal of the implant but also possess natural characteristics of being an antimicrobial agent for long-term release of antibiotics and guarantee biocompatibility with native cells. Besides, an immune modulatory effect of electrospun hydrogels with antimicrobial activities encourages both tissue growth and differentiation in tissue culture [14,94].

The increased awareness of antimicrobial activity for medical applications has promoted the growing trend of using antimicrobial polymers for implantation to reduce the outbreak of infectious diseases. In 2007, Townsend Polymer Services stated in their global plastics polymer additives market report that there was a worldwide consumption of 15,500 tons of polymer/biocides formulation [95]. It is the major reason why biocide additives, either inorganic (metallic nanoparticles: silver and copper) or organic (natural and synthetic compounds) are incorporated into polymers [96]. However, certain medical practitioners tend to avoid inorganic and synthetic biocides for bone conduction. So, new tissue engineering approach is addressed for infection control by using electrospinning method to cross-link natural biocides with polymers, which in many cases do not inhibit the biocompatibility of scaffolds. These engineered scaffolds can give limited cytotoxicity and a much longer duration of the antimicrobial activity as well as provide diverse pharmacological and therapeutic activities such as antiinflammatory and antioxidant effects, while simultaneously initiating bone regeneration [97].

Natural biocides have originated from animals (e.g., chitosan, propolis) and plants (e.g., *Aloe vera*) [98–100]. For example, Paipitak et al. had characterized the synthetic PVA polymer with chitosan (animal-derived polysaccharides) nanofibers prepared by electrospinning [99]. Chitosan has been considered as an excellent material for hydrogel production in bone tissue engineering because chitosan has cellular binding capability [101] and is able to induce migration of cells [102], as well as confer considerable antibacterial activity against a broad spectrum of bacteria [73]. All these characteristics can accelerate the bone regeneration process after a surgical intervention of PVA/chitosan. In addition, Selvakumar et al. produced a promising scaffold which has antimicrobial activity against various human pathogens while retaining the function of enhancing cell proliferation for bone regeneration using segmented polyurethane (SPU) and *Aloe vera* wrapped mesoporous hydroxyapatite (Al-mHA) nanorods [100]. Both studies have proved that the use of naturally derived synergistic compounds can tackle biofilms of multidrug-resistant microorganisms, such as Methicillin-resistant *Staphylococcus aureus* (MRSA) and extended spectrum beta-lactamases (ESBL) *E. coli* and *Klebsiella pneumoniae*.

2.4 Impact of various parameters on the electrospinning process for nanofiber morphology

Many experiments have been conducted to determine the effects of various parameters on morphological structures and diameters of electrospun nanofibers. Through appropriate adjustment of different variables from processing parameters, polymer solution parameters, and ambient parameters (Table 2.4), different properties of nanofibers can be generated [8,22,74,103]. However, not all the variables listed in Table 2.4 are fundamental control parameters or are they independent of each other. For example, applied voltage, target distance, and electric field are all interconnected [104]. Owing to the fact that one variable can influence one or more parameters, it is ideal to change the setting of one variable at a time.

2.4.1 Polymer solution parameters

Chain entanglement is considered to be one of many parameters that can significantly influence fiber formation during polymer electrospinning [104] and it can be varied depending on fundamental variables: the polymer molecular weight and concentration. The chain entanglement plays a major role in stabilizing the fibrous structure. If the chain entanglements in the solution are insufficient to stabilize the solution jet, only beads will form during electrospinning [92]. The establishment of the optimum ranges for concentration and molecular weight is desirable to ensure stable nanofiber formation. It is well known that for a given molecular weight (M), the entanglement

Table 2.4 Polymer solution, electrospinning process, and ambient parameters that affect the characteristics of electrospun nanofibers

Electrospinning		
Polymer solution parameters	**Processing parameters**	**Ambient parameters**
• Molecular weight • Architecture (branched, linear etc.) of the polymer • Concentration • Viscosity • Conductivity • Surface tension • Elasticity • Solvent vapor pressure • Solvent characteristics (poor, good)	• Electric potential at the metal needle tip • Diameter of needle tip • Solution slow rate and feed rate • Distance between the capillary tip and collection screen • Hydrostatic pressure in the capillary tube • Type of collector used	• Solution temperature • Humidity • Air velocity in the electrospinning chamber

density varies exponentially with concentration [105]. Alternatively, the same result is achieved at a fixed polymer concentration by increasing molecular weight [104]. It is also known that the polymer concentration and molecular weight may have a significant effect on electrical conductivity and fibers diameter [74].

Furthermore, the increase in both the polymer concentration and the molecular weight also result in a corresponding increase in solution viscosity $[\eta]$ [106]. The Mark-Houwink-Sakurada equation relating the intrinsic viscosity to the molecular weight of a linear polymer is

$$[\eta]=KM^{a}$$

where the empirical constants K and α depend on the nature of polymer, solvent, and temperature [107]. For example, Tacx et al. have obtained the Mark-Houwink relationship for PVA in different solvents and temperatures as: $[\eta] = 1.51 \times 1024\ M_w^{0.804}$ for PVA in DMSO at 65°C, $[\eta] = 3.54 \times 1024\ M_w^{0.692}$ for PVA in ethyleneglycol at 140°C, and $[\eta] = 6.51 \times 1024\ M_w^{0.628}$ for PVA in water at 30°C [108]. This equation has provided a simple approach for characterizing the intrinsic viscosity of a polymer and can be used for determining the Berry number (Be). The relationship between intrinsic viscosity and the concentration can be normalized with respect to the Berry number as

$$Be = [\eta]C$$

where $[\eta]$ is the intrinsic viscosity and C is the solution concentration [109]. The Berry number is generally used by the researchers as a processing index for controlling the diameter of electrospun nanofibers [110].

At a constant concentration, the morphological transition is defined from beads, to beaded fibers, to complete fibers, and to flat fibers with the increase of molecular weight. A fibrous structure cannot be stabilized when the Berry number is $[\eta]C < 4$. As a minimum degree of chain entanglement is needed for producing fibrous structures, it required a $[\eta]C$ between 5 and 12 for producing a stable fibrous structure. Therefore, when $[\eta]C > 4$, where the solution is under a semidilute entangled regime, the polymer chains in the solution begin to entangle with each other and the solution viscosity increases significantly [74], while the gradual shift from circular fibers to flat fibers starts at a Berry number of 12 ($[\eta]C \geq 12$) [111].

2.4.2 Processing parameters

By changing the distance between the capillary tip and collection screen, the applied electric field between the tips and collector can be altered, which affects the formation of the fibrous membranes. Owing to the distance increase, the fibers are continually stretched and thinned within the whipping region, resulting in smaller fiber diameters [111]. Bead generation will appear when the distance is too large or too small [30]. Moreover, there is a direct impact on the fiber diameter of injection needle tip diameter. Wang et al. reported that the fiber diameter becomes smaller with a smaller injection needle tip diameter and increased working distance [112]. Another process parameter is the type of collector used. It can vary the arrangement of electrospun fibers as random nonwoven fibrous mats, or as uniaxially aligned arrays [87].

2.4.3 Ambient parameters

In addition to the effects of controlled processing and solution parameters on fiber morphology and properties, the influence of ambient parameters such as humidity, temperatures, and air velocity in the electrospinning chamber were investigated. Nezarati et al. reported that fiber breakage occurs at low humidity due to decreased electrostatic discharge from the jet. However, a high humidity level did not guarantee the fiber formation, it was dependent on the polymer hydrophobicity, solvent volatility, and miscibility with water. Furthermore, the humidity directly influences the number of pores on the fiber. The surface pores formed via vapor-induced phase separation increased with a high evaporation rate of the highly volatile solvent at a high humidity level [30,113].

2.5 Inventions related to electrospun hydrogels for bone tissue engineering

There are numerous patents on hydrogels for producing contact lenses, hygiene products, and wound dressings and these have well-established roles in the markets. However, the commercial hydrogel-based products (e.g., coating of scaffolds) in tissue engineering market are still limited. Many hydrogel-based scaffolds have been designed, studied, and published in research papers and in some cases, these are even patented, but not many have been commercialized [114]. Therefore, there will be even less numbers of patents granted specifically for electrospun hydrogels in bone tissue engineering. Limited patents and commercial products with electrospun hydrogels in tissue engineering are related to some extent to their low mass production (further discussion to follow in Section 3.6). Further research is required to elucidate the influence of production rate on the electrospun hydrogels.

It is generally supposed that the interest in electrospinning started with the significant contributions of Anton Formhals in the 1930s who filed 22 patents on different aspects of electrospinning processes carried out in several countries, such as United States, France, the United Kingdom, and Germany between 1931 and 1944 [5]. Although this technology was invented in the early 20th century, there are still many opportunities to use electrospinning in various forms of organic and inorganic materials for different types of applications. Their potential in bone tissue engineering can be widely explored. According to the European Patent Office, the patents describing these electrospun hydrogel scaffolds exclusively for bone tissue engineering were first granted in the 21st century (Table 2.5). The electrospun hydrogels scaffolds have experienced an increasing global interest during last few years. It is expected that there would be a significant increase in the number of patents granted for bone tissue repair and regeneration in order to increase the entrepreneurial activities and investments, consequently expanding the size of the market opportunity for new uses in the future [121].

2.6 Future applications of electrospun hydrogels

The work in the past regarding electrospinning has mainly focused on determining suitable settings for electrospinning of various polymers and on understanding the important features of the preparation processes with the purpose of gaining control of electrospun nanofiber morphology, configuration, porosity, etc. [8]. In contrast, recent investigations have focused on introducing different types of organic materials (i.e., plant-based materials, biologics, and nanoformulated vitamins) into hydrogels via electrospinning. In addition, it is also worth to focus on investigating surface functionalization with aminoalkyl groups, aligned electrospun nanofibers, and the methods to scale-up nanofibers production.

Animal-based, plant-based, and natural nanofibers have been used for therapeutic approaches, yet plant-based materials are rarely used for electrospinning. Zhang

Table 2.5 Patents applications of electrospun hydrogels, exclusively for the use of bone tissue engineering

Patent #	Subject	Applicant (s)	Inventor(s)	Priority application date	Publication date	Reference
WO2012048188	Electrospun mineralized chitosan nanofibers cross-linked with genipin for bone tissue engineering	Lelkes Peter I; Frohbergh Michael; Drexel University	Lelkes Peter I; Frohbergh Michael	07-10-2011	12-04-2012	[115]
KR20130091824A	The fabrication method of porous hyaluronic acid-gelatin hydrogel scaffolds for bone tissue engineering and the hydrogel scaffolds fabricated thereby	Soonchunhyang University Industry Academy Cooperation Foundation	Lee, Byong Taek; Jang, Dong Woo; Nguyen Thuy Ba Linh	09-02-2012	20-08-2013	[116]
KR20130091822	A manufacturing method of novel fibrous scaffold composed of electrospun porous poly([epsilon]-caprolactone) (PCL) fibers for bone tissue engineering	Soonchunhyang University Industry Academy Cooperation Foundation	Lee, Byong Taek; Jang, Dong Woo; Son, So Ra	09-02-2012	20-08-2013	[117]
KR101428514B1	A hybrid manufacturing method of tissue scaffolds for bone regeneration using electrospinning and freeze drying	Yonsei University Industry-Academic Cooperation Foundation	Ryu, Won Hyoung; Yang, Sung Yeun; Hwang, Tae Heon; Kim, Hyun Ryung	31-05-2013	12-08-2014	[118]
KR20140147916A	Method for fabricating composite bone hemostatic material composed of chitosan hydrogel and electrospun gelatin/BCP	Soonchunhyang University Industry Academy Cooperation Foundation	Lee, Byong Taek; Lee, Byoung Youl; Jang, Dong Woo; Kim, Bo Ram; Padalhin Andrew	19-06-2013	31-12-2014	[119]
KR101723193B1	Method of fabricating hydrogel scaffold using electrospinning and hydrogel scaffold manufactured by the method	Gwangju Institute of Science and Technology	Yoon, Myung Han; Kim, Dong Yoon	24-08-2015	06-03-2017	[120]

et al. reported that there were only around a hundred publications on plant extract among all the journal articles (i.e., *Cissus quadrangularis* and Asian *Panax ginseng* root) from 2008 to 2016 when compared with the total amount of papers published annually on electrospinning. The plant extracts have been found to give a positive effect on the hydrophilicity and mechanical properties of nanofibers. Besides, these are important for inducing osteogenic differentiation of MSCs, giving a significant osteocalcin gene expression in human osteoblast cells. The osteocalcin is fundamental for bone formation, as it involves bone mineralization and calcium ion homeostasis. It is expected that additional research in the field of plant-based nanomaterials would be advantageous with respect to cost, accessibility, and other commercial issues [92].

Biopharmaceuticals are also known as biologics, which include not only recombinant therapeutic proteins but also naturally sourced proteins and peptides, live virus vaccines, and blood components [122]. In the pharmaceutical industry, biopharmaceuticals have become one of the nascent and rapid growing sectors for drug delivery due to ongoing technological advancements and multidisciplinary efforts of researchers in different fields, such as tissue engineering, molecular medicines, and therapeutic field. As of 2014, >300 biopharmaceutical molecules have been approved for marketing with monoclonal antibodies (mAbs) leading the market growth, followed by recombinant proteins [123]. To date, there are currently >900 biologics in the developmental stage for the treatment and prevention of >100 diseases [124]. However, the major drawback in the development of biologics is in their formulation, as it is difficult to formulate into a suitable drug delivery system. Hence, the researchers have shown increased interest in pursuing methodologies that can shorten the window for both process development and manufacturing of biologics [125].

One of the successful examples is the invention of alginate slow-release antibacterial peptides microspheres for bone tissue engineering which was patented by Fei and Yu in 2009. The peptides tend to have more effective potential to confront drug-resistant bacteria and biological membranes than any other antimicrobial agents. In detail, the peptides, beta-alexin, and lactoferrin are attached on micropores of a bone grafting material to improve the antiinfection capability and to maintain antibiotic concentration for a long time in the scaffold [126]. Despite the fact that the micro-scale biologics interact well in bone, the development of nanostructure-mediated transport of biologics with specific controlled diameter and physiochemical properties is recently one aspect of particular interest in the biopharmaceutical field. Electrospraying is emerging as the most efficient technique for the preparation of nanoparticles/nanospheres biologics, which allows specific tailoring for drug delivery applications. Nanoparticles are ideal candidates for these advanced requirements, and one of the easiest technique that can produce such nanostructured materials for delivering the biologics is the electrospinning process.

Furthermore, a potential approach to expand the effectiveness of electrospun hydrogels to bone tissue engineering is nanoformulated vitamins which are commonly used for delivery applications via oral, pulmonary, transdermal, and ocular routes of administration. However, vitamins have not yet been developed to undergo electrospinning or electrospraying, which might provoke a surge in research efforts to optimize nanoparticulate vitamin formulations in the near future for both tissue regeneration and pharmaceutical delivery [24]. A recent work by Li et al. had incorporated vitamins A and E into electrospun gelatine nanofibers and cross-linked through a glutaraldehyde treatment as antibacterial wound dressing materials. It was found that both vitamins A and E can continually be released for >60 h and greatly inhibited the growth of microorganisms, *E. coli* and *S. aureus* as well as promote the adhesion and proliferation of L929 fibroblasts cells [127].

Despite successful incorporation of antimicrobial agents in electrospun hydrogels (discussed in Section 3.3.3), significant efforts have been made in the development of antibacterial electrospun hydrogels without the use of antibacterial agents to enhance the antimicrobial effect of scaffolds. Attempts have been made to obtain the antibacterial activities of scaffolds by surface functionalization with aminoalkyl groups [128]. Roemhild et al. and Fernandes et al. have studied the electrospun PVA nanofibers containing amino-modified cellulose nanofibrils and nanostructured amino-modified bacterial cellulose membranes, respectively. Both studies have reported that the chemical grafting of aminoalkyl groups onto the surface of nanomaterial scaffolds can mimic the antimicrobial property of antimicrobial agents, exhibiting a high antimicrobial activity against *S. aureus*, *K. pneumoniae*, *E. coli*, etc. [129,130].

The natural compact bone tissue consists of highly oriented osteons which are aligned parallel to the long axis of the bone. Therefore, some researchers have started to investigate the effect of organization of cells along the aligned electrospun nanofibers in a directional manner typified by the orientation of the osteons, suggesting that nanofiber orientation can impart a functional development on the cells [131]. Doustgani et al. and Jose et al. have reported that the mechanical response of uniaxially aligned electrospun nanofibers was significant, providing higher tensile strength and elastic modulus than randomly oriented electrospun nanofibers. In the biological site, the alignment and orientation of nanofibers provide a higher surface-to-volume ratio, which enhances the osteogenic differentiation of mesenchymal stem cells and drug release rate [132,133]. Hence, the aligned electrospun nanofibers worked effectively than randomly oriented electrospun nanofibers for tissue-engineered scaffolds, especially in the field of artificial bone implant.

Electrospinning is one of the applications of the nanotechnology with the potential of industrial processing. The ability of electrospinning to process up to several liters of polymer solution under continuous operations led electrospinning research toward commercialization. The efforts to scale-up electrospinning from a laboratory scale to an industrial production level are needed. The researchers at North Carolina State University in the United States have modified the traditional electrospinning setup, called "bowl edge electrospinning." This apparatus can trigger multiple jets to deposit nanofibers onto a collector placed around the outside of the bowl. The results of this study demonstrated the orders of magnitude to be approximately 40 times higher than that of traditional needle electrospinning (TNE) [25]. Due to the advancement in the scaling-up technologies on electrospinning, it is projected that the existing nanofiber market worldwide may be around 400 million USD and will increase up to 1 billion USD by 2020 [134].

2.7 Conclusion

Nanomedicine and nanotechnology are transforming society's approach to medicine and biomedical science. However, as in every material science technology, there are three distinct interrelated aspects: the applications, the process, and the material. The applications are diverse for electrospinning and include wound healing, bone regeneration, and tissue engineering. Electrospun materials have remarkable properties in terms of biocompatibility and this combined with the developments in the electrospinning process will lead to a rapidly growing area. In particular, the reconstruction of defected bone tissues is supported by the polymer composites formed by the electrospinning process. As carriers for bone tissue regeneration, these nanofibrous hydrogel composites are made by ionic complexation in the absence of toxic cross-linking agents. The scaffolds and the by-products of these composites degradation are nontoxic to tissues and cells in vivo. Therefore, electrospinning has emerged as a favorable nanostructuring methodology.

The polymer blends are of particular significance for improving or modifying the physicochemical properties of constructed scaffolds. The current scientific approaches in developing electrospun nanofiber-reinforced hydrogels, electrospun hydrogels with biological electrospray cells, and electrospun hydrogels with antimicrobial activity are likely to enable even closer matching of scaffolds to the in vivo environment. The electrospun nanofiber-reinforced hydrogels have the ability to resist the contractile force during the expansion process of newly grown tissues; the electrospun hydrogels with biological electrospray cells can promote cells evenly distributed along the nanofibers at the beginning of cultivation stage of the scaffolds to remain higher cell viability; and the electrospun hydrogels with antimicrobial activity can offer antiinflammatory and antioxidant activity for implanted scaffolds.

However, further development cannot be carried out unless the challenges of electrospinning dealing with different parameters (discussed in Section 3.4) for different types of polymers are resolved because different types of polymer blends have different kinds of requirements during electrospinning. It has been projected that the majority of the issues will be addressed in the near future. In addition, the ability to mass-produce hydrogels nanofibers with the bowl edge electrospinning method could possibly bring electrospun hydrogels into a new era.

References

[1] S. Ramakrishna, K. Fujihara, W. Teo, T. Lim, Z. Ma, An Introduction to Electrospinning and Nanofibers, World Scientific, New Jersey, 2005.

[2] K. Gkioni, S.C.G. Leeuwenburgh, T.E.L. Douglas, A.G. Mikos, J.A. Jansen, Mineralization of hydrogels for bone regeneration, Tissue Eng. Part B Rev. 16 (6) (2010) 577–585.

[3] I. Armentano, M. Dottori, E. Fortunati, S. Mattioli, J.M. Kenny, Biodegradable polymer matrix nanocomposites for tissue engineering: a review, Polym. Degrad. Stab. 95 (11) (2010) 2126–2146.

[4] E. Kostakova, E. Zemanova, P. Mikes, J. Soukupova, H. Matheisova, K. Klouda, in: Electrospinning and Electrospraying of polymer solutions with spherical fullerenes, Nanocon 2012, 4th Int. Conf, 2012, pp. 435–439.

[5] N. Tucker, J. Stanger, M. Staiger, H. Razzaq, K. Hofman, The history of the science and technology of electrospinning from 1600 to 1995, J. Eng. Fiber Fabr. 7 (2012) 63–73.

[6] W.J. Morton, Method of Dispersing Fluids, 1902691. United State patent 705, 691.

[7] M. Zafar, S. Najeeb, Z. Khurshid, M. Vazirzadeh, S. Zohaib, Potential of electrospun nanofibers for biomedical and dental applications, Materials 9 (2), (2016) 1–21.

[8] A. Frenot, I.S. Chronakis, Polymer nanofibers assembled by electrospinning, Curr. Opin. Colloid Interface Sci. 8 (1) (2003) 64–75.

[9] S.J. Reinholt, A. Sonnenfeldt, A. Naik, M.W. Frey, A.J. Baeumner, Developing new materials for paper-based diagnostics using electrospun nanofibers, Anal. Bioanal. Chem. (2013) 3297–3304.

[10] D. Rodoplu, M. Mutlu, Effects of electrospinning setup and process parameters on nanofiber morphology intended for the modification of quartz crystal microbalance surfaces, J. Eng. Fiber Fabr. 7 (2) (2012) 118–123.

[11] X. Li, M.A. Kanjwal, L. Lin, I.S. Chronakis, Electrospun polyvinyl-alcohol nanofibers as oral fast-dissolving delivery system of caffeine and riboflavin, Colloids Surf. B: Biointerfaces 103 (2013) 182–188.

[12] S. Gupta, S. Goswami, A. Sinha, A combined effect of freeze--thaw cycles and polymer concentration on the structure and mechanical properties of transparent PVA gels, Biomed. Mater. 7 (1) (2012) 15006.

[13] M. Yunmin, L. Yuanyuan, C. Haiping, H. Qingxi, Application and analysis of biological electrospray in tissue engineering, Open Biomed. Eng. J. 9 (2015) 133–137.

[14] G.B.C. Cardoso, A.B. Machado-Silva, M. Sabino, A.R. Santos, C.A.C. Zavaglia, Novel hybrid membrane of chitosan/poly (ε-caprolactone) for tissue engineering, Biomatter 4 (2014) 1–9.

[15] O. Karaman, C. Celik, A.S. Urkmez, Self-assembled biomimetic scaffolds for bone tissue engineering, in: M. Bououdina (Ed.), Emerging Research on Bioinspired Materials Engineering, first ed., IGI Global, Hershey, 2016, pp. 104–132.

[16] M. Raucci, K. Donato, D. Giugliano, R. Donato, A. Soriente, H. Schrekker, L. Ambrosio, Antimicrobial ionic liquids in bioactive sol-gel hydroxyapatite for bone tissue engineering, Front. Bioeng. Biotechnol. 4 (2016).

[17] G. Ghersi, et al., PLLA scaffold via TIPS for bone tissue engineering, Chem. Eng. Trans. 49 (2016) 301–306.

[18] J. Erben, et al., The combination of meltblown and electrospinning for bone tissue engineering, Mater. Lett. 143 (2015) 172–176.

[19] L. Cui, N. Zhang, W. Cui, P. Zhang, X. Chen, A novel nano/micro-fibrous scaffold by melt-spinning method for bone tissue engineering, J. Bionic Eng. 12 (1) (2015) 117–128.

[20] J. Wang, G. Liu, J. Chen, B. Zhao, P. Zhu, Synthesis of biocompatible hydroxyapatite using chitosan oligosaccharide as a template, Materials (Basel) 8 (12) (2015) 8097–8105.

[21] G.G. De Lima, et al., in: Synthesis and characterization of electrospun PVA scaffolds for bone tissue engineering, Fifth International Symposium Frontiers in Polymer Science, 2017.

[22] T. Subbiah, G.S. Bhat, R.W. Tock, S. Parameswaran, S.S. Ramkumar, Electrospinning of nanofibers, J. Appl. Polym. Sci. 96 (2) (2005) 557–569.

[23] S.J. Lee, S.H. Oh, J. Liu, S. Soker, A. Atala, J.J. Yoo, The use of thermal treatments to enhance the mechanical properties of electrospun poly(epsilon-caprolactone) scaffolds, Biomaterials 29 (10) (2008) 1422–1430.

[24] R. Sridhar, R. Lakshminarayanan, K. Madhaiyan, V. Amutha Barathi, K.H.C. Lim, S. Ramakrishna, Electrosprayed nanoparticles and electrospun nanofibers based on natural materials: applications in tissue regeneration, drug delivery and pharmaceuticals, Chem. Soc. Rev. 44 (2015) 790–814.

[25] N.M. Thoppey, J.R. Bochinski, L.I. Clarke, R.E. Gorga, Edge electrospinning for high throughput production of quality nanofibers, Nanotechnology 22 (34) (2011) 345301.

[26] X. Wang, T. Lin, X. Wang, Scaling up the production rate of nanofibers by needleless electrospinning from multiple ring, Fibers Polym. 15 (5) (2014) 961–965.

[27] A.N.A. Shukry, K.A. Sekak, M.R. Ahmad, T.J.B. Effendi, in: Characteristics of electrospun PVA-*Aloe vera* nanofibres produced via electrospinning, Proceedings of the International Colloquium in Textile Engineering, Fashion, Apparel and Design 2014 (ICTEFAD 2014), vol. 2014, Ictefad, 2014, pp. 7–11.

[28] R.S. Barhate, C.K. Loong, S. Ramakrishna, Preparation and characterization of nanofibrous filtering media, J. Membr. Sci. 283 (1–2) (2006) 209–218.

[29] S. Bahrainian, M. Abbaspour, M. Kouchak, P. Taghavi Moghadam, A review on fast dissolving systems: from tablets to nanofibers, Jundishapur J. Nat. Pharm. Prod. 12 (2) (2017) e34267.

[30] M. Zhu, et al., Electrospun nanofibers membranes for effective air filtration, Macromol. Mater. Eng. 302 (1) (2016) 1600353.

[31] A.K. Haghi, Electrospinning of Nanofibers in Textiles, first ed., Apple Academic Press, Oakville, 2011.

[32] K.H. Kim, H. Yun, New drug delivery system using electrospinning of biodegradable polymers, 2010.

[33] M.M. Stevens, Biomaterials for bone tissue engineering, Mater. Today 11 (5) (2008) 18–25.

[34] G.M. Calori, E. Mazza, M. Colombo, C. Ripamonti, The use of bone-graft substitutes in large bone defects: any specific needs? Injury 42 (2011) S56–S63.

[35] N. Bhardwaj, S.C. Kundu, Electrospinning: a fascinating fiber fabrication technique, Biotechnol. Adv. 28 (3) (2010) 325–347.

[36] R. Cancedda, P. Giannoni, M. Mastrogiacomo, A tissue engineering approach to bone repair in large animal models and in clinical practice, Biomaterials 28 (29) (2007) 4240–4250.

[37] L. Nagarajan, N. Gayathri, Production of nanofibres using rotary jet spinning method for tissue engineering, Int. J. Sci. Res. 5 (3) (2016) 858–864.

[38] D.G.T. Strange, K. Tonsomboon, M.L. Oyen, Mechanical behaviour of electrospun fibre-reinforced hydrogels, J. Mater. Sci. Mater. Med. 25 (3) (2014) 681–690.

[39] T. Albrektsson, C. Johansson, Osteoinduction, osteoconduction and osseointegration, Eur. Spine J. 10 (2001) 96–101.

[40] M. Vallet-Regi, I. Izquierdo-Barba, M. Colilla, Structure and functionalization of mesoporous bioceramics for bone tissue regeneration and local drug delivery, Philos. Trans. R. Soc. A Math. Phys. Eng. Sci. 370 (2012) 1400–1421.

[41] A.S. Greenwald, et al., Bone-Graft Substitutes: Facts, Fictions & Applications, in: American Academy of Orthopaedic Surgeons, 73rd Annual Meeting, 2006.

[42] S. Minardi, et al., Evaluation of the osteoinductive potential of a bio-inspired scaffold mimicking the osteogenic niche for bone augmentation, Biomaterials 62 (2015) 128–137.

[43] H. Frost, The biology of fracture healing. An overview for clinicians. Part I, Clin. Orthop. Relat. Res. (248) (1989) 283–293.

[44] H. Frost, The biology of fracture healing. An overview for clinicians. Part II, Clin. Orthop. Relat. Res. (248) (1989) 294–309.

[45] L. Bosworth, S. Downes (Eds.), Electrospinning for Tissue Regeneration, Woodhead Publishing, Cambridge, 2011.
[46] K. Lee, E.A. Silva, D.J. Mooney, Growth factor delivery-based tissue engineering: general approaches and a review of recent developments, J. R. Soc. Interface 8 (55) (2011) 153–170.
[47] G.T.S. Kirby, et al., PLGA-based microparticles for the sustained release of BMP-2, Polymers (Basel) 3 (1) (2011) 571–586.
[48] F. Tao Meng, G. Hui Ma, W. Qiu, Z. Guo Su, W/O/W double emulsion technique using ethyl acetate as organic solvent: effects of its diffusion rate on the characteristics of microparticles, J. Control. Release 91 (2003) 407–416.
[49] E.O. Hollinger, T.A. Einhorn, B. Doll, C. Sfeir, Bone Tissue Engineering, CRC Press, Boca Raton, 2004.
[50] T. Kokubo, H.M. Kim, M. Kawashita, Novel bioactive materials with different mechanical properties, Biomaterials 24 (13) (2003) 2161–2175.
[51] C. Li, C. Vepari, H.-J. Jin, H.J. Kim, D.L. Kaplan, Electrospun silk-BMP-2 scaffolds for bone tissue engineering, Biomaterials 27 (16) (2006) 3115–3124.
[52] B.D. Smith, D.A. Grande, The current state of scaffolds for musculoskeletal regenerative applications, Nat. Rev. Rheumatol. 11 (2015) 213–222.
[53] W.J. King, P.H. Krebsbach, Growth factor delivery: how surface interactions modulate release in vitro and in vivo, Adv. Drug Deliv. Rev. 64 (12) (2012) 1239–1256.
[54] H.S. Mansur, C.M. Sadahira, A.N. Souza, A.A.P. Mansur, FTIR spectroscopy characterization of poly (vinyl alcohol) hydrogel with different hydrolysis degree and chemically crosslinked with glutaraldehyde, Mater. Sci. Eng. 28 (4) (2008) 539–548.
[55] M.D. Schofer, et al., Electrospun plla nanofiber scaffolds and their use in combination with bmp-2 for reconstruction of bone defects, PLoS ONE 6 (9) (2011).
[56] A.L. Butcher, G.S. Offeddu, M.L. Oyen, Nanofibrous hydrogel composites as mechanically robust tissue engineering scaffolds, Trends Biotechnol. 32 (11) (2014) 564–570.
[57] S.I. Jeong, M.D. Krebs, C.A. Bonino, J.E. Samorezov, S.A. Khan, E. Alsberg, Electrospun chitosan–alginate nanofibers with *in situ* polyelectrolyte complexation for use as tissue engineering scaffolds, Tissue Eng. Part A 17 (1–2) (2011) 59–70.
[58] J. Venugopal, et al., Electrosprayed hydroxyapatite on polymer nanofibers to differentiate mesenchymal stem cells to osteogenesis, J. Biomater. Sci. Polym. Ed. 24 (2) (2013) 170–184.
[59] A. Ferrand, et al., Osteogenetic properties of electrospun nanofibrous PCL scaffolds equipped with chitosan-based nanoreservoirs of growth factors, Macromol. Biosci. 14 (1) (2013) 45–55.
[60] A.K. Ekaputra, G.D. Prestwich, S.M. Cool, D.W. Hutmacher, Combining electrospun scaffolds with electrosprayed hydrogels leads to three-dimensional cellularization of hybrid constructs, Biomacromolecules 9 (8) (2008) 2097–2103.
[61] S.-R. Son, N.-T.B. Linh, H.-M. Yang, B.-T. Lee, *In vitro* and *in vivo* evaluation of electrospun PCL/PMMA fibrous scaffolds for bone regeneration, Sci. Technol. Adv. Mater. 14 (1) (2013) 15009.
[62] J.R. Venugopal, S. Low, A.T. Choon, A.B. Kumar, S. Ramakrishna, Nanobioengineered electrospun composite nanofibers and osteoblasts for bone regeneration, Artif. Organs 32 (5) (2008) 388–397.
[63] M.P. Prabhakaran, J. Venugopal, S. Ramakrishna, Electrospun nanostructured scaffolds for bone tissue engineering, Acta Biomater. 5 (8) (2009) 2884–2893.
[64] M.D. Schofer, et al., Characterization of a PLLA-collagen I blend nanofiber scaffold with respect to growth and osteogenic differentiation of human mesenchymal stem cells, Sci. World J. 9 (2009) 118–129.

[65] D. Rong, et al., Fabrication of gelatin/PCL electrospun fiber mat with bone powder and the study of its biocompatibility, J. Funct. Biomater. 7 (1) (2016) 6.
[66] L. Ren, et al., Fabrication of gelatin-siloxane fibrous mats via sol-gel and electrospinning procedure and its application for bone tissue engineering, Mater. Sci. Eng. C 30 (3) (2010) 437–444.
[67] N.T.B. Linh, B.T. Lee, Electrospinning of polyvinyl alcohol/gelatin nanofiber composites and cross-linking for bone tissue engineering application, J. Biomater. Appl. 27 (3) (2012) 255–266.
[68] P. Agrawal, K. Pramanik, Chitosan-poly(vinyl alcohol) nanofibers by free surface electrospinning for tissue engineering applications, Tissue Eng. Regen. Med. 13 (5) (2016) 485–497.
[69] B. Dhandayuthapani, Y. Yoshida, T. Maekawa, D.S. Kumar, Polymeric scaffolds in tissue engineering application: a review, Int. J. Polym. Sci. 2011 (ii) (2011).
[70] C.T. Johnson, A.J. García, Scaffold-based anti-infection strategies in bone repair, Ann. Biomed. Eng. 43 (3) (2015) 515–528.
[71] L. Francis, J. Venugopal, M.P. Prabhakaran, V. Thavasi, E. Marsano, S. Ramakrishna, Simultaneous electrospin-electrosprayed biocomposite nanofibrous scaffolds for bone tissue regeneration, Acta Biomater. 6 (10) (2010) 4100–4109.
[72] P. Jayaraman, C. Gandhimathi, J.R. Venugopal, D.L. Becker, S. Ramakrishna, D.K. Srinivasan, Controlled release of drugs in electrosprayed nanoparticles for bone tissue engineering, Adv. Drug Deliv. Rev. 94 (2015) 77–95.
[73] E. Yang, X. Qin, S. Wang, Electrospun crosslinked polyvinyl alcohol membrane, Mater. Lett. 62 (20) (2008) 3555–3557.
[74] A. Koski, K. Yim, S. Shivkumar, Effect of molecular weight on fibrous PVA produced by electrospinning, Mater. Lett. 58 (3–4) (2004) 493–497.
[75] I. Uslu, H. Daştan, A. Altaş, A. Yayli, O. Atakol, M.L. Aksu, Preparation and characterization of PVA/boron polymer produced by an electrospinning technique, E-Polymers 133 (2007) 1–6.
[76] S.J. Bryant, K.A. Davis-Arehart, N. Luo, R.K. Shoemaker, J.A. Arthur, K.S. Anseth, Synthesis and characterization of photopolymerised multifunctional hydrogels: water-soluble poly(vinyl alcohol) and chondroitin sulfate macromers for chondrocyte encapsulation, Macromolecules 37 (18) (2004) 6726–6733.
[77] V. Giménez, A. Mantecón, V. Cádiz, Crosslinking of poly(vinyl alcohol) using dianhydrides as hardeners, Appl. Polym. Sci. 59 (3) (1996) 425–431.
[78] B. Allan, Closer to nature: new biomaterials and tissue engineering in ophthalmology, Br. J. Ophthalmol. 83 (1999) 1235–1240.
[79] M.C. Lensen, V.A. Schulte, M. Diez, Cell adhesion and spreading on an intrinsically anti-adhesive PEG biomaterial, INTECH (2011) 397–414.
[80] L.A. Bosworth, L.A. Turner, S.H. Cartmell, State of the art composites comprising electrospun fibres coupled with hydrogels: a review, Nanomed. Nanotechnol. Biol. Med. 9 (3) (2013) 322–335.
[81] Y.M. Kolambkar, et al., An alginate-based hybrid system for growth factor delivery in the functional repair of large bone defects, Biomaterials 32 (1) (2011) 65–74.
[82] L. Keller, Q. Wagner, P. Schwinte, Bi-layered nano active implant with hybrid stem cell microtissues for tuned cartilage hypertrophy, J. Stem Cell Res. Ther. 1 (1) (2015) 1–9.
[83] N. Han, J. Johnson, J.J. Lannutti, J.O. Winter, Hydrogel-electrospun fiber composite materials for hydrophilic protein release, J. Control. Release 158 (1) (2012) 165–170.
[84] W. Xu, J. Ma, E. Jabbari, Material properties and osteogenic differentiation of marrow stromal cells on fiber-reinforced laminated hydrogel nanocomposites, Acta Biomater. 6 (6) (2010) 1992–2002.

[85] B. Guo, P.X. MA, Synthetic biodegradable functional polymers for tissue engineering: a brief review, Sci. China Chem. 57 (4) (2014) 490–500.
[86] S. Sakai, Y. Takagi, Y. Yamada, T. Yamaguchi, K. Kawakami, Reinforcement of porous alginate scaffolds by incorporating electrospun fibres, Biomed. Mater. 3 (3) (2008) 34102.
[87] W. Cui, Y. Zhou, J. Chang, Electrospun nanofibrous materials for tissue engineering and drug delivery, Sci. Technol. Adv. Mater. 11 (1) (2010) 14108.
[88] W.G. Koh, H.J. Lee, Y.M. Kook, Multi-layered electrospun fiber incorporated hydrogel, KR20150084519, 2015.
[89] D.I. Braghirolli, F. Zamboni, G.A.X. Acasigua, P. Pranke, Association of electrospinning with electrospraying: a strategy to produce 3D scaffolds with incorporated stem cells for use in tissue engineering, Int. J. Nanomed. 10 (2015) 5159–5169.
[90] J.J. Stankus, J. Guan, K. Fujimoto, W.R. Wagner, Microintegrating smooth muscle cells into a biodegradable, elastomeric fiber matrix, Biomaterials 27 (5) (2006) 735–744.
[91] S. Sahoo, W.C. Lee, J.C.H. Goh, S.L. Toh, Bio-electrospraying: a potentially safe technique for delivering progenitor cells, Biotechnol. Bioeng. 106 (4) (2010) 690–698.
[92] W. Zhang, S. Ronca, E. Mele, Electrospun nanofibres containing antimicrobial plant extracts, Nano 7 (2) (2017) 42.
[93] L.G. Harris, S.J. Foster, R.G. Richards, P. Lambert, D. Stickler, A. Eley, An introduction to *Staphylococcus aureus*, and techniques for identifying and quantifying *S. aureus* adhesins in relation to adhesion to biomaterials: review, Eur. Cells Mater. 4 (2002) 39–60.
[94] W.W. Thein-Han, Y. Kitiyanant, R.D.K. Misra, Chitosan as scaffold matrix for tissue engineering, Mater. Sci. Technol. 24 (9) (2008) 1062–1075.
[95] J. Markarian, Antimicrobials find new healthcare applications, Plast. Addit. Compd. 11 (1) (2009) 18–22.
[96] M.C. Chifiriuc, et al., Soft tissue engineering and microbial infections: challenges and perspectives, in: Nanobiomaterials in Soft Tissue Engineering, William Andrew Publishing, 2016, pp. 1–29. (chapter 1).
[97] M.B. Nair, J.D. Kretlow, A.G. Mikos, F.K. Kasper, Infection and tissue engineering in segmental bone defects – a mini review, Curr. Opin. Biotechnol. 22 (5) (2011) 721–725.
[98] G.G. De Lima, R.O. De Souza, A.D. Bozzi, M.A. Poplawska, D.M. Devine, M.J.D. Nugent, Extraction method plays critical role in antibacterial activity of propolis-loaded hydrogels, J. Pharm. Sci. 105 (3) (2016) 1248–1257.
[99] K. Paipitak, T. Pornpra, P. Mongkontalang, W. Techitdheer, W. Pecharapa, Characterization of PVA-chitosan nanofibers prepared by electrospinning, Procedia Eng. 8 (2011) 101–105.
[100] M. Selvakumar, H.S. Pawar, N.K. Francis, B. Das, S. Dhara, S. Chattopadhyay, Excavating the role of aloe vera wrapped mesoporous hydroxyapatite frame ornamentation in newly architectured polyurethane scaffolds for osteogenesis and guided bone regeneration with microbial protection, ACS Appl. Mater. Interfaces 8 (9) (2016) 5941–5960.
[101] K.Y. Lee, L. Jeong, Y.O. Kang, S.J. Lee, W.H. Park, Electrospinning of polysaccharides for regenerative medicine, Adv. Drug Deliv. Rev. 61 (12) (2009) 1020–1032.
[102] Y. Usami, et al., Chitin and chitosan induce migration of bovine polymorpho-nuclear cells, J. Vet. Med. Sci. 56 (4) (1994) 761–762.
[103] Z.M. Huang, Y.Z. Zhang, M. Kotaki, S. Ramakrishna, A review on polymer nanofibers by electrospinning and their applications in nanocomposites, Compos. Sci. Technol. 63 (15) (2003) 2223–2253.

[104] S.L. Shenoy, W.D. Bates, H.L. Frisch, G.E. Wnek, Role of chain entanglements on fiber formation during electrospinning of polymer solutions: good solvent, non-specific polymer-polymer interaction limit, Polymer (Guildf) 46 (10) (2005) 3372–3384.
[105] W. Graessley, The entanglement concept in polymer rheology, Adv. Polym. Sci. 16 (1974) 179.
[106] J.D. Ferry, Viscoelastic Properties of Polymers, third ed., Wiley, New York, 1980.
[107] C. Tanford, Physical chemistry of macromolecules, J. Polym. Sci. 62 (173) (1961) S22–S23.
[108] J.C.J.F. Tacx, H.M. Schoffeleers, A.G.M. Brands, L. Teuwen, Dissolution behavior and solution properties of polyvinylalcohol as determined by viscometry and light scattering in DMSO, ethyleneglycol and water, Polymer (Guildf) 41 (3) (2000) 947–957.
[109] P. Gupta, C. Elkins, T.E. Long, G.L. Wilkes, Electrospinning of linear homopolymers of poly(methyl methacrylate): exploring relationships between fiber formation, viscosity, molecular weight and concentration in a good solvent, Polymer (Guildf) 46 (13) (2005) 4799–4810.
[110] K. Nasouri, A.M. Shoushtari, A. Kaflou, Investigation of polyacrylonitrile electrospun nanofibres morphology as a function of polymer concentration, viscosity and Berry number, Micro Nano Lett. 7 (5) (2012) 423.
[111] M. Cells, H. Hall, Tuning electrospinning parameters for production of 3D-fiber-fleeces with increased porosity for soft tissue engineering applications, Eur. Cell Mater. 21 (2011) 286–303.
[112] M. Wang, H.W. Tong, Effects of processing parameters on the morphology and size of electrospun PHBV micro- and nano-fibers, Key Eng. Mater. 334–335 (II) (2007) 1233.
[113] R.M. Nezarati, M.B. Eifert, E. Cosgriff-Hernandez, Effects of humidity and solution viscosity on electrospun fiber morphology, Tissue Eng. Part C Methods 19 (10) (2013) 810–819.
[114] N. Chirani, H. Yahia, L. Gritsch, F.L. Motta, S. Chirani, S. Faré, History and applications of hydrogels, J. Biomed. Sci. 4 (2) (2015) 13–23.
[115] I. Lelkes Peter, F. Michael, Electrospun mineralized chitosan nanofibers crosslinked with genipin for bone tissue engineering, US Patent 20130274892 A1, 2012.
[116] B.T. Lee, D.W. Jang, N.T.B. Linh, The fabrication method of porous hyaluronic acid-gelatin hydrogel scaffolds for bone tissue engineering and the hydrogel scaffolds fabricated thereby, Korean patent 20130091824A, 2013.
[117] B.T. Lee, D.W. Jang, S.R. Son, A manufacturing method of novel fibrous scaffold composed of electrospun porous poly([epsilon]-caprolactone) (PCL) fibers for bone tissue engineering, Korea patent 20130091822, 2013.
[118] W.H. Ryu, S.Y. Yang, T.H. Hwang, H.R. Kim, A hybrid manufacturing method of tissue scaffolds for bone regeneration using electrospinning and freeze drying, Korea patent 101428514, 2014.
[119] B.T. Lee, B.Y. Lee, D.W. Jang, B.R. Kim, P. Andrew, Method for fabricating composite bone hemostatic material composed of chitosan hydrogel and electrospun gelatin/BCP, Korean patent 101507589, 2014.
[120] M.H. Yoon, D.Y. Kim, Method of fabricating hydrogel scaffold using electrospinning and hydrogel scaffold manufactured by the method, Korean patent 101723193, 2017.
[121] C.J. Luo, S.D. Stoyanov, E. Stride, E. Pelan, M. Edirisinghe, Electrospinning versus fibre production methods: from specifics to technological convergence, Chem. Soc. Rev. 41 (13) (2012) 4708–4735.
[122] H. Yang, Emerging Non-Clinical Biostatistics in Biopharmaceutical Development and Manufacturing, CRC Press, Boca Raton, 2016.

[123] G. Mahler, Advancing biologics development and manufacturing, Pharma's Alm. (2016) 72–74.
[124] The Economist Group Limited, Going large, The Economist (2015). London.
[125] G. Walsh, Biopharmaceutical benchmarks, Nat. Biotechnol. 28 (9) (2010) 917–924.
[126] J. Fei, H. Yu, Tissue engineering bone carrying sustained-release antimicrobial peptide and preparation method thereof, China patent 101401975A, 2009.
[127] H. Li, et al., Electrospun gelatin nanofibers loaded with vitamins A and E as antibacterial wound dressing materials, RSC Adv. 6 (55) (2016) 50267–50277.
[128] N. Lin, A. Dufresne, Nanocellulose in biomedicine: current status and future prospect, Eur. Polym. J. 59 (2014) 302–325.
[129] S.C.M. Fernandes, et al., Bioinspired antimicrobial and biocompatible bacterial cellulose membranes obtained by surface functionalization with aminoalkyl groups, ACS Appl. Mater. Interfaces 5 (8) (2013) 3290–3297.
[130] K. Roemhild, C. Wiegand, U. Hipler, T. Heinze, Novel bioactive amino-functionalized cellulose nanofibers, Macromol. J. (2013) 1767–1771.
[131] C.Y. Xu, R. Inai, M. Kotaki, S. Ramakrishna, Aligned biodegradable nanofibrous structure: a potential scaffold for blood vessel engineering, Biomaterials 25 (5) (2004) 877–886.
[132] A. Doustgani, E. Vasheghani-Farahani, M. Soleimani, Aligned and random nanofibrous nanocomposite scaffolds for bone tissue engineering nanofibrous scaffolds for bone tissue engineering, Nanomed. J. 1 (1) (2013) 20–27.
[133] M.V. Jose, V. Thomas, K.T. Johnson, D.R. Dean, E. Nyairo, Aligned PLGA/HA nanofibrous nanocomposite scaffolds for bone tissue engineering, Acta Biomater. 5 (1) (2009) 305–315.
[134] G. Bhat, Polymeric nanofibers: recent technology advancements stimulating their growth, J. Text. Sci. Eng. 5 (1) (2015) 1–2.

Further Reading

[1] N.M. Thoppey, R.E. Gorga, L.I. Clarke, J.R. Bochinski, Control of the electric field-polymer solution interaction by utilizing ultra-conductive fluids, Polymer (United Kingdom) 55 (24) (2014) 6390–6398.
[2] T. Dvir, B.P. Timko, D.S. Kohane, R. Langer, Nanotechnological strategies for engineering complex tissues, Nat. Nanotechnol. 6 (1) (2011) 13–22.
[3] J. Tao, S. Shivkumar, Molecular weight dependent structural regimes during the electrospinning of PVA, Mater. Lett. 61 (11–12) (2007) 2325–2328.

Fabrication and applications of hydroxyapatite-based nanocomposites coating for bone tissue engineering

3

Jignesh P. Raval, Parth Joshi*, Dharmesh R. Chejara*, Imad A. Disher†*
*Uka Tarsadia University, Bardoli, India, †University of Babylon, Al-Hilla, Iraq

3.1 Introduction

Bone is an organic-inorganic nanocomposite made of (i) collagen, which makes it flexible and tough, (ii) carbonated apatite that gives structural reinforcement and stiffness, and (iii) bone matrix that supports cellular functions [1,2]. Maximum strength and toughness is along the lines of the applied stress [3] as its microstructure is organized three dimensionally along multiple length scales.

The number and sizes of flaws are minimized to yield unusual mechanical properties because the fundamental structure of the bone involves the coordination of bone mineral nanocrystals and collagen. The bone's microstructure further acts as a scaffold for cell regulation. The nanometer length scale is critical in regulating cell behavior as the surface features that these cells encounter are mostly of nanometer dimensions.

3.2 Hydroxyapatite: Structure and properties

Hydroxyapatite is a well-known bioactive ceramic material used in medicine [4]. The strength to the skeleton is provided by biological apatites that is the inorganic constituent of bone and acts as a storehouse for calcium, phosphorus, sodium, and magnesium. These biological apatites are structurally similar to hydroxyapatite ($Ca_{10}(PO_4)_6(OH)_2$) and brushite [B, ($CaHPO_4{\cdot}2H_2O$)]. These two forms of calcium phosphates are stable at body temperature and in body fluid. These materials, along with fluorapatite [FAp, ($Ca_5(PO_4)_3F$)], monetite [M, ($CaHPO_4$)], tricalcium phosphate [TCP, ($Ca_3(PO_4)_2$)], tetracalcium phosphate [TTCP, ($Ca_4(PO_4)_2$)], and octacalcium phosphate [OCP, ($Ca_8H_2(PO_4)_6{\cdot}5H_2O$)] belong to a family of minerals known as apatites. These materials demonstrate similar structures, that is, hexagonal system, $P6_3/m$, space group, and possess the structural formula, $X_3Y_2(TO_4)Z$. This structure allows easy substitution. In nature, apatite compositions include X and Y = Ca, Sr, Ba, Re, Pb, U, or Mn (rarely Na, K, Y, Cu); T = P, As, V, Si, S, or C (as CO_3); and Z = F, Cl, OH, or O. In medicine, apatites of interest possess X=Y=Ca, T = P, and Z = F or OH. The apatite is called hydroxyapatite when T = P and Z = OH [5].

Applications of Nanocomposite Materials in Orthopedics. https://doi.org/10.1016/B978-0-12-813740-6.00004-1

Hydroxyapatite forms crystals that are best described as hexagonal rhombic prisms. The lattice parameters for hydroxyapatite are $a=9.432$ Å and $c=6.881$ Å. Hydroxyl ions (OH^-) occur at the corners of the basal plane. These ions are positioned at every 3.44 Å (one-half the unit cell), parallel to the *c*-axis, and perpendicular to the basal plane. Thus, 60% of calcium ions in the unit cell are associated with the hydroxyl ions. The density of this material is 3.219 g/cm^3 [5].

3.3 Conventional orthopedic implants

3.3.1 Metallic implants

PMMA is the most common bone cement used in implants. It provides mechanical fixation by penetrating the interstices of the surrounding bone and also by adapting to the surface features of the metal stem [6].

The PMMA bone cement presents many problems. On one hand, bone cement can be loaded with antibiotics [7] and also assists in distributing stresses between the bone and the metal. On the other hand, micromotion at the implant-bone interface may lead to the release of large amounts of bone cement particles. Also, PMMA microfracture can occur due to stress concentrations at the implant/PMMA. Third-body wear of the metal and polymer components can be generated by metals, polymers, and bone cement fragments. Finally, inflammation in the surrounding tissues, bone breakdown (osteolysis), and implant loosening [8] is induced by these wear particles.

The in-situ polymerization reaction of acrylic bone cement is exothermic reaction with a significant temperature peak ranging between 80°C and 124°C in the cement and between 48°C and 105°C at the bone/cement interface. It has been reported that the polymerization temperature of the acrylic bone cement causes the necrosis of the tissues in the surrounding areas of the prosthesis [9]. Another disadvantage is that after exothermic polymerization reaction, monomers may be released from the cement structure which is itself toxic [10].

The cementless technique is an alternative that may prove to be more reliable. This technique depends on biological fixation provided by insertion and initial press fit followed by bone ingrowth into a textured or porous implant surface. However, it has been shown that within 10 years after surgery, in up to 40% of cases, osteolysis [11] is a common problem which leads to pain in the thigh. This is caused by the active resorption of the bone around the implant. A few studies have also shown that instead of bone ingrowths the majority of fixation takes place by the fibrous tissue [4]. Assistant treatments become necessary in order to enhance bony implant fixation and with the intention of improving the survival rate of cementless joint prostheses, calcium phosphate-based coatings were introduced clinically.

As mentioned earlier, wear of orthopedic implant materials is another serious issue. A large number of polyethylene wear particles are generated by the metal/polyethylene implant. In fact, it is suggested by some investigators that polyethylene molecules are released in each step [5]. Wear of the metal component of the joint also occurs. For example, Co-Cr-Mo alloy degrades at an average rate of 0.02–0.06 mm in 10 years [12]. Moreover, greater tissue toxicity is caused by metal particles than polyethylene particles [13,14].

Another phenomenon that affects the use of implants is stress shielding. This term refers to an uneven distribution of load at the bone-implant interface that can lead to implant loosening [15]. Every current metal implant component is affected by this problem. For example, Co-Cr-Mo alloy exhibits a modulus of elasticity of 220 GPa. This value is 10 times higher than that of the surrounding bone (17 GPa); this results in significant stress shielding. Ti-Al-V a biocompatible, highly corrosion resistant alloy exhibiting a modulus of elasticity of 110 GPa is an alternative metal component material. Unfortunately, when fixed using PMMA bone cement Ti-Al-V alloy exhibits crevice corrosion and demonstrates poor wear resistance [16].

3.3.2 Nonmetallic implants

To achieve mechanical resilience and bone bonding, different implant architectures and biomaterials have been attempted. Alumina and zirconia are examined as the first approach for possible substitutes for metallic alloys, but it is indicated clinically that they induce stress shielding and do not bond with the bone [17].

The second approach focuses on enhancing mechanical interlocking by increasing the surface area available to bone bonding by the use of porous implant architectures. Although increasing the interfacial area is expected to reduce implant movement, pore sizes must be greater than 100–150 μm to maintain healthy and viable tissue. The overall mechanical integrity of the implant is compromised by these large pores. Furthermore, the ingrown tissue is likely to be damaged, if micromovement does occur. This induces inflammation and destroys interfacial stability. Thus, porous implants have been limited to nonload-bearing clinical applications [16].

The bioactive materials that draw out a specific biological response at the material interface is focused on in the third approach. This causes the formation of a bond between the material and its surrounding tissue. The property of the bioactive materials is similar to that of the natural bone, that the bone-dissolving cells (osteoclasts) tear down these materials and replace them with natural bone. The degree of bioactivity is measured in terms of the rate of bone formation, bonding strength, and thickness of the bonding layer. Bioglass, glass ceramics, β-tricalcium phosphate (β-TCP), calcium sulfate, and hydroxyapatite are the common bioactive orthopedic materials. Bioglass is limited in strength, although it can rapidly bond with the bone. The use of glass-ceramics as materials for spinal fusion is limited as it has a low bending strength. Because the surrounding host tissue replaces the implant, β-TCP and calcium sulfate are designed to be resorbed. There are many problems associated with these resorbable systems including the low initial strength, maintenance of strength as the biomaterial degrades instability of the interface, causing mismatch between the resorption rate and the host tissue regeneration rate [18–20].

Based on their initial fixation, the stable bone ingrowth and remodeling around the stem, the HA coatings have been tested. Studies have revealed that the micromotion of HA-coated implants were comparable to that of cemented implants [21]. Additionally, comparative studies have also shown that HA-coated implants had a higher survival rate and less bone resorption than the cemented implants [21]. Thus, there is a trend in favor of HA-coated implants in terms of improved fixation and durability with respect to uncoated porous implants.

3.4 Composites of hydroxyapatite with ceramics

3.4.1 *Hydroxyapatite-Al_2O_3 composites*

Adolfsson and Hermansson [22] studied the thermal stability of Al_2O_3/HA composites in a closed system to overcome the decomposition of HA to tricalcium phosphate which is initiated in air at temperatures below 1000°C and completed at 1200°C. Composites of Al_2O_3 and various apatites (hydroxyapatite, fluorapatite, and chlorapatite) were subjected to hot isostatic pressing at 1200°C for 1 h at a pressure of 160 MPa. They found that it was possible to reduce the fraction of vacancies formed in the microstructure of HA with the use of a closed system and thereby the thermal stability improved. The temperature at which the decomposition reaction of HA is initiated in an Al_2O_3/HA composite can be increased and after hot isostatic pressing at 1200°C no decomposition of the HA was detected. The thermal stability was improved and no phase changes were detected in the hot isostatically pressed Al_2O_3/hydroxyapatite composites due to the closed system. The apatite reacted with the moisture in air and partly converted to oxyhydroxyapatite and decomposition of the latter phase was initiated in the fluorapatite and chlorapatite-based composites.

Shaping and implantation becomes difficult as the porous HA bodies are mechanically weak and brittle. One way to solve this problem is by introducing a strong porous network onto which the hydroxyapatite coating is applied. Narulkar et al. [23] prepared porous zirconia and alumina-added zirconia ceramics by ceramic slurry infiltration of expanded polystyrene bead compacts, followed by firing at 1500°C. Then the porous ceramics was coated with a slurry of hydroxyapatite-borosilicate glass-mixed powder, followed by firing at 1200°C. The porous structures without the coating had high pore interconnectivity, high porosities of 51%–69%, and sufficiently large pore window sizes (300–500 μm). The porous ceramics had compressive strengths of 5.3–36.8 MPa, which was favorably comparable to the mechanical properties of cancellous bones. In addition, a borosilicate glass layer was found on the interface, whereas the porous hydroxyapatite surface was formed on the top of the composite coating. Thus, modification of the porous zirconia-based ceramics with a bioactive composite coating was done for biomedical applications.

3.4.2 *Hydroxyapatite-glass nanocomposites*

The mechanical properties of porous glass-reinforced hydroxyapatite bioceramics including microhardness, bending and compression tests and fracture toughness determination were studied by Silva et al. [24]. Porous disks were produced by a dry method using wax spheres as pore formers. Green bodies were sintered and due to the reaction between the glassy phase and the hydroxyapatite matrix, the final microstructure of the composites consists of hydroxyapatite, α- and β-tricalcium phosphate (α- and β-$Ca_3(PO_4)_2$). When compared with the literature data for single hydroxyapatite, they found that the glassy phase yielded higher fracture toughness and bending strength. However, the achieved value is still less than that for the natural bone. They noticed a compromise between mechanical properties and the porosity level for bioceramics: for example, according to the Weibull statistics for composites with 65% porosity the maximum bending stress level is 0.2 MPa for 100% survival probability, whereas this stress level increases to 2.5 MPa for composites with 40% survival probability. However, only the 65% porosity composite samples seem to have the complete adequate morphology for bone ingrowths.

Seo and Lee [25] prepared HA ceramics with $30P_2O_5$-30CaO-$40Na_2O$ glass (1.0 and 2.5 wt%) to improve the resistance of monophase HA because the poor mechanical properties of HA induced by severe dissolution in biological milieu limit medical applications and lead to clinical failure. Due to the grain boundary dissolution in buffered water the monophase HA sintered body showed microstructural degradation. However, HA/glass composites showed no apparent evidence of dissolution, suggesting that a less soluble glass phase should be placed at grain boundaries to protect HA from dissolution.

Zhu et al. [26] made an attempt to prepare near net-shape fabrication of HA/glass composites by infiltrating a glass into porous HA performs. Their main aim was to develop glasses that are chemically compatible with HA at elevated temperatures. After extensive investigations in the phosphate and borosilicate systems, glasses of (50–55)SiO_2-(20–25)B_2O_3-(1−20)Li_2O-(0–6)CaO (wt%) composition were successfully developed. Good chemical compatibility with HA at elevated temperatures was shown by glass. The melt infiltration process was used to fabricate dense HA/glass composites at 850–95°C. Though the investigations demonstrated a good near net-shape capability of the process, the linear shrinkage induced by the infiltration process is less than 0.1%; the preliminary mechanical tests showed that the fracture toughness of the infiltrated HA/glass composite are incomparable with that for natural bone.

3.4.3 Hydroxyapatite-mullite composites

The result of a transmission electron microscopy (TEM) study of HA-mullite composites was reported by Nath et al. [27]. TEM analysis confirms the decomposition of HA to tricalcium phosphate in combination with selected-area diffraction pattern analysis, irrespective of the mullite content. Importantly, it has been observed in the investigated HA composites that there is presence of sintering liquid residue of tetragonal gehlenite as well as cubic CaO and rhombohedral Al_2O_3.

3.4.4 Hydroxyapatite-YSZ nanocomposites

Highly dispersed YSZ grains were obtained by Ahn et al. [28], in a dense HA matrix; the nanocomposite powders were densified via pressure-assisted sintering by 1000°C and 50 MPa load. High-resolution TEM showed that the interfaces between HA and YSZ grains are highly crystalline in nature. They found that with the addition of 3 wt% YSZ, the fracture toughness of an HA-based nanocomposite could be increased to 2.0 MPa/m$^{1/2}$, which was very significant. Their study showed that to provide HA with significantly enhanced sinterability and stability, particle size and morphology, stoichiometry, and phase homogeneity and purity can be controlled.

3.5 Composites of hydroxyapatite with metals

3.5.1 Hydroxyapatite-Pt nanocomposites

In order to improve fracture toughness, Clegg and Paterson [29] described a method for incorporating ductile platinum particles into a hydroxyapatite matrix. The effect of volume fraction of platinum particles on fracture toughness was determined by them. They compared the results with predictions based on models in the literature. They also measured the fracture toughness of the composite using the Vickers indentation technique. The results indicated that the incorporation of the particles improved the measured fracture toughness of the composite in a manner consistent with that predicted in the literature.

3.5.2 Hydroxyapatite-Ti nanocomposites

Estrada et al. [30] tried to improve the mechanical properties of HA by coating its surface with Ti nanoparticles in sizes from 8 to 70 nm through the deposition of Ti by laser ablation. A comparative study of the hydroxyapatite hardness before and after deposition was performed using scanning probe microscopy (SPM) nanoindentation. Based on TEM observations, it was shown that the Ti nanoparticles obtained were covered by an oxygen shell and the nanoparticles were deposited on the irregular surface of HA which produced a smoother morphology and totally covered the ceramic. The characterization methods allowed determining a structure of Ti core cluster with a thin titanium oxide shell for the particles passivating the surface and generating spherical shapes in the material. The hardness of the material was reduced by up to 62% by the coating on the HA surface, and in consequence the damage produced on the substrate was reduced significantly. They reported that the surface modifications of the covered HA resulted in better mechanical properties, reducing the possible damage to the HA ceramic by friction or impacts as it often happens in meniscus, bone junctions, and in prosthetic implants in humans.

A new system of functionally graded materials consisting of titanium with HA was considered by Simon et al. [31]. The samples were obtained by sintering titanium with HA powders. They tested the bioactivity in simulated body fluid (SBF). The sintering conditions lead to convenient values of sample density as compared with bone density. The size of HA developed at the sample surface after soaking for a week in SBF is about hundreds of micrometers, depending on the pressing force applied during the sintering process, and tends to form a continuous network of the new developed bioactive layer indicating that these materials proved to have bioactive behavior.

3.6 Composites of hydroxyapatite with polymers

3.6.1 Hydroxyapatite-epoxy composites

Fu et al. [32] chose the thermosetting epoxy resin for its simple forming processing to make HA/epoxy composites because of the complicated processing techniques of thermoplastic polymer matrix composites. They suggested to use a three-dimensional carbon fiber fabric to reinforce HA/epoxy composite through resin transfer molding (RTM) processing as the epoxy has poor mechanical properties that are far lower than those of the human cortical bone. The wet method was used to synthesize HA powder with a grain size of 3–4 μm. HA powder with or without silane treatment was added into the epoxy under stirring. The mixtures were casted into a mold after heating it at 130°C for 1 h. It was found that the HA particles were distributed gradually from the surface to the center of the fiber-reinforced composite and the HA-containing epoxy resin impregnated carbon fiber fabric except small defects. The flexural strength of the fiber-reinforced hydroxyapatite/epoxy composite was found to be much higher than that of the human cortical bone, and also the elastic modulus was close to that of cortical bone. The cytotoxicity test with L929 cells showed that the toxicity of the epoxy resin had diminished by the addition of HA powder.

3.6.2 Hydroxyapatite-PVA nanocomposites

Fenglan et al. [33] used HA nanoparticles to make a hydrogel biocomposite with polyvinyl alcohol (PVA) by a unique technique, where HA and PVA composite solutions were exposed to repeated cycles of freezing at −20°C and thawing at room temperature for crystallization and cross-linking of PVA molecules. It was shown that the composite had good homogeneity and thermal stability. Because of the formation of chemical bonding between HA and PVA, the HA crystals were uniformly distributed in the polymer matrix. The solution-blending technique also contributed to the uniform dispersion and reduced the aggregation of HA.

3.6.3 Hydroxyapatite-polyamide nanocomposites

Jie et al. [34] prepared nanocomposites of HA and polyamide under normal atmospheric pressure directly using nano-hydroxyapatite slurry and co-solution method. It was shown in the results that the HA content in the composite can reach up to 65 wt%. Interface chemical bonding formed between HA and polyamide and the HA kept the original morphological structure with a crystal size of 10–30 nm in diameter by 50–90 nm in length and distributed uniformly in the composite. "One of the best bioactive materials for load-bearing bone repair or substitute may be the synthetic nanocomposite." However, the mechanical properties of the prepared material which is an essential demand in this field was not covered in this study.

3.6.4 Hydroxyapatite-PMMA composites

HA/PMMA and calcium silicate/PMMA (CS/PMMA) composites were prepared by Monvisade et al. [35] by interpenetrating bulk polymerization of methyl methacrylate

(MMA) monomer in porous structures of HA and CS. The calcined powders of HA and CS were mixed with PVA solution, given shape by uniaxial pressing, and then fired at 1100°C for HA and 900°C for CS to form porous HA and CS. The solution mixture of MMA monomer and 0.1 mol% of benzoyl peroxide (BPO) was used to soak the templates for 24 h. Then bulk polymerization of precomposites at 85°C for 24 h under nitrogen atmosphere occurred. The interpenetrating of PMMA into the porous HAp and CS structures was shown by the microstructures of the composites. It was indicated by the thermogravimetric analysis that the PMMA content in the HA/PMMA and CS/PMMA composites were 13 and 26 wt%, respectively. The average molecular weights (Mw) of PMMA were about 491,000 for HA/PMMA composites and about 348,000 for CS/PMMA composites. Compressive strengths of these composites were significantly higher than their starting porous templates of about 90–131 MPa.

Singh et al. [36] reported a freeze-granulation technique to prepare a novel nanocomposite of PMMA/modified HA with multiwalled carbon nanotubes (MWCNTs) as reinforcement for a new-generation biomedical bone cement and implant coatings. This technique is also used to increase material homogeneity and also enhance the dispersion of MWCNTs in the composite matrix. The phase composition and the surface morphology of the nanocomposite material were studied using X-ray diffraction, field-emission scanning electron microscopy, and micro-Raman spectroscopy. It was indicated that a concentration of 0.1 wt% MWCNTs in the PMMA/HA nanocomposite material gives the best mechanical properties by performing a nanoindentation technique on different concentrations of MWCNT-reinforced nanocomposites.

3.6.5 Hydroxyapatite-polylactide composites

A novel technique whereby hydroxyapatite powder is encapsulated in polylactide-based microspheres, processed by an emulsion-solvent evaporation method, and then used as the building blocks to produce dense and uniform composites through a hot pressing route was reported by Russias et al. [37] They found that the mechanical properties, specifically, the modulus, strength and the fracture toughness, of these materials were comparable to those of human cortical bone. This suggested that the hot pressing of hydroxyapatite/polylactide microspheres can be a viable route for the synthesis of load-bearing bone-replacement materials. However, both strength and fracture toughness degrade with immersion in a simulated environment due to the degradation of the polymer phase though the elastic modulus was relatively unaffected by in vitro degradation.

3.6.6 Hydroxyapatite-PS composites

The potential biomaterials for bone replacements are high-impact polystyrene (HIPS) composites due to their good biocompatibility and adequate mechanical properties. Gong et al. [38] studied the effect of the modification of the surface of the micron-sized HA particles by in situ polymerization of styrene (St) which then compounds with HIPS. The modification effects of the HA surface on the morphology and mechanical properties of HIPS/HA composites were investigated. The results showed that the polymerization of St is not inhibited by HA particles. The compatibility between HA and HIPS is enhanced by the PS segments coated on the HA surface by in situ polymerization of St. This also improves the dispersion of HA particles in HIPS matrix, and enhances the interfacial adhesion between HA and the matrix. Thereby, the stiffness, tensile strength and notch impact strength of HIPS/HA composites are improved at the same time. And for the optimum mechanical properties of HIPS/HA composites there is a critical coating thickness of PS on the HA surface.

3.6.7 Hydroxyapatite-PE nanocomposites

Two composite systems composed of high-density polyethylene (HDPE) were prepared by Sousa et al. [39], filled with hydroxyapatite (HA) and carbon fiber. In order to test bars with a sandwich-like morphology, these composites were compounded in a co-rotating twin screw extruder and subsequently molded in a two-component injection molding machine. These moldings are based on an HDPE/HA composite outer layer and an HDPE/C fiber composite core. The tensile and impact testing were used to assess the mechanical performance of the obtained specimens. As a result of the C fiber reinforcement present in the molding core these moldings present a high stiffness. Furthermore, the sandwich moldings exhibit a clear in vitro bioactive behavior under simulated physiological conditions as a result of the HA loading, which indicates that an in vivo bone-bonding behavior can be expected for these materials. However, the sandwich moldings exhibited low strength which is attributed to the existence of the heavily filled skin with HA.

Pandey et al. [40] synthesized HA-reinforced high-density polyethylene (HDPE) composites using the hot rolling technique that facilitated uniform dispersion and blending of the reinforcements in the matrix. The composites were processed with up to 50 wt% HA particles. They found that a crystalline, uniformly reinforced composite having chemical affinity between the matrix and reinforcement can be synthesized using the new blending process. An increase in volume fraction of reinforcement from 10 to 50 wt% resulted in a 150% increase in elastic modulus (531 MPa) and 20% increase in tensile strength (25 MPa).

Eniwumide et al. [41] prepared HA/polyethylene composites. Hydroxyapatite composite reinforced with 40 vol% was made using sintered and unsintered grades of HA and two grades of high-density polyethylene. Compact tension testing was performed at room temperature (37°C) and also at three strain rates. To increase fracture toughness, the loading rate was increased from 2 to 200 μm/s. By increasing the testing temperature or decreasing the surface area of the reinforcing particles, the fracture toughness can be increased. Thus, a low surface area sintered HA in a high-molecular weight polyethylene is required for higher fracture toughness.

3.6.8 Hydroxyapatite-collagen nanocomposites

The mineralization of HA nanoparticles in the presence of collagen fibrils was studied by Verde-Carvallo et al. [42] Two different methods to produce the HA/collagen composite were used. In the first method (method A), a $Ca(OH)_2$ aqueous suspension and a solution $(NH_4)_2HPO_4$ with collagen were gradually added into a reaction vessel with a pH ~10.5. The weight ratio of HAp/collagen was fixed at 80/20. Then, the precipitate was aged at 40°C for 24h. The second reaction (method B) was made with H_3PO_4 aqueous solution in the presence of collagen and $Ca(OH)_2$ suspension with a final pH of 8.5 following the same procedure of method (A). They found that method (A) is not exactly a biomimetic synthesis, but the crystal size corresponds with that reported in the literature for natural bone, and method (B) which was made with an almost neutral pH, produced crystals of size larger than those obtained with method (A). Hence it was proved that an interaction exists between collagen and hydroxyapatite, and this was possible without the presence of noncollagenous proteins.

It was noticed that the two different XRD patterns shown in this work give rise to the question regarding the phase purity of the prepared HA which have low crystallinity for both methods.

3.6.9 Hydroxyapatite-PEEK nanocomposites

Relative to powder reinforcement, HA whisker reinforcement resulted in composites with increased elastic modulus, tensile strength, and toughness. Like the human cortical bone, whisker-reinforced composites also exhibited anisotropic mechanical properties. In this regard, the effects of HA whisker reinforcement in poly-ether-ether-ketone (PEEK) was investigated by Converse and Roeder [43]. HA whiskers and PEEK powder were mixed in amounts ranging from 0 to 40vol% HA. The mixture was compression molded into a composite bar using an open-channel dye by uniaxially pressing it into a composite pellet. They found that increased elastic modulus of up to 17.2GPa for PEEK reinforced with 40vol% HA whiskers was a result of increased HA whisker reinforcement. However, the ultimate tensile strength was only 56.3MPa, whereas the human cortical bone has an elastic modulus of 17.4GPa in the longitudinal direction and an ultimate tensile strength in the range 80–150MPa.

3.7 Conclusion

Significant development has been achieved in the construction of hydroxyapatite nanocomposites with ceramics, polymers, and metals. HA-based nanocomposites have been explored in this chapter for bone tissue engineering. The HA-based nanocomposite biomaterials proved to be promising biomaterials for bone tissue engineering. The combination of HA with polymer nanocomposites is the best approach to bone tissue engineering. Combination of polymer with HA nanocomposites can result in the expected 3D scaffold as bone graft substitute with sufficient bone behavioral properties such as mechanical strength, pore size, and osteoconductivity.

References

[1] A. Ravagliodi, A. Krajewcki, Bioceramics: Materials, Properties, Applications, Chapman & Hall, New York, 1992.

[2] L. Kuhn-Spearing, C. Rey, H.M. Kim, M.J. Glimcher, in: Carbonated apatite nanocrystals from bone, TMS Annual Meeting, 1996.

[3] P. Fratzl, N. Fratzl-Zelman, K. Klaushofer, Biophys. J. 64 (1993) 260.
[4] J.P. Collier, T.W. Bauer, R.D. Bloebaum, J.D. Bobyn, S.D. Cook, J.O. Galante, W.H. Harris, W.C. Head, M.J. Jasty, M.B. Mayor, et al., Clin. Orthop. 274 (1992) 97–112.
[5] T.P. Schmalzried, P. Campbell, A.K. Schmitt, I.C. Brown, H.C. Amstutz, J. Biomed. Mater. Res. 38 (1997) 203–210.
[6] M. Spector, Prostheses: materials, design, and strategies for implant fixation, orthopedic knowledge update 3, Am. Acad. Orthop. Surg. (1990) 185–205.
[7] A.D. Hanssen, M.J. Spangehl, Clin. Orthop. Relat. Res. 427 (2004) 79–85.
[8] D. Taylor, M. Merlo, R. Pegley, M.P. Cavatorta, Mater. Sci. Eng. A 382 (2004) 288–294.
[9] B. Pascual, B. Vzquez, M. Gurruchaga, I. Goni, M.P. Ginebra, F.J. Gil, J.A. Planell, B. Levenfeld, J.S. Romn, Biomaterials 17 (1996) 509–516.
[10] H.C. Park, Y.K. Liu, R.S. Lakes, J. Biomech. Eng. 108 (1986) 141–148.
[11] D.J. Berry, W.S. Harmsen, D. Ilstrup, D.G. Lewallen, M.E. Cabanela, Clin. Orthop. 319 (1995) 168–177.
[12] P.F. Doorn, P.A. Campbell, J. Worrall, P.D. Benya, H.A. McKellop, H.C. Amstutz, J. Biomed. Mater. Res. 42 (1998) 103–111.
[13] N. MacKinnon, K.J. Crowell, A.K. Udit, P.M. Macdonald, Chem. Phys. Lipids 132 (2004) 23–36.
[14] J.J. Rodriguez-Mercado, E. Roldan-Reyes, M. Altamirano-Lozano, Toxicol. Lett. 144 (2003) 359–369.
[15] T. Karachalios, C. Tsatsaronis, G. Efraimis, P. Papadelis, G. Lyritis, G. Diakoumopoulos, J. Arthroplasty 19 (2004) 469–475.
[16] B.D. Ratner, A.S. Hoffman, F.J. Schoen, J.E. Lemons, Biomaterials Science: An Introduction to Materials in Medicine, first ed., Elsevier Science, San Diego, CA, 1996.
[17] J.H. Uang, S.M. Best, in: J.E. Gough, A.R. Boccaccini (Eds.), Tissue Engineering Using Ceramics and Polymers, Wood-Head Publishing Limited, Cambridge, United Kingdom, 2007, pp. 3–31.
[18] L.L. Hench, E.C. Etheridge, Biomaterials: An Interfacial Approach, Academic Press, New York, 1982.
[19] L.L. Hench, J. Am. Ceram. Soc. 74 (1991) 1487.
[20] W. Suchanek, M. Yoshimura, J. Mater. Res. 13 (1998) 94.
[21] l. Sun, C.C. Berndt, A. Kucuk, R.S. Lima, K.A. Khor, J. Biomed. Mater. Res. 58 (5) (2001) 570–592.
[22] E. Adolfsson, L. Hermansson, J. Mater. Sci. 35 (2000) 5719–5723.
[23] V.V. Narulkar, S. Prakash, K. Chandra, Bull. Mater. Sci. 30 (4) (2007) 309–314.
[24] M.H.P.D. Silva, A.F. Lemo, J.M.D.F. Ferreirab, J.D. Santos, Mater. Res. 6 (3) (2003) 321–325.
[25] D.S. Seo, J.K. Lee, Met. Mater. Int. 15 (2) (2009) 265–271.
[26] Q. Zhu, G.D. With, L.J.M.G. Dortmans, F. Feenstra, J. Mater. Sci. Mater. Med. 15 (2004) 1187–1191.
[27] S. Nath, K. Biswas, B. Basu, Scr. Mater. 58 (2008) 1054–1057.
[28] E.S. Ahn, N.J. Gleason, A. Nakahira, J.Y. Ying, Nano Lett. 1 (3) (2001) 149–153.
[29] R.E. Clegg, G.D. Paterson, SIF2004 Structural Integrity and Fracture, 2004.
[30] J.R.D. Estrada, E. Camps, L.E. Alarco n, J.A. Ascencio, J. Mater. Sci. (2006), https://doi.org/10.1007/s10853-006-1216-z.
[31] V. Simon, D. Muresan, C. Popaa, S. Simon, J. Optoelectron. Adv. Mater. 7 (6) (2005) 2823–2826.
[32] T. Fu, J.L. Zhao, J.H. Wei, Y. Han, K.W. Xu, J. Mater. Sci. 39 (2004) 1411–1413.
[33] X. Fenglan, L. Yubao, W. Xuejiang, W. Jie, Y. Aiping, J. Mater. Sci. 39 (2004) 5669–5672.

[34] W. Jie, L. Yubao, C. Weiqun, Z. Yi, J. Mater. Sci. 38 (2003) 3303–3306.
[35] P. Monvisade, P. Siriphannon, R. Jermsungnern, S. Rattanabodee, J. Mater. Sci. Mater. Med. 18 (2007) 1955–1959.
[36] M.K. Singh, T. Shokuhfar, J.J.A. Gracio, A. Carlos, M. Sousa, J.M.F. Fereira, H. Garmestani, S. Ahzi, Adv. Funct. Mater. 9999 (2008) 1–7.
[37] J. Russias, E. Saiz, R.K. Nalla, A.P. Tomsia, J. Mater. Sci. 41 (2006) 5127–5133.
[38] X.H. Gong, C.Y. Tang, H.C. Hu, X.P. Zhou, X.L. Xie, J. Mater. Sci. Mater. Med. 15 (2004) 1141–1146.
[39] R.A. Sousa, A.L. Oliveira, R.L. Reis, A.M. Cunha, M.J. Bevis, J. Mater. Sci. Mater. Med. 14 (2003) 385–397.
[40] A. Pandey, E. Jan, P.B. Aswath, J. Mater. Sci. 41 (2006) 3369–3376.
[41] J.O. Eniwumide, R. Joseph, K.E. Tanner, J. Mater. Sci. Mater. Med. 15 (2004) 1147–1152.
[42] G. Verde-Carvallo, A. Guarino, G. González, Eur. Cells Mater. 7 (Suppl. 2) (2004) 58–59.
[43] G.L. Converse, R.K. Roeder, Mater. Res. Soc. Symp. Proc. 898E (2006). 0898-L05-07.1-0898-L05-07.6.

Magnesium-based alloys and nanocomposites for biomedical application

4

Magesh Sankar, Jithin Vishnu†, Manoj Gupta‡, Geetha Manivasagam*,†*
*School of Mechanical Engineering, Vellore Institute of Technology, Vellore, India, †Centre for Biomaterials Cellular and Molecular Theranostics, Vellore Institute of Technology, Vellore, India, ‡Department of Mechanical Engineering, National University of Singapore, Singapore, Singapore

4.1 Introduction

Biomaterial is defined as "a material intended to interface with biological systems to evaluate, treat, augment or replace any tissue, organ or function of the body" and biocompatibility has been defined as "the study and knowledge of the interactions between living and nonliving materials" [1]. These biomaterials have been in use for the past one century and in the last three decades the need for these materials has become intense. The four major types of materials used as biomaterials are metals, ceramics, polymers, and their composites. Among these, metallic materials tend to have major advantage as biomaterials due to their high mechanical strength and integrity. In the case of load-bearing applications, metallic materials are more suitable compared to ceramics or polymers due to their high fracture toughness. The commercially available biomaterials possess high strength and corrosion as well as wear resistance properties and hence they are useful in load-bearing applications such as hip, knee, spinal, and dental implants. If in the case of minor fractures, stainless steel or titanium is used as a temporary implant, there is a need for revision surgery which is costly and painful. Also when these implants are exposed to body solutions for long duration may undergo leaching/corrosion or wear and tend to lead to aseptic loosening and eventually causes implant failure. Hence, this prompted the development of biodegradable materials that could serve as a potential replacement as they have the desired property of naturally getting adsorbed in the human body after the healing process of the fractured bone is completed. Polymers either natural such as collagen, chitosan or synthetic like polycaprolactone (PCL), poly-L-glutamic acid (PLGA), poly-L-lactic acid, etc., are the key examples of these biodegradable materials.

However, polymers possess poor mechanical properties which are comparatively inferior to metallic implants and human bone. Thus, they are unable to withstand the forces and hence, tend to disintegrate faster than the required remodeling period. To overcome this drawback, metallic materials that possess biodegradable property and at the same time possess strength equivalent to that of the human bone have been examined with great interest for implant applications. Thus, current research focuses

Applications of Nanocomposite Materials in Orthopedics. https://doi.org/10.1016/B978-0-12-813740-6.00005-3

on the development of high-strength biodegradable metallic materials that are highly biocompatible. In this chapter, an insight on the significance of Mg as a biodegradable material has been discussed along with the corrosion mechanism, role of Mg in cardiovascular and orthopedic applications, development of Mg-based nanocomposites and their electrochemical behavior, and finally the surface modification of Mg alloys for the effective utilization as biomaterial.

4.2 Magnesium-based biomaterials

Most orthopedic implants have been used in fracture management devices and joint replacement [2]. Fracture management devices include wires, screws, pins, plates, spinal fixation devices, and other artificial ligaments. In spite of having several advantages, these orthopedic implants tend to carry severe limitations and possess risk for the human body. The most important limitation of these implants is the degradation products produced due to corrosion and wear and exposed to the human body environment [3]. Polymers like poly(lactic-*co*-glycolic acid) (PLGA), poly-L-lactide (PLLA), collagen, and chitosan have also been used as scaffolds reinforced with either HAp or some growth factors, to promote bone healing. However, these polymers lack mechanical strength which limits their application for load-bearing purposes. Magnesium on the other hand has mechanical properties similar to that of the human bone. Only concern with the usage of Mg implants is its poor corrosion resistance.

4.2.1 Why magnesium and magnesium alloys?

Among different metallic elements, Mg is mostly preferred as biodegradable metallic material owing to its low density (1.78 g/cm^3) and elastic modulus (45 GPa), which are closer to that of the human bone (density 1.8 g/cm^3; elastic modulus: 2–20 GPa). Apart from the mechanical properties, the other main advantage of magnesium is its superior biocompatibility property. Mg is the fourth richest element present in the human body. It weighs about 21–28 g on an average 70 kg human being. The distribution of Mg in the human body is mainly concentrated in the bone (60%–70%) and the remaining in cells and blood vessels. It helps to maintain normal muscle (contraction of muscles), steady heart rhythm, healthy immune system, strong teeth and bones, and transmits nerve impulses (neurological) [4–6]. The blood sugar level, blood pressure, energy metabolism, and protein synthesis in the human body are managed mainly by magnesium. The Mg that gets degraded in the body will be first absorbed by ileum and colon and will be flushed out of the body through the regular functioning of the kidneys. Thus for the above said reasons, Mg is considered as one of the potential candidate as metallic biomaterial.

Apart from Mg, iron is an alternative candidate for use as biodegradable implant material owing to its high strength and superior biocompatibility. The mechanical properties of iron can be compared to that of conventional and permanent implant material such as stainless steel as its density is 7.87 g/cm^3 and elastic modulus 200 GPa. However, unlike the permanent implants, iron degrades with time, but at much slower phase when compared to other metallic implants. Potential for iron to be a degradable medical implant has been shown primarily through some in vivo: stents developed using pure iron was implanted in the aorta of porcine and it did not show any local or systemic toxicity [7]. However, the degradation rate of Fe is very slow when exposed to physiological environment and hence they perform in a manner similar to that of permanent implants. To overcome this limitation iron is alloyed with manganese in an effort to increase iron's corrosion rate by creating microgalvanic corrosion sites, in turn reducing its magnetic susceptibility. The daily exposure limits of Fe and Mn for an average 60-kg adult are 2.55 mg and 0.42 mg, respectively, which is much lower than Mg (~300–350 mg/day) [8]. Thus, Mg is studied with more interest than Fe-based biodegradable materials.

4.2.2 Corrosion behavior of medical implants

Corrosion is defined as a destructive attack on the surface of a material as a result of an electrochemical reaction with its environment. Electrochemical deterioration occurs as positive ions are released from the anode as electrons flow toward the cathode. Typical anodic and cathodic reactions are shown below.

$$M \rightarrow M^{n+} + ne^- \tag{4.1}$$

$$2H_2O + O_2 + 4e^- \rightarrow 4OH^- (pH \geq 7) \tag{4.2}$$

$$O_2 + 4H^+ + 4e^- \rightarrow 2H_2O(pH < 7) \tag{4.3}$$

Eq. (4.1) represents the oxidation of a metal; Eq. (4.2) is the reduction process in neutral or basic conditions; and Eq. (4.3) [9] is also a reduction process in acidic conditions. Pitting corrosion is one of the most common forms of corrosion in implants, where intense attack occurs at localized sites while the remainder of the surface corrodes at a much lower rate, either because of the formation of a protective oxide layer or due to some physiological conditions. Some other contributing factors to pitting are caused by the presence of reactive sites on the surface that are more anodic or cathodic.

4.2.2.1 Magnesium—Corrosion mechanism

Equilibria between metal and solution can be illustrated via the Eh-pH (Pourbaix) diagrams [10]. The first extensive use of such diagrams was made by Marcel Pourbaix while describing the thermodynamics of metallic corrosion. The Eh-pH diagram primarily describes the equilibrium conditions in terms of two variables, a single electrode (reduction) potential and pH, under which dissolved ions are in equilibrium with the metal/metallic phases and is shown in Fig. 4.1 [10]. Diagrams are drawn with contour lines depicting solution composition, such as metal ion concentration at 25°C or some higher temperature.

In fatigue analysis, there is no endurance limit for Mg and its alloys. Under corrosive conditions, this phenomenon is more pronounced and the slope of the fatigue curve depends on the environment and the alloy composition. Corrosion is generally unwanted in engineering and science applications, whereas in the case of biodegradable implants, this phenomenon could revolutionize the biomedical industry if the

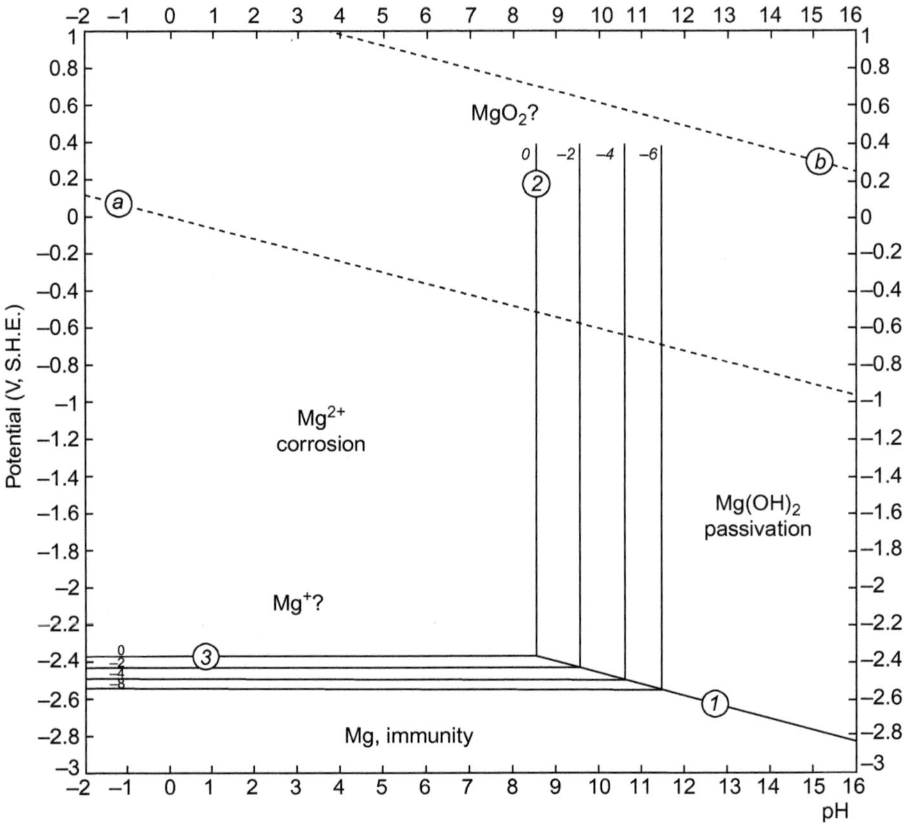

Fig. 4.1 EH-pH diagram for magnesium [10].

degradation rate could be controlled. Mg is susceptible to oxidation at room temperature, which produces a passivation oxide layer. Mg standard electrode potential is −2.37V as compared with the standard hydrogen electrode. In the presence of moisture, Mg oxide is converted to Mg hydroxide. The immunity region of Mg oxide in the Eh-pH diagram falls below the region of water stability, which is indicative of its vulnerability to corrosion. At potentials above −2.37V and pH value below 11, Mg corrodes producing Mg^{2+} and H_2. At pH values between 8.5 and 12.5, a protective layer [MgO and $Mg(OH_2)$] is formed. It has been reported that these protective or passivating layers can promote osteoinductivity and osteoconductivity.

4.2.3 Current research to overcome the challenges in Mg-based biomaterials

4.2.3.1 Corrosion

Metallic implant materials that are implanted in the human body should withstand aggressive environment with several ions present in the body fluid with pH 7 and temperature of 37°C [10a]. Hence the need for high corrosion resistance of these materials is high. To substantiate this titanium and titanium alloys exhibit elevated corrosion resistance in biological environment when compared with other conventional biomaterials like 316L grade stainless steel and cobalt-chromium alloys. The reason for this performance of titanium is the formation of thin adherent passivating oxide layer formed on the surface of Ti when exposed to body fluid environment. Apart from corrosion, wear is yet another phenomenon which affects the performance of implant materials. Ti implants undergo wear due to accelerated corrosion and result in the formation of Ti debris near the implant region, which in turn results in the blackening of the tissues, leading to catastrophic failure of the implant. Magnesium implants used in in vivo to secure fracture in early 1907 was observed to undergo rapid corrosion due to the segregation of excess gas under the tissue [11]. As a continuation of this work, several Mg alloys like Mg-Cd, Mg-Al, and Mg-Al-Mn were developed as an alternate for pure Mg for fracture fixation. The time taken for wound healing is approximately 12 weeks, however, all these alloys failed to survive this time period and underwent rapid corrosion. Electrochemical studies performed in simulated body fluid (SBF) on pure Mg are influenced mainly by pH of the solution, temperature in which the test is performed, and the presence of blood plasma and proteins, which mimics the real composition of the human body condition. Mg undergoes a complex corrosion process in the human body and results in the formation of magnesium particles, $Mg(OH)_2$ and H_2 gas in presence of water. This H_2 gas produced at the implant site will form H_2 bubbles and delays the healing process by restricting the flow of oxygen to the surrounding tissue which in turn leads to necrosis. Corrosion in the presence of serum solution leads to release of gas bubbles when Mg stents are employed. This leads to the blockage of blood vessel causing death. The presence of secondary phase and impurities in van lead to localized corrosion and pitting, which are the most predominant types of corrosion that occur in Mg and its alloys. In a primary study on 31 Mg alloys tested for corrosion almost 29 underwent pitting and localized corrosion whereas only 2 samples underwent uniform corrosion. To enhance the mechanical strength of Mg alloys intermetallics (secondary phases) are of high importance. Majority of the alloying elements and impurities are nobler compared to Mg, and hence, the intermetallic particles act as cathode and the Mg matrix will act as an anode and hence when exposed to biological environment the potential difference between them results in a galvanic coupling and rapidly increases the corrosion rate of Mg. In addition to exposure to aggressive environments, if mechanical loading also takes place then these kinds of nonuniform corrosion is highly dangerous. Low volume-to-surface area ratio results in high pH. According to ISO 10993-12, the ideal surface to volume ratio that is recommended is $3\,cm^2\,mL^{-1}$. In the case of simulating the degradation of Mg stent in the artery, the SV/SA ratio has to be 6.7 whereas when the testing of bone screws in cortical bone is done in Hank's solution, then the ratio should be maintained at 0.67. It is important to note that temperature at which the corrosion testing is performed is a key to analyze the exact corrosion behavior of Mg and its alloys. For example, Kirkland et al. studied the corrosion behavior of Mg alloys in minimum essential medium at 20°C and 37°C and found that the samples tested at 37°C exhibited 100% higher corrosion. Hence, it becomes evident that the body temperature alters the electrochemical reactions and accelerates the process resulting in the variation of corrosion mechanism of these Mg alloys compared to that of room temperature. It becomes evident that corrosion tests carried out at room temperature underrate the corrosion behavior of Mg alloys.

4.2.3.2 Effect of alloying elements on corrosion behavior of Mg materials

Several alloying elements like Al, Zn, Mn, La, Ca, Zr, and RE (rare earth) have been alloyed with Mg to improve its corrosion resistance and mechanical strength. Among numerous combinations of Mg alloys the most commonly studied alloys for biomedical application are WE43, LAE442, Mg-Gd, Mg-Dy, and Mg-Nd-Zn-Zr alloys. Interesting part is that, among these alloys, WE43 has already been tested for clinical trials. The Mg substrate that undergoes corrosion has a poor oxide film protection and hence the pH of the solution goes down (<7) and in turn accelerates the corrosion process. To overcome this limitation, alloying addition is done using Y, Li, Nd, and Zr. Mg-Dy alloy poses very high corrosion resistance and also Dy has high solubility in Mg with up to 10Dy addition to the Mg matrix. Apart from this the main advantage of Dy is that it is one of the highly compatible RE element for human body compared to others. Another advantage of adding Dy is that it does not allow any secondary phase formations and hence a homogeneous and stable single phase matrix could be obtained. The main constituents of WE43 are Mg-3.5Y-2.3Nd-0.5Zr. When Li is added to Mg it helps in increasing its corrosion resistance and also restricts the pH variation of the system. In yet another work two extruded Mg-11.3Gd-2.5Zn-0.7Zr and Mg-10.2Gd-3.3Y-0.6Zr alloys were tested electrochemically and the sample with high Gd exhibited better corrosion resistance in Hank's SBF solution. It resulted in the formation of magnesium, calcium, phosphorus, and oxygen solid film on the surface of the sample containing 11.3Gd. Pitting corrosion is predominant in Mg and hence the sample with 10.2Gd resulted in heavy pitting compared to 11.3Gd.

4.3 Magnesium for cardiovascular application

Coronary artery disease, one of the most common types of cardiovascular diseases, occurs by the deposition of cholesterol and other fatty substances (plaque) on the interior walls of coronary arteries. This will lead to thickening and narrowing of arteries which is termed as stenosis. Consequently the blood flow and oxygen supply to cardiac muscles will be reduced and finally results in fatal medical conditions such as angina or myocardial infarction or even stroke. A stent is a mesh-like tubular structure which is used to prop open the diseased artery by a minimally invasive procedure. Depending on the method of deployment, stents can be of balloon expandable type which are expanded plastically by a balloon and compressed against arterial lumen or can be self-expanding type which expand from a constraining sheath.

4.3.1 Limitations of bare metal stents and drug eluting stents

One of the successful candidates for manufacturing stent tubes are metals, as they possess better radial strength, tensile properties, high radiopacity for better visibility during stenting procedure, enhanced corrosion resistance, and better biocompatible properties [12]. Traditional coronary stent materials include stainless steel (316L), cobalt-chromium alloys, nickel-titanium alloy (Nitinol), platinum, and tantalum alloys. However, several problems are associated with bare metal stents viz. thrombus formation, in-stent restenosis, chronic inflammatory reactions, mechanical mismatch between stented and nonstented vessel regions, interference with reinterventional procedures, prolonged antiplatelet therapy, and inability to adapt with vessel growth in the case of congenital heart diseases involving pediatric surgery [13].

Two novel concepts developed to rectify the problems associated with bare metal stents are drug eluting stents and biodegradable stents. The drug eluting stents work on the concept of using a metallic stent as a platform for controlled drug delivery with the aid of a drug loaded polymer matrix covering the metallic platform to assist in arterial remodeling. However, the long-term success of drug eluting stents is still hindered by the occurrence of late stent thrombosis.

4.3.2 Biodegradable stents

The innovative concept of biodegradable stent is to use a material to provide temporary mechanical scaffolding. After that it will be bioabsorbed within a reasonable period, leaving behind the healed and remodcled artery preventing the long-term problems associated with permanent stents. An ideal material for biodegradable stent should possess a compromise between degradation rate and mechanical integrity. Initially the degradation rates should be lower to maintain mechanical support for the artery to remodel and after complete remodeling, when there is a loss of mechanical integrity, the degradation should proceed rapidly so as to prevent the retention of degradation products near the implant site. Currently biodegradable stents are manufactured from two classes of materials—polymers and metals.

4.3.2.1 Magnesium alloy biodegradable stents

The metals proposed for biodegradable stent applications are magnesium alloys, iron and its alloys, and most recently zinc alloys. Magnesium alloys have become the primary focus of investigation for biodegradable implant applications [14]. The use of magnesium as a biodegradable material dates back to 1878, when Edward C. Huse used ligatures made of magnesium to successfully heal bleeding vessels. Clinical studies have proved that magnesium is an excellent candidate for cardiovascular applications as Mg ions inhibit platelet activation, relaxes vascular smooth muscle cells, and prevents vasoconstriction and subsequent increase in blood pressure by inhibiting hormones such as angiotensin and norepinephrine [15,16]. Also magnesium due to its high electronegativity exhibits excellent hypothrombogenic property. Above all excess Mg from the human body is efficiently eliminated by kidney.

The first reported study that investigated local response to biocorrosion of magnesium in porcine arteries was done by Heublein and colleagues using AE21 alloy [17]. Clinical trials demonstrated low thrombogenicity and improved endothelialization which are essential for successful stent applications. The first magnesium corrodible stent that was used for human clinical trial was Lekton Magic coronary stent made from WE43 alloy [18]. The stents implanted in 20 patients for critical lower limb ischemia developed no adverse reactions. One of the practical limitations associated with Lekton Magic stent was its radiolucency which made it difficult for precise placements. Magnesium alloy stent fabricated by Biotronik company was the first successfully implanted biodegradable coronary stent implanted in the left pulmonary artery of a preterm baby [19]. The results showed total degradation of the stent during a 5-month follow-up, developed no neointimal hypertrophy, and proved to be efficient for pediatric stent applications. The AMS stent (Biotronik company) made from a bioabsorbable alloy containing magnesium (90%) and RE elements was implanted in 20 patients by Peeters et al. This laser-cut tubular stent achieved angiographic procedural success. Waksman et al. have compared the clinical efficacy of bioabsorbable magnesium stents and stainless steel stents in porcine coronary arteries [20]. These WE43 Mg alloy stents were found to be safe and feasible and resulted in lower neointimal formation compared to stainless steel stents. Even though it was associated with low increase in lumen area which can be attributed to factors such as underexpansion or recoil of stents. PROGRESS-AMS is the first human trial which used Lekton Magic bioabsorbable coronary stent by Biotronik made of WE43. The PROGRESS-AMS (Clinical Performance and angiographic Results of Coronary stenting with Absorbable Metal Stents) was a nonrandomized multicenter clinical trial that involved implantation of magnesium alloy stent in 63 patients having coronary artery disease [21]. Complete degradation occurred at 4 months and early recoil was found to be a major contributor for restenosis. In order to improve the biodegradation kinetics to improve the corrosion resistance and drug delivery rate of magnesium alloys, a micro-arc oxidation/polymer composite coating was developed on AZ81 alloy [22]. The famous BIOSOLVE I and II in human clinical trials have used a drug-eluting absorbable metal scaffold (DREAMS stent by Biotronik) [23]. These tailor-made magnesium alloys are found to be competent as an alternative to commercially available polymer counterparts in terms of better scaffolding, and safe degradation without neointimal formation.

Despite these clinical advances, the rapid corrosion rate of Mg alloys is still challenging for biodegradable coronary stent applications. This limitation of short-term support of <6 months can lead to premature loss of mechanical strength and structural integrity which can eventually result in stent failure. Recently, Lin Mao et al. reported an Mg alloy Mg-2.2Nd-0.1Zn-0.4Zr (JDBM-2) with excellent biocompatibility, mechanical integrity, and degradation kinetics favorable for biodegradable stents.

4.4 Magnesium for orthopedic application

Metallic implants made of stainless steel, Ti alloys, and Co-Cr alloys are widely used in orthopedic applications to assist with the repair or replacement of the bone tissue, for example, plates, screws, and pins to fix fractures. These alloys possess good mechanical strength, which is required for load-bearing applications; however, their mechanical properties are not well matched with those of natural bone tissue, leading to stress shielding effects that reduce stimulation of new bone formation and remodeling, eventually leading to implant loosening [24]. In addition, removal of permanent implants after the tissue has healed requires a second surgery, which increases health-care costs and patient morbidity.

Magnesium has mechanical properties comparable to those of human bone and has shown stimulatory effects on the growth of new bone tissue, which makes it an attractive material for orthopedic applications [25]. Magnesium has a Young's modulus and density closer to the natural bone compared to other metallic materials, while its fracture toughness is greater than that of ceramic biomaterials such as hydroxyapatite. Regarding fracture treatment, Mg alloys possess the advantage of providing an adapted stabilization depending on its degradation state, thus improving the physiological stimulation of the bone tissue and the healing process. Mg biodegradable implants are expected to maintain their mechanical properties for a time period of 12–18 weeks, eventually being replaced by natural bone tissue. Recent studies suggest a positive role of Mg on the bone cell attachment and tissue growth, probably by promoting precipitation of biologically equivalent apatite ($(Ca_{1-x}Mg_x)_{10}(PO_4)_6OH_2$). Although several studies have shown the potential of Mg alloys for orthopedic applications, control of the implant corrosion rate in the physiological environment, optimization of the implant osseointegration (implant ability to integrate with bone tissue), and thorough evaluation of the Mg alloys biocompatibility are still under investigations. Interestingly, in the last one century Mg has effectively been taken up for human trials and the schematic of the same is represented in Fig. 4.2. It can be seen that Mg has been used for human trials since 1900; however, it is surprising to note that in the last five decades, that is, between 1948 and 2010 there exist no clinical trials of Mg for orthopedic applications.

4.4.1 Current status of Mg-based materials for orthopedic application

4.4.1.1 In vitro testing of Mg-based orthopedic biomaterials

Medical devices assessed for their biocompatibility and cytotoxicity adheres to the ISO 10993 Part 5 and 12 documents. Biometallic implants such as titanium, stainless steel are inert in nature whereas polymers hydrolyze at the later stage of implantation. Hence, it is essential to determine the health risk of metals and polymers in the human body. In most of the Mg materials, when ISO cytotoxicity test standards were applied, it was found to exhibit cellular toxicity of <75%. The microenvironment in the in vivo and in vitro condition is very different; this explains the observed difference between in vitro and in vivo biocompatibility test results. Under in vivo conditions, the Mg^{2+} and OH^- ions released from Mg implants gets diluted by the surrounding body fluid and diffuse into the circulatory system prior to excretion from the human body via urine and feces. Therefore, dilution of extracts of Mg-based metals is one of the recommended approaches to mimic the levels of in vivo degradation products prior to cytotoxicity testing. Some studies found that Mg ion level of <15 mM in the diluted extracts of Mg materials can be considered nontoxic (above 75% cell viability) while 10 mM is the critical dose without deteriorating cell viability. Therefore, the revised testing protocol for biocompatibility assessment can guide us on how to develop a "safe" Mg-based implant prior to bio-efficacy evaluation.

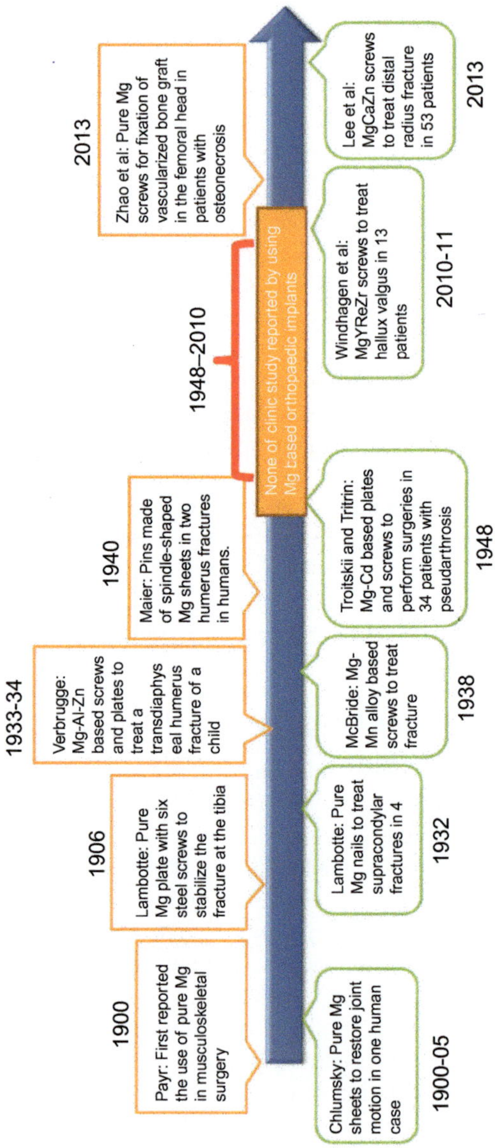

Fig. 4.2 Human case studies using Mg-based orthopedic implant materials [26].

4.4.1.2 Preclinical studies of Mg or its alloys for orthopedic application

In addition to in vitro assessments, in vivo studies are highly essential for the confirmation of biosafety and effectiveness before the using potential speculative medical devices for clinical trials. In general, tiny animals, for example, mice, rats, or rabbits, can be used for biosafety analysis of implantable devices to look for the potential health risks in patients through a series of biological tests, together with irritation and skin sensitization tests, general toxicity tests, tests for native effects after implantation, and toxicokinetic study in keeping with the ISO standards [26]. In terms of the analysis on bio-efficacy, large animals, for example, sheep/goats, pigs, monkeys, and dogs, are needed to conduct clinical indication-oriented experiments to check the treatment outcomes of the novel medical devices compared to their counterparts. Implant materials based on pure Mg, binary Mg series (e.g., Mg-Sr, Mg-Ag, Mg- La/Nd/Ce, Mg-Ca and Mg-Zn), and ternary and quaternary series (Mg-Al-Zn, Mg-Ca-Zn, Mg-Zn-Mn, Mg-Zn-Zr, Mg-Nd-Zn-Zr, etc.) are tested for their biocompatibilities, and repair the effectiveness on fracture healing and other orthopedic traumas. Even though large variety of animals are subjected to in vivo testing of Mg alloys, the major limitation is that the Mg-based devices are inserted within the femoral cavity or shaft which is not the exact reciprocation of the clinical indication-oriented surgeries. As a result, these might cause discrepancies in healing outcomes between in vivo experimental studies and clinical trials. To overcome this limitation, in recent times, four specially developed disease models, that is, anterior cruciate ligament reconstruction (ACLR), avulsion fracture in distal femur, ulna fracture, and femur fracture, have been developed in animals to impersonate clinical signs, in turn aiming to prove the screening of Mg-based implant materials which are used for clinical trials. Mg plates and screws developed by Chaya et al. were implanted on rabbit ulna fracture [27]. Fracture healing took place and also significant bone growth was observed in 8 weeks which in turn matured or remodeled after 16 weeks. The healed ulna with the Mg implant was subjected to flexural strength testing using bending test. The flexural strength of the healed ulnae with the Mg implant was evaluated using bending test and it was observed that the Mg ions degraded to stimulate new bone formation and as a result significant stability was retained. Femoral fracture fixation in mice model was carried out using intramedullary Mg-Ag pins. These pins were used to repair the augmented callus formed in mice. The in vitro results showcased that the degradation products can be partially ascribed to impaired activity of osteoclasts. The intracondylar fracture fixation, which is a low load-bearing implant site, was modified with high-purity Mg screws and its performance was compared with that of PLLA screws. The bone mass and bone mineral density was evaluated and it was observed that Mg screws possessed increased bone mass and better osseointegration. The implanted screws had stable fixation and resulted in enhanced fracture healing with controlled degradation. Cheng et al. evaluated the in vivo behavior of pure Mg screws in comparison with Ti screws for reconstruction of semitendinosus autograft to the femoral tunnel in ACL reconstruction in rabbit [28]. Table 4.1 presents the details of the different Mg alloys subjected to in vivo analysis and the gas formation and degradation behavior has been explained [26].

Table 4.1 The biological response of Mg alloys as orthopedic materials with coatings [26]

Implants	Treatment/surface modification	Implant site	Period time	Gas formation	New bone	Bone contact	Degradation rate or residual implant%
LAE442 rods, extruded	MgF_2 coating	The medial femoral condyle of rabbits	12 weeks	–	+	+	4%
AZ31B screws	Fluoride coating	Femur of rabbits	3 months	–	+	++	NA
AZ31B screws	Si-containing coating	Femoral mid-diaphyseal region in rabbits	21 weeks	NA	++	++	NA
AZ91 rods	Diopside/micro arc oxidation (MAO) coating	Greater trochanter of rabbits	2 months	–	+++	NA	$7\,mg/cm^2$
AZ91 rods	Plasma treated	Lateral epicondyle in rats	2 months	NA	++	++	98%
ZX50 pins	MAO	Femur of the rats	16 weeks	+	++	++	0.25 mm/year
WE43 rivet-screws	Plasma electrolytic coating (PEO)	Mandibular bone of mini-pigs	24 weeks	+	++	++	NA
ZK60 rod	MAO	Femoral shaft of rabbit	12 weeks	+	+	+	NA
Mg-3Zn-0.8Zr rod	MgF_2 coating	Femora of rabbits	3 months	–	++	+	23.85%

AZ91 rod	Fluoridated hydroxyapatite (F-HA)/MAO	Greater trochanter of rabbits	2 months	–	++	NA	4 mg/cm^2
AZ31 rod	Beta-tri calcium phosphate (β-TCP) coating	Femora of rats	12 weeks	NA	++	++	37%
AZ31 disk	Octa calcium phosphate (OCP) + HA coating	Back of mice	16 weeks	+	NA	NA	NA
Mg-Zn-Ca alloy rod	Calcium deficient (Ca-def) HA coating	Femur of rabbits	24 weeks	NA	++	++	0.15 mm/year
High purity Mg screw	HA coating	Tibial defect of rabbits	12 weeks	NA	++	++	33.60%
Mg-Nd-Zn-Zr alloy plates and screws	Ca-P coating [$CaHPO_4 \cdot 2H_2O$]	Tibiae of rabbits	18 weeks	+	++	NA	NA

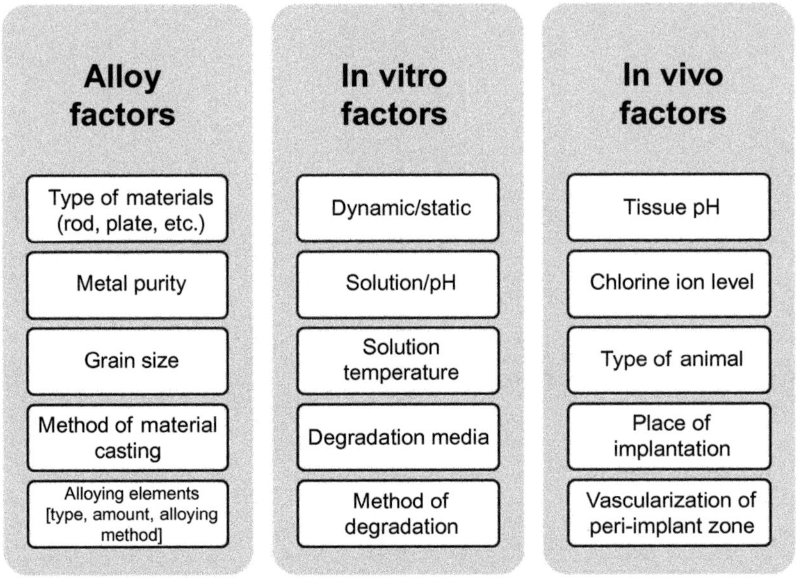

Fig. 4.3 Factors influencing Mg-based biodegradable implants.

Overall, to summarize, the factors influencing Mg and Mg alloys for biomedical applications has been depicted in Fig. 4.3. It can be clearly understood that these factors are interdependent and hence, it has to be well planned to develop the Mg-based alloys for the effective utilization of its properties and to obtain a commercially viable biodegradable implant material.

4.5 Magnesium-based nanocomposites

Nanocomposites are composites in which at least one of the phases shows dimensions in the nanometer range. Nanocomposites are the materials which tend to have nanoparticulates dispersed in the matrix. Nanocomposite materials have emerged as suitable alternatives to overcome limitations of microcomposites and monolithics, while posing preparation challenges related to the control of elemental composition and stoichiometry in the nanocluster phase. They are reported to be the materials of 21st century in the view of possessing design uniqueness and property combinations that are not found in conventional composites. The general understanding of these properties is yet to be reached, even though the first inference on them was reported as early as 1992.

In the last two decades several processing techniques have been developed to fabricate these nanocomposites and they are widely used for almost all the applications in the world. The several methods that are currently available to develop these nanocomposites are shown in Fig. 4.4. Development of magnesium-based nanocomposites have been initiated as a part of improving the mechanical properties of pure Mg and at the

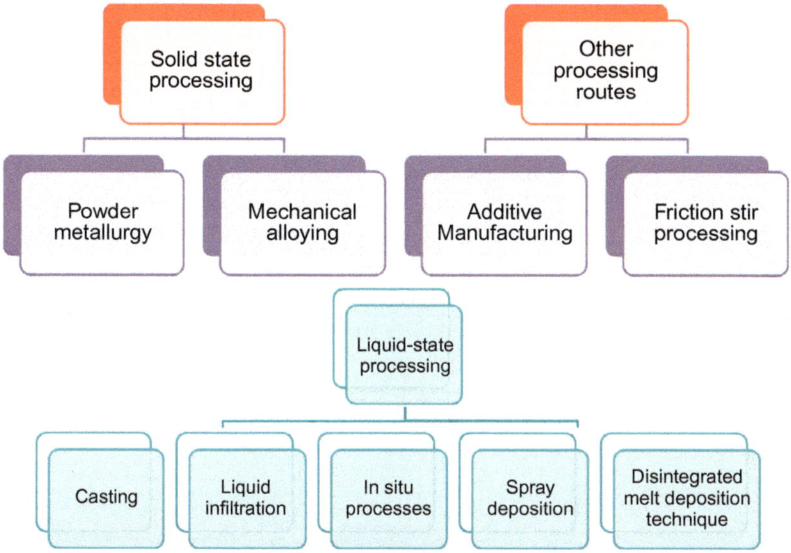

Fig. 4.4 Schematic representation of the different processing routes for the development of nanocomposites.

same time use the advantage of the surface-to-volume ratio enhancement and simpler processes to develop these materials. The most prominent methods developed using these composites are as follows: powder metallurgy (PM), microwave sintering, and disintegrative melt deposition. The PM technique typically involves mixing the metal powders and nano-reinforcements in a desired composition by simple blending or by high-energy mechanical alloying. The blended powders are pressed to form green compacts and then sintered using resistance heating. The authors have developed a sintering setup which made use of microwave energy to rapidly sinter the magnesium nanocomposites without the need for an inert protective atmosphere. Details of the hybrid heating microwave sintering setup can be referenced from Ref. [29].

4.5.1 Disintegrated melt deposition (DMD) technique

The DMD is a unique technique developed under the spray forming processes, which is a highly cost-effective technique similar to that of conventional casting process, and it was designed to develop metallic nanomaterials and hence can be extensively used for the development of Mg-based nanocomposites as shown by Gupta et al. [29]. This DMD process uses low impinging gas jet velocity at high temperature to develop the nanocomposites. The synthesized nanomaterials through DMD is transferred to a graphite crucible along with alloying elements or reinforcements and heated to 750°C under controlled environment in a heat-resistance furnace. To achieve homogeneity, the superheated slurry is stirred at 465 rpm for 5–10 min using a twin blade with a pitch of 45 degrees. The impeller is coated with Zirtex 25 (86% ZrO_2, 8.8% Y_2O_3, 3.6% SiO_2, 1.2% K_2O and Na_2O, and 0.3% trace inorganic) to avoid iron contamination of the molten metal. The melt is then released through a 10-mm diameter orifice at the base of the crucible. The composite melt is disintegrated by two jets of argon gas orientated normal to the melt stream before being deposited onto a metallic substrate. Preform of 40 mm diameter is obtained following the deposition stage. A schematic diagram of the DMD setup is shown in Fig. 4.5 [29]. From the literature it can be seen that this DMD process can be used for the development of metallic materials with alloying inclusions, bulk metallic glass, and composites with micro- and nano-reinforcements.

Mg-5.6 Ti and Mg-9.6Ti (wt%) composites were developed using DMD process followed by hot extrusion by Hassan et al. [30]. The Ti particulates were in the size range of ~19 ± 10 μm, the 0.2% yield strength and the fracture strain of the Mg-5.6 wt% composite increased by 63 MPa and 44%, conversely the ultimate tensile strength decreased by 4 MPa when compared with pure Mg. In the case of Mg-9.6Ti the tensile properties showed detrimental effect when compared with the Mg-5.6Ti. Graphene nanoplatelets (GNPs) was added to Mg-Ti microcomposite developed through PM

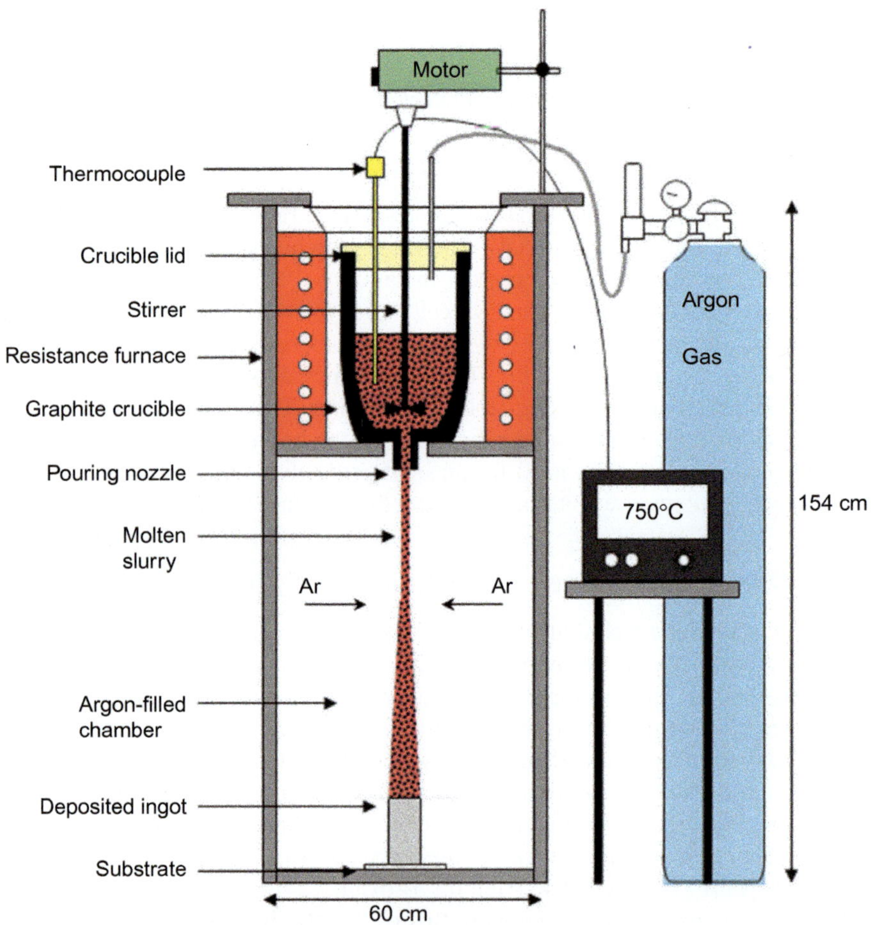

Fig. 4.5 Schematic representation of the DMD process [29].

route and the effect of strength and ductility was observed by Rashad et al. [31]. The tensile properties of this composite was compared with that of pure Mg and it showed a drastic increase in the yield strength, ultimate tensile strength, and fracture strain by 8 MPa, 30 MPa, and 243% respectively. The bubble chart representations of the comparison of microhardness with grain size and yield strength with ductility are shown in Figs. 4.6 and 4.7. From this it can be clearly seen that ceramic composites developed using DMD process covers a wide range of mechanical properties and can be utilized effectively for biomedical applications.

4.5.2 Electrochemical behavior of Mg nanocomposites

The electrochemical behavior of three different Mg nanocomposites, viz., Mg-TiN, Mg-TiO_2, and Mg-TiC developed using DMD process has been evaluated in Kokubo's SBF [32]. The three nanoparticles were dispersed in the range of 0.58, 0.97, 1.98, and 2.5 wt% in the magnesium matrix to evaluate which combination is resulting in improved mechanical and electrochemical properties compared to pure-Mg. A three-electrode potentiostat (ACM Gill) was used to carry out the corrosion testing with the Mg substrates acting as working electrode, platinum as counter electrode to measure the variation of current in the system, and the standard calomel electrode (SCE) as reference electrode to measure the variation in the potential due to the ionic interaction between the substrate surface and the electrolyte (SBF). The test was carried out with pH of 7.4 and initially the open circuit potential was performed for duration of 3600 s to develop the potential stability of the electrolyte and the substrate. Further the potentiodynamic polarization studies were carried out with a potential range of −2 to −1 V with sweep rate of 1 mV/min.

4.5.2.1 Potentiodynamic polarization

The Tafel graph refers to the graph plotted between rest potential and corrosion current density and it provides information about the rest potential, that is, the potential in which the system comes to equilibrium where no cathodic or anodic reactions will take place. Further this graph depicts the cathodic and anodic plot and the slopes measured from the anodic and cathodic region of the plot is used to calculate the corrosion rate along with the corrosion current density. From the polarization results obtained it becomes evident that all the combinations of nanocomposites exhibited increased rest potential compared to that of the pure Mg. The rest potential of pure Mg was −1.68 V, which is about 12% and 15% less compared to Mg-0.58 and Mg-2.5% TiN, respectively. Similarly, the Mg-0.58 TiO_2 exhibited an improved potential of 0.3 V compared with pure Mg. In the case of Mg-0.97 and 1.98% combination of TiN, TiO_2, and TiC, a similar range of potential ranging from −1.45 to −1.55 V was observed. However, Mg-2.5 TiO_2 (−1.63 V) alone showed no significant variation when compared with pure Mg. In the case of TiC nanocomposite, Mg-2.5% TiC resulted in increased rest potential, however, the current density of this composition was also higher and hence resulted in increased corrosion rate. The passivation trend of 2.5% TiC was not appreciable compared with that of 0.58% TiC. The corrosion rate of the samples was evaluated with the help of the corrosion current density and is shown in

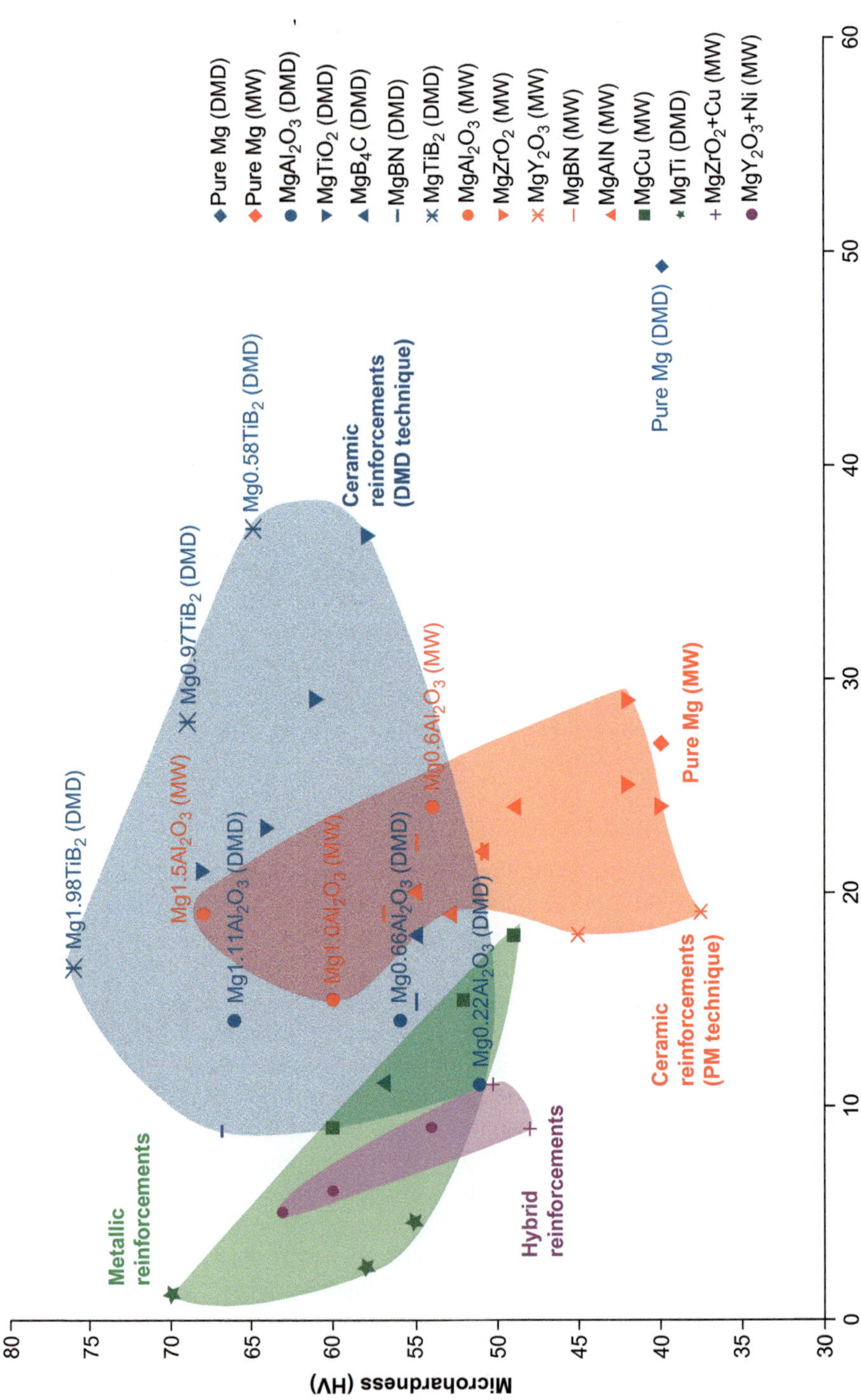

Fig. 4.6 Bubble chart showing the microhardness and grain size of magnesium nanocomposites [29].

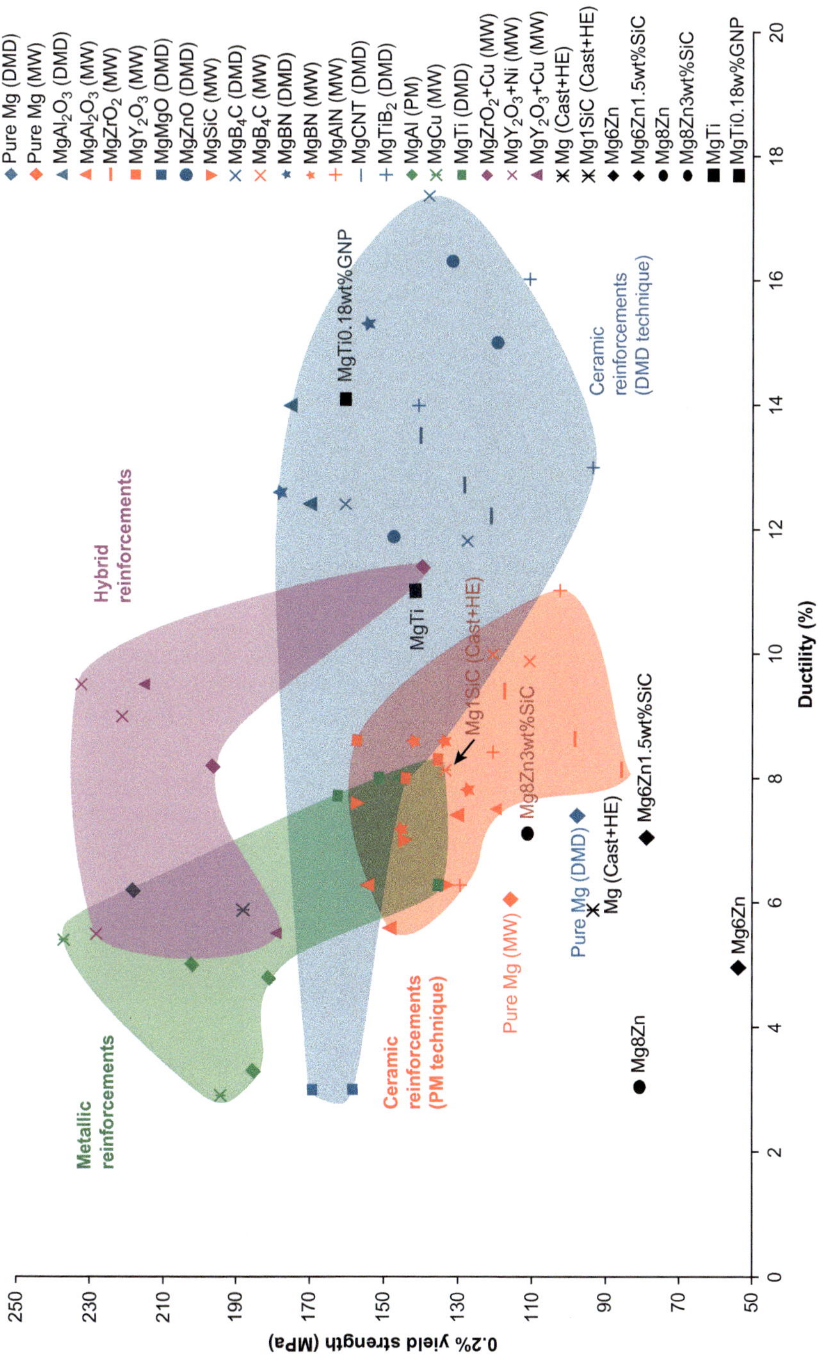

Fig. 4.7 Bubble chart showing the yield strength and density of magnesium nanocomposites [29].

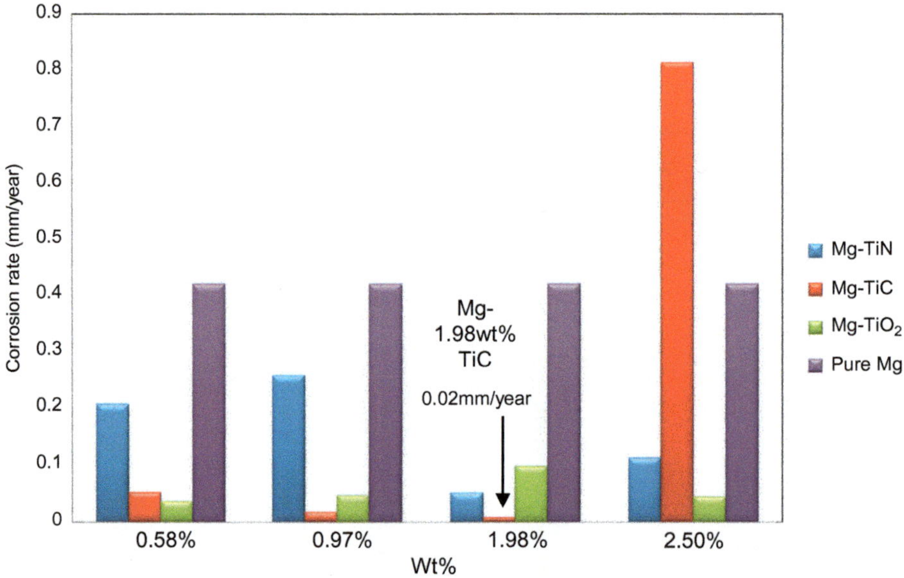

Fig. 4.8 Corrosion rate comparison of pure Mg and Mg-based nanocomposites in simulated body fluid (SBF).

Fig. 4.8. Overall all the samples underwent pitting corrosion. In comparison, 1.98 wt% TiO_2, TiN, and TiC resulted in low corrosion rate compared to other combinations. Among them Mg-1.98 wt% TiC resulted in the lowest corrosion rate of 0.02 mm/year. Mg-0.58 wt% combination of the all three composites exhibited higher corrosion rate compared to other combinations. Overall, Mg-2.5 wt% TiC exhibited highest corrosion rate (0.8 mm/year) which is greater than pure Mg (0.45 mm/year). Mg-2.5 wt% TiC underwent faster diffusion past the rest potential, which resulted in the increased corrosion rate compared to the other compositions.

4.6 Surface modification of Mg alloys

4.6.1 Effect of surface modification

Only concern with the usage of Mg implants is the poor corrosion resistance. Rapid degradation of Mg alloys leads to the evolution of the gas and subsequent pit formation. Only possibility to improve corrosion resistance while still retaining the bulk properties is surface modification. In terms of corrosion, which is the major concern for Mg implant material, even though alloying additions have resulted in improved corrosion resistance, the controlled degradation rate desired to use it for orthopedic applications (3–6 months) is not achieved. According to the requirements for diverse applications, suitable permanence is necessary for medical implants. Further, the alloying additions have the limitations of toxicity and the degradation product could pose threat to the human body. It is, therefore, essential to build a negotiation between the metal surface

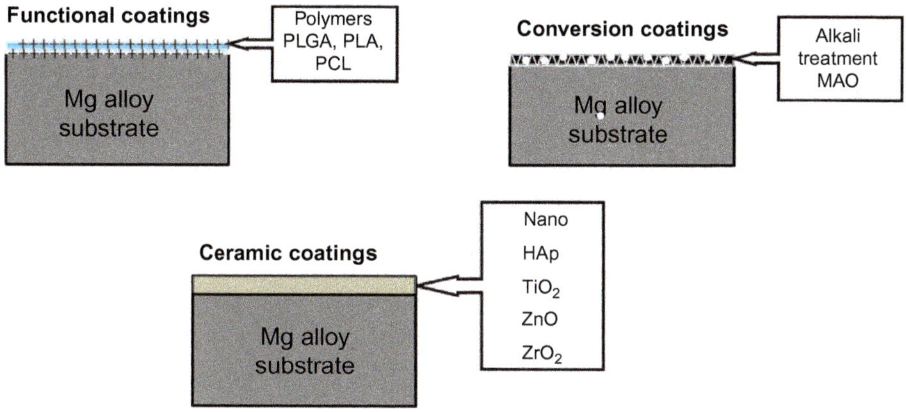

Fig. 4.9 Schematic representation of the coating processes on Mg alloy substrates.

and bulk properties. Hence, to overcome this limitation, surface modification will be an apt solution in which several processes have been studied and developed.

Till date for Mg and Mg alloys, different surface modification technologies that are under research can be classified into (i) functional coatings; (ii) conversion coatings; and (iii) ceramic coating processes as shown schematically in Fig. 4.9.

4.6.1.1 Functional coatings

First, the functional coatings are one in which polymers are coated on the Mg alloy and this coated surface helps in improving the bioactivity as well as reduces the degradation rate of the Mg alloy. The most commonly used functional coatings are PLA, PLGA, PCL, and chitosan [33]. Even though the bioactivity of the Mg alloy is enhanced with these polymers, the main limitation is their mechanical integrity to the surface. Hydrogen gas evolution is the major limitation in the case of Mg alloys when exposed to body fluid environment. This H_2 gas formation underneath the deposited polymer coating layer results in the bulging of the coating and as polymers poses poor mechanical properties it eventually losses the integrity with the Mg surface. Another limitation for polymer coatings is the identification of suitable dispersion medium and deposition system to achieve uniform coating on the Mg substrates.

4.6.1.2 Conversion coatings

On the other hand, conversion coatings were extensively researched on Mg alloys to enhance the corrosion resistance as this coating is obtained by a simple cost-effective process. In the case of conversion coating, the Mg substrate is exposed to alkali treatment in which a uniform thick oxide layer is formed on the surface.

Several surface modification techniques are used to enhance the corrosion resistance of Mg alloys: (i) chemical method—acid etching, (ii) alkaline passivation, (iii) conversion coating, (iv) microarc oxidation (MAO), and (v) biomimetic coating (solution immersion in supersaturated calcification solution).

Mg alloy exposed to atmospheric condition results in the formation of Mg oxide layer on the surface which is nonuniform and highly porous. This native oxide layer is several micrometers thick and due to its porous and nonuniform structure it is not sufficient to protect the Mg from undergoing corrosion. To obtain a protective layer, a biomimetic coating of around 10 μm thicknesses is formed by immersing in supersaturated calcification solution (SCS), which helps in enhancing the corrosion resistance. The Ca-P crystals are developed on the surface of the Mg alloy substrates. In this process the native oxide layer or inner layer is subjected to dissolution due to the environment in which the Mg is exposed and reprecipitation of the Ca-P crystalline phase takes place from the SCS medium. The SCS formation is followed by alkali heat treatment to improve the adhesion strength of the Ca-P layer. The apatite forming ability is highly governed by the surface hydration by alkali treatment and the subsequent heat treatment resulting in the improved crystallinity of the same.

The MAO is a high-voltage plasma-assisted anodic oxidation process, which is most widely used to modify the surface of Mg and its alloys. It is also termed as plasma electrolytic oxidation (PEO) and microplasma oxidation (MPO). The main objective of PEO is to develop highly adherent ceramic oxide coating on the surface of Mg alloys with the help of plasma discharges generated during the process, which in turn causes fractional short-term melting of the native oxide layer formed on the surface of Mg. The main advantage of MAO coating [34] is that it helps in improving the hardness of the Mg substrate which in turn improves the wear and corrosion resistance and provides better thermal stability and dielectric properties. By imparting hardness the surface roughness of the Mg surface is increased post MAO and hence, it increases the adhesion of organic or polymeric topcoats developed through several other coating processes. The main limitation in the case of MAO is that, the dense oxide layer developed in this process is porous and hence enhancement in corrosion resistance cannot be obtained. A secondary coating has to be made over the MAO using conventional coating processes to seal the porous surface and reduce the degradation of the Mg substrate.

4.6.1.3 Surface coating processes

In spite of the fact that conversion coatings reduced the degradation of Mg substrates, the main limitation of this process is that the coating formed on the surface is porous and needs secondary treatments to further control the degradation. To overcome this limitation several other coating processes were developed viz., (i) electrodeposition and electroless deposition; (ii) electrophoretic deposition; (iii) sol-gel coatings; (iv) chemical vapor deposition; (v) physical vapor deposition; (vi) sputtering; (vii) pulsed laser deposition (PLD); and (viii) thermal spraying.

Using the above-mentioned coating processes, the coatings with materials that are biocompatible, biomimetic, nontoxic, and corrosion resistant were developed. Several nanopowders such as TiO_2 [35], ZnO [36], ZrO_2 [37], and hydroxyapatite (HAp) [38] are used to coat metallic implants to improve surface-related properties. TiO_2 is widely used either alone or in combination with HAp as a coating to various metallic implant materials to improve corrosion resistance. However, earlier results have showed that

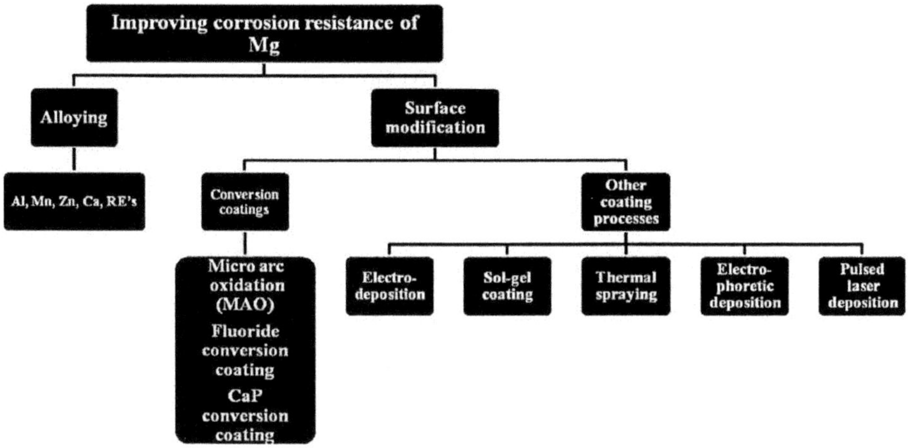

Fig. 4.10 Schematic illustration of the various processes for improving corrosion resistance of Mg.

TiO_2 nanoparticles have the possibility to develop tumor growth by immunomodulation of B and T lymphocytes, macrophages, and natural killer (NK) cells [39,40]. Hence, it would be most suitable to coat the surface only with HAp because, in addition to being biocompatible, HAp coating can mimic inorganic component of bone. In addition, it also provides an osteophilic surface to initiate bone bonding.

The most commonly studied coating processes to develop HAp coating on Mg alloys are shown in Fig. 4.10.

Thermal spraying is one of the most commonly used techniques to coat ceramics such as HAp on implant surfaces to make the surface bioactive. However, it is a high-temperature technique and suffers from drawback such as spallation of HAp coating which can possibly occur at three locations: (1) between lamellae of the coating, (2) coating-metal interface, and (3) coating-bone interface. The temperature at which HAp coating is performed is much higher than the melting point of magnesium (~600°C).

The most common coating processes that have drawn the attention of researchers for the development of thin and adherent coatings on Mg are electrodeposition and sol-gel coating process. Electrodeposition is performed potentiostatically by varying voltages with the Mg substrate as cathode and calomel electrode as anode. In the case of sol-gel coating process, the HAp deposition is obtained by using calcium and phosphate-based precursor and immersing the metal substrate in the solution and by varying the voltage between 100 and 300V DC and temperature between 100°C and 400°C. The main limitation of these two coating processes is the high temperature at which the deposition is performed to attain crystalline HAp deposition. In spite of the fact that these processes are simple and cost effective, the lower melting point of Mg alloys (~600°C) hinder their application. Hence, it is of great importance to opt for a coating technique that could be carried out at room temperature (RT) and also can retain the crystallinity of the HAp powder coated on the substrate. Crystalline HAp is highly preferred because it exhibits slower degradation in vivo compared to amorphous HAp [41]. Electrophoretic deposition (EPD) and PLD fit well to the desired process of depositing the HAp on Mg alloy under ambient conditions [42]. The main advantage of EPD is that the particles suspended in the colloidal solution tend to undergo agglomeration which can be removed by prefiltering and the insulating properties of depositing material can influence the deposition and the defect regions can be covered with the coating by increasing the electric field, that is, increasing the voltage which in turn will result in the efficient packing and uniform deposition of the material [43]. Similarly, using PLD technique, the atomic and molecular species are condensed using high kinetic energy which results in the production of uniform thin films of the target material. The PLD coating technique effectively produces crystalline and highly adherent films [44]. The PLD provides special advantages for the development of chemically complex materials. Various advantages such as fast deposition time, higher flexibility, maintenance of stoichiometry, energetic evaporants, and the simplicity for growth of multilayered structures.

Apart from these coatings several other coating processes have been developed and also under development with a concern of improving the performance of the Mg alloys as an effective and safe biodegradable implant material.

4.7 Future aspects

In spite of the development of several magnesium alloys and nanocomposites for biomedical applications, in vivo studies are still scarce. Also the in vitro and in vivo results have large variation in many cases and hence development of protocols and standards for the study of Mg samples is required. Recent studies have proved that the Mg alloys and nanocomposites are mechanically stable even after in vivo testing, however, their complete degradation hasn't been evaluated. Hence, the complete degradation study of Mg has to be evaluated which will help in better understanding the degradation kinetics so that the studies can be extended to big animals and human trials effectively.

References

[1] D.F. Williams, Definitions in Biomaterials: Proceedings of a Consensus Conference of the European Society for Biomaterials, 1987.

[2] M. Marlowe, J.E. Parr (Eds.), Modularity of Orthopedic Implants, American Society for Testing and Materials (ASTM), Pennsylvania, 1997.

[3] M.P. Staiger, A.M. Pietak, J. Huadmai, G. Dias, Magnesium and its alloys as orthopedic biomaterials: a review, Biomaterials 27 (2006) 1728–1734, https://doi.org/10.1016/j.biomaterials.2005.10.003.

[4] F. Pi-Sunyer, Obesity. Mod. Nutr. Heal. Dis. (1999) 1395–1414, https://doi.org/10.1016/0738-3991(95)90004-7.

[5] R.D. Grubbs, M.E. Maguire, Magnesium as a regulatory cation: criteria and evaluation, Magnesium 6 (1987) 113–127. http://www.ncbi.nlm.nih.gov/pubmed/3306178, (Accessed 21 November 2016).

[6] G.A. Quamme, Renal magnesium handling: new insights in understanding old problems. Kidney Int. 52 (1997) 1180–1195, https://doi.org/10.1038/ki.1997.443.

[7] M. Schinhammer, A.C. Hänzi, J.F. Löffler, P.J. Uggowitzer, Design strategy for biodegradable Fe-based alloys for medical applications, Acta Biomater. 6 (2010) 1705–1713, https://doi.org/10.1016/j.actbio.2009.07.039.

[8] C.K. Yuen, W.Y. Ip, Theoretical risk assessment of magnesium alloys as degradable biomedical implants, Acta Biomater. 6 (2010) 1808–1812, https://doi.org/10.1016/j.actbio.2009.11.036.

[9] G.L. Song, A. Atrens, Corrosion mechanisms of magnesium alloys, Adv. Eng. Mater. 1 (1999) 11–33, https://doi.org/10.1002/(SICI)1527-2648(199909)1:1<11::AID-ADEM11>3.0.CO;2-N.

[10] M. Pourbaix, Atlas of Electrochemical Equilibria in Aqueous Solutions, second ed., National Association of Corrosion Engineers, Houston, TX, 1974.

[10a] G. Manivasagam, S. Suwas, Biodegradable Mg and Mg based alloys for biomedical implants. Mater. Sci. Technol. 30 (5) (2014) 515–520, https://doi.org/10.1179/1743284713Y.0000000500.

[11] J.J. Jacobs, J.L. Gilbert, R.M. Urban, Current concepts review—corrosion of metal orthopaedic implants corrosion of metal orthopaedic implants, J. Bone Joint Surg. 80 (2) (1998) 268–282.

[12] G. Mani, M.D. Feldman, D. Patel, C.M. Agrawal, Coronary stents: a materials perspective, Biomaterials 28 (2007) 1689–1710, https://doi.org/10.1016/j.biomaterials.2006.11.042.

[13] M. Moravej, D. Mantovani, Biodegradable metals for cardiovascular stent application: interests and new opportunities, Int. J. Mol. Sci. 12 (2011) 4250–4270, https://doi.org/10.3390/ijms12074250.

[14] F. Witte, Reprint of: The history of biodegradable magnesium implants: a review, Acta Biomater. 23 (2015) S28–S40, https://doi.org/10.1016/j.actbio.2015.07.017.

[15] J.L. Nadler, S. Goodson, R.K. Rude, Evidence that prostacyclin mediates the vascular action of magnesium in humans, Hypertension 9 (1987) 379–383.

[16] M. Gawaz, I. Ott, A.J. Reininger, F.J. Neumann, Effects of magnesium on platelet aggregation and adhesion. Magnesium modulates surface expression of glycoproteins on platelets in vitro and ex vivo, Thromb. Haemost. 72 (1994) 912–918.

[17] B. Heublein, R. Rohde, V. Kaese, M. Niemeyer, W. Hartung, A. Haverich, Biocorrosion of magnesium alloys: a new principle in cardiovascular implant technology? Heart 89 (2003) 651–656, https://doi.org/10.1136/heart.89.6.651.

[18] C. Di Mario, H. Griffiths, O. Goktekin, N. Peeters, J. Verbist, M. Bosiers, K. Deloose, B. Heublein, R. Rohde, V. Kasese, C. Ilsley, R. Erbel, Drug-eluting bioabsorbable magnesium stent, J. Interv. Cardiol. 17 (2004) 391–395, https://doi.org/10.1111/j.1540-8183.2004.04081.x.

[19] P. Zartner, R. Cesnjevar, H. Singer, M. Weyand, First successful implantation of a biodegradable metal stent into the left pulmonary artery of a preterm baby, Catheter. Cardiovasc. Interv. 66 (2005) 590–594, https://doi.org/10.1002/ccd.20520.

[20] R. Waksman, R. Pakala, P.K. Kuchulakanti, R. Baffour, D. Hellinga, R. Seabron, F.O. Tio, E. Wittchow, S. Hartwig, C. Harder, R. Rohde, B. Heublein, A. Andreae, K.-H. Waldmann, A. Haverich, Safety and efficacy of bioabsorbable magnesium alloy stents in porcine coronary arteries, Catheter. Cardiovasc. Interv. 68 (2006) 607–617, https://doi.org/10.1002/ccd.20727.

[21] R. Waksman, R. Erbel, C. Di Mario, J. Bartunek, B. De Bruyne, F.R. Eberli, P. Erne, M. Haude, M. Horrigan, C. Ilsley, D. Böse, Early- and long-term intravascular ultrasound and angiographic findings after bioabsorbable magnesium stent implantation in human coronary arteries, JCIN 2 (2009) 312–320, https://doi.org/10.1016/j.jcin.2008.09.015.

[22] P. Lu, H. Fan, Y. Liu, L. Cao, X. Wu, X. Xu, Controllable biodegradability, drug release behavior and hemocompatibility of PTX-eluting magnesium stents, Colloids Surf. B Biointerfaces 83 (2011) 23–28, https://doi.org/10.1016/j.colsurfb.2010.10.038.

[23] M. Haude, R. Erbel, P. Erne, S. Verheye, H. Degen, D. Böse, I. Vermeersch, P. Wijnbergen, N. Weissman, F. Prati, R. Waksman, J. Koolen, Safety and performance of the drug-eluting absorbable metal scaffold (DREAMS) in patients with de-novo Bouchi and Gogas. Global Cardiology Science and Practice 2015:63 coronary lesions: 12 month results of the prospective, multicentre, first-in-man BIOSOL, Lancet 381 (2013) 836–844.

[24] J. Nagels, M. Stokdijk, P.M. Rozing, Stress shielding and bone resorption in shoulder arthroplasty, J. Shoulder Elb. Surg. 12 (2003) 35–39, https://doi.org/10.1067/mse.2003.22.

[25] P. Gill, N. Munroe, Review on magnesium alloys as biodegradable implant materials, Int. J. Biomed. Eng. Technol. 10 (2014) 383–398, https://doi.org/10.1504/IJBET.2012.052400.

[26] D. Zhao, F. Witte, F. Lu, J. Wang, J. Li, L. Qin, Biomaterials current status on clinical applications of magnesium-based orthopaedic implants: a review from clinical translational perspective, Biomaterials 112 (2017) 287–302, https://doi.org/10.1016/j.biomaterials.2016.10.017.

[27] A. Chaya, S. Yoshizawa, K. Verdelis, N. Myers, B.J. Costello, D.-T. Chou, S. Pal, S. Maiti, P.N. Kumta, C. Sfeir, In vivo study of magnesium plate and screw degradation and bone fracture healing, Acta Biomater. 18 (2015) 262–269, https://doi.org/10.1016/j.actbio.2015.02.010.

[28] P.-Y. Cheng, W.-R. Liao, The relationship between test anxiety and achievement in accounting students with different cognitive styles: the mediating roles of self-regulation, Int. Res. Educ. 4 (2016) 14, https://doi.org/10.5296/ire.v4i2.9177.

[29] M. Gupta, W.L.E. Wong, Materials characterization magnesium-based nanocomposites: lightweight materials of the future, Mater. Charact. 105 (2015) 30–46, https://doi.org/10.1016/j.matchar.2015.04.015.

[30] S.F. Hassan, M. Gupta, Development of ductile magnesium composite materials using titanium as reinforcement, J. Alloys Compd. 345 (2002) 246–251, https://doi.org/10.1016/S0925-8388(02)00413-9.

[31] M. Rashad, F. Pan, A. Tang, Y. Lu, M. Asif, S. Hussain, J. She, J. Gou, J. Mao, Effect of graphene nanoplatelets (GNPs) addition on strength and ductility of magnesium-titanium alloys, J. Magnesium Alloys 1 (2013) 242–248, https://doi.org/10.1016/j.jma.2013.09.004.

[32] T. Kokubo, H. Takadama, How useful is SBF in predicting in vivo bone bioactivity? Biomaterials 27 (2006) 2907–2915, https://doi.org/10.1016/j.biomaterials.2006.01.017.

[33] S. Agarwal, J. Curtin, B. Duffy, S. Jaiswal, Biodegradable magnesium alloys for orthopaedic applications: a review on corrosion, biocompatibility and surface modifications, Mater. Sci. Eng. C 68 (2016) 948–963, https://doi.org/10.1016/j.msec.2016.06.020.

[34] T.S.N. Sankara Narayanan, I.S. Park, M.H. Lee, Strategies to improve the corrosion resistance of microarc oxidation (MAO) coated magnesium alloys for degradable implants: prospects and challenges, Prog. Mater. Sci. 60 (2014) 1–71, https://doi.org/10.1016/j.pmatsci.2013.08.002.

[35] X. Nie, A. Leyland, A. Matthews, Deposition of layered bioceramic hydroxyapatite/TiO_2 coatings on titanium alloys using a hybrid technique of micro-arc oxidation and electrophoresis, Surf. Coat. Technol. 125 (2000) 407–414, https://doi.org/10.1016/S0257-8972(99)00612-X.

[36] D. Gopi, N. Murugan, S. Ramya, L. Kavitha, Electrodeposition of a porous strontium-substituted hydroxyapatite/zinc oxide duplex layer on AZ91 magnesium alloy for orthopedic applications, J. Mater. Chem. B 2 (2014) 5531–5540, https://doi.org/10.1039/c4tb00960f.

[37] Z. Yao, H. Gao, Z. Jiang, F. Wang, Structure and properties of ZrO_2 ceramic coatings on AZ91D Mg alloy by plasma electrolytic oxidation, J. Am. Ceram. Soc. 91 (2008) 555–558, https://doi.org/10.1111/j.1551-2916.2007.02155.x.

[38] R. Rojaee, M. Fathi, K. Raeissi, Electrophoretic deposition of nanostructured hydroxyapatite coating on AZ91 magnesium alloy implants with different surface treatments, Appl. Surf. Sci. 285 ((2013) 664–673, https://doi.org/10.1016/j.apsusc.2013.08.108.

[39] E.-Y. Moon, G.-H. Yi, J.-S. Kang, J.-S. Lim, H.-M. Kim, S. Pyo, An increase in mouse tumor growth by an in vivo immunomodulating effect of titanium dioxide nanoparticles, J. Immunotoxicol. 8 (2011) 56–67, https://doi.org/10.3109/1547691X.2010.543995.

[40] E. Demir, D. Burgucu, F. Turna, S. Aksakal, B. Kaya, Determination of TiO_2, ZrO_2, and Al_2O_3 nanoparticles on genotoxic responses in human peripheral blood lymphocytes and cultured embyronic kidney cells, J. Toxicol. Environ. Health A 76 (2013) 990–1002, https://doi.org/10.1080/15287394.2013.830584.

[41] L. Sun, C.C. Berndt, K.A. Khor, H.N. Cheang, K.A. Gross, Surface characteristics and dissolution behavior of plasma-sprayed hydroxyapatite coating, J. Biomed. Mater. Res. 62 (2002) 228–236, https://doi.org/10.1002/jbm.10315.

[42] M. Sankar, S. Suwas, S. Balasubramanian, G. Manivasagam, Comparison of electrochemical behavior of hydroxyapatite coated onto WE43 Mg alloy by electrophoretic and pulsed laser deposition, Surf. Coat. Technol. (2016) https://doi.org/10.1016/j.surfcoat.2016.10.077.

[43] A. Tahmasbi Rad, M. Solati-Hashjin, N.A.A. Osman, S. Faghihi, Improved biophysical performance of hydroxyapatite coatings obtained by electrophoretic deposition at dynamic voltage, Ceram. Int. 40 ((2014) 12681–12691, https://doi.org/10.1016/j.ceramint.2014.04.116.

[44] V. Nelea, C. Ristoscu, C. Chiritescu, C. Ghica, I.N. Mihailescu, H. Pelletier, P. Mille, A. Cornet, Pulsed laser deposition of hydroxyapatite thin films on Ti-5Al-2.5Fe substrates with and without buffer layers, Appl. Surf. Sci. 168 (2000) 127–131, https://doi.org/10.1016/S0169-4332(00)00616-4.

Multiwalled carbon nanotube-based nanocomposites for artificial bone grafting

5

Ruma Perveen, Inamuddin, Abu Nasar
Advanced Functional Materials Laboratory, Department of Applied Chemistry, Faculty of Engineering and Technology, Aligarh Muslim University, Aligarh, India

5.1 Introduction

With the great progress in medical science and allied areas, artificial bone grafting—an advanced technology in the domain of medical research has also been established over the recent years. A wide range of valuable artificial biomaterials has been developed that have been found successful to repair the defects, or to treat the complete failure of the bones. The primitive autografting and allografting methods of bone replacement suffer from inevitable drawbacks of ruptures and instability along with the risks of transmission of dangerous diseases [1,2].

This has led to a major concern on the development as well as advancement in the discipline of artificial bone grafting. In this context, polymers, ceramics, and metals have drawn significant attention of researchers for the preparation of artificial bones which may overcome the limitations of conventional materials and methods accompanied with long-term stability. Although, metals are extensively used to replace and heal the defects or failure of bone owing to their outstanding mechanical properties, but in order to circumvent the issues of toxicity of metals [3], biocompatible materials like bioceramics and biopolymers are being explored nowadays as an efficient and green alternative of metals for artificial bone grafting [4–7]. Since this is the era of green revolution, the emergence of these fledgling biomaterials is envisaged to give new dimensions to the field of artificial bone grafting. The primary issue associated with these biomaterials is low mechanical strength which is a serious topic of conflict in the current research. Carbon nanotube (CNT) especially multiwalled carbon nanotubes (MWCNTs) are receiving appreciable attention to address this issue as a consequence of its immensely strong mechanical properties. Thus, MWCNT might be a promising candidate to serve as artificial bone grafts by incorporating it with other biomaterials [8].

5.2 Artificial bone grafting

Artificial bone grafting is a surgical operation that involves replacement or treatment of the damaged bone with patient's own bone tissue incorporated with natural or synthetic materials. Since, bone is highly dynamic and vascularized tissue, it is capable

Applications of Nanocomposite Materials in Orthopedics. https://doi.org/10.1016/B978-0-12-813740-6.00006-5

of complete regeneration and remodeling following the mechanism of bone deposition by osteoblasts and bone resorption by osteoclasts [9]. This mechanism facilitates the formation of a completely new natural bone with the removal of the graft material completely. Bone grafts are bioresorbable and biocompatible so as to avoid any antigen-antibody reaction. These can be implanted in the form of fillers or scaffolds which serve as a mineral reservoir for bone regeneration. Bone is composed of a bioceramic termed as hydroxyapatite (HA) which is responsible for the rigidity of the bone and the biopolymeric component (collagen) which is an elastic protein, provides fracture resistance to it. The basic function of bone is to provide structural support to the body and to protect the internal organs such as brain, heart, lungs, etc. Therefore, it is essential to look for such biomaterials which can effectively treat bone diseases or bone failures that comply with osteogenesis, osteoconduction, and osteoinduction processes of bone grafting [10]. Biopolymers such as collagen, chitin, chitosan, polycaprolactone, poly(methyl methacrylate), polydopamine, polyglycolic acid, alginate, etc. and bioceramics like HA have gained significant attention as valuable alternatives to solve the issues prevailing in artificial bone grafting [11–13]. But, the instability and short-term degradation of these biomaterials pose a serious challenge to the researchers. Hence, attempts are being made to develop improved bone tissue scaffolds combining the unique properties of these biomaterials with the excellent mechanical strength of MWCNT. The schematic procedure of artificial bone grafting has been demonstrated in Fig. 5.1.

5.2.1 Strategies for artificial bone grafting

The current state of the art for the development of artificial bone implants is the introduction of cells, biomaterials, and signaling molecules in association into the body, in the form of bioactive constructs called scaffolds (Fig. 5.2). This technique has greatly evolved as reflected by the achievements in the following objectives [2]:

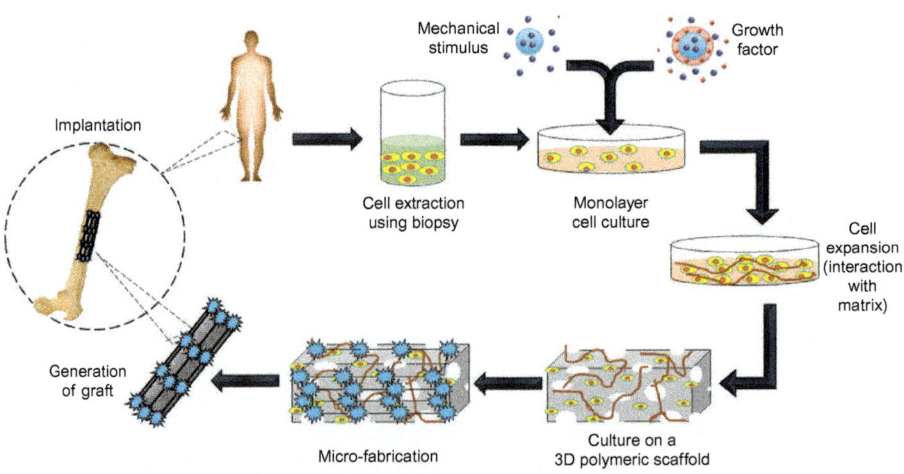

Fig. 5.1 Schematic showing the basic artificial bone grafting methodology.

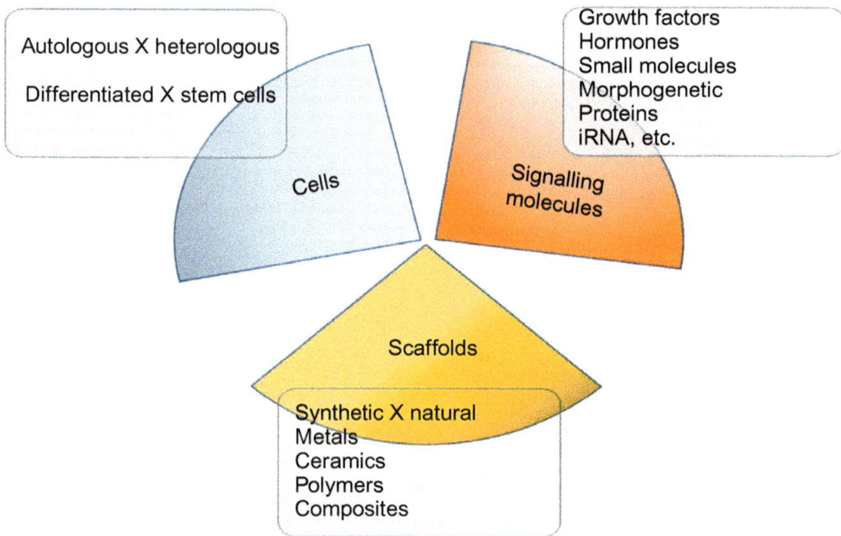

Fig. 5.2 Current strategies for artificial bone grafting.

(a) promotion of cell adhesion and fixation
(b) direct cell distribution and differentiation, however restricting their distribution in the tissue
(c) sustenance of large tissue losses while new tissue is formed
(d) ultimately, new tissue formation

5.3 Carbon nanotube

Carbon is abundantly available in nature in the form of different allotropes among which graphite, graphene, and CNTs have emerged as fascinating biomaterials recently. CNTs possess cylindrical nanoarchitecture with unusual properties worthwhile for a wide spectrum of applications. Among the biocompatible composites, single-walled carbon nanotube (SWCNT) and MWCNT biocomposites have gained extensive attention for multiple applications [14,15]. The classification of CNTs as well as their applications is pictorially represented in Figs. 5.3 and 5.4, respectively.

CNTs are considered as the strongest and stiffest biomaterial yet discovered attributed to their high tensile strength and elastic modulus, respectively. Such mechanical strength arises from the covalent sp^2 bonding between the arrays of carbon atoms. Although, CNTs are benefited with unique combination of mechanical, electrical, and thermal properties but, the toxicity of CNTs is a matter of concern. Therefore, researchers are trying to circumvent this limitation of CNTs by synthesizing biocomposites to withstand its toxic nature as well as to broaden its application in implantable medical devices including bone tissue engineering.

A study reported the improved tracking of cells, distribution of transfection agents, microenvironments sensing, and scaffold incorporation in the host's body by using CNTs. A comparative study of the toxicities of SWCNT and MWCNT was

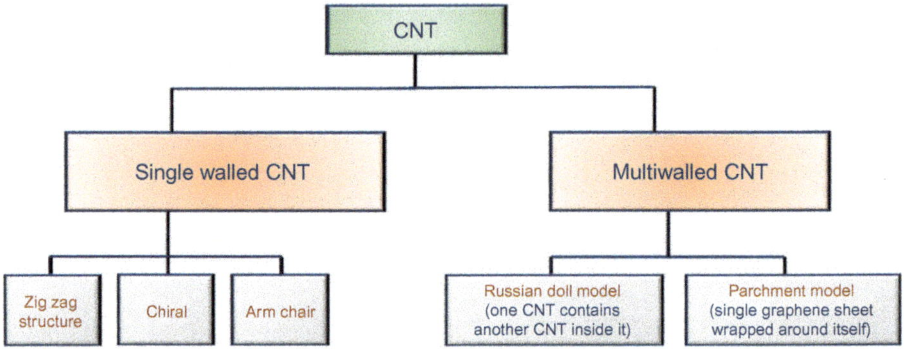

Fig. 5.3 Types of CNTs.

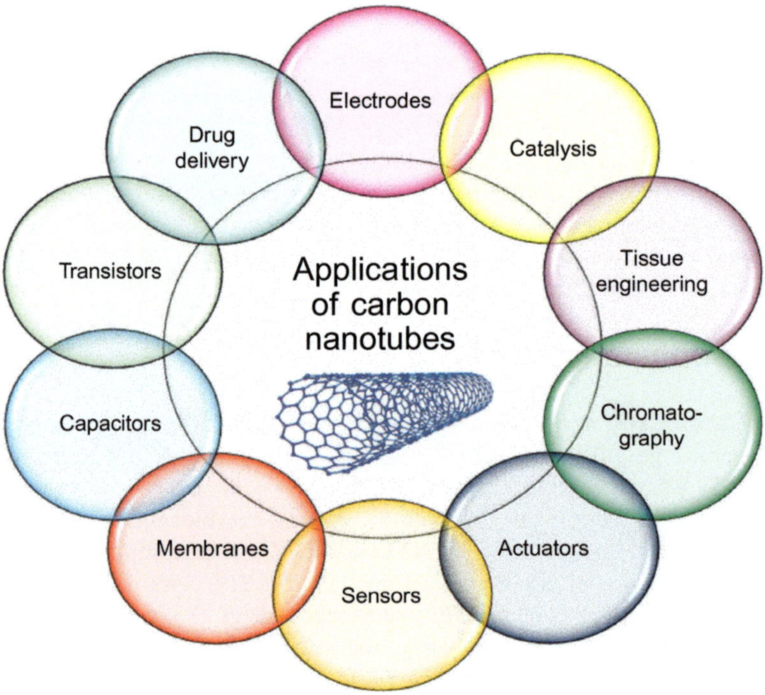

Fig. 5.4 Applications of CNTs.

also carried out at in vivo condition and it was discovered that the CNTs' toxicity is governed by length, shape, size, aspect ratio, and chemical surface. The successful implantation of CNT in an animal femur model was resulted with no toxicity and no severe inflammatory response which suggests its scope in the regeneration of damaged bones. The dependence of toxicity on the orientation of CNT was confirmed by many investigators. They proposed that the proper orientation of CNT networks can greatly enhance the proliferation and osteogenic differentiation than randomly aligned CNT arrays [16–19].

As far as MWCNT is concerned, it integrates into new bone accompanied with bone repairment, also exhibits excellent bone tissue compatibility with little local inflammation and significant acceleration in bone formation stimulated by recombinant human bone morphogenetic protein-2 (rhBMP-2). However, few reports reveal that carboxyl-functionalized CNTs (COOH-SWNTs and COOH-MWNTs) restrict the proliferation, adipogenic, osteogenic differentiation, and mineralization of mesenchymal stem cells (MSCs) [20,21]. Incorporation of MWCNTs in the composite materials is anticipated to be a promising and effective measure to tackle this issue. The MWCNT-based biocomposites can be used to develop high load-bearing orthopedic implants that are not only capable to reduce the toxicity of the CNT but also have the potential to mimic the natural activity of the bone. This chapter particularly reports on the composite scaffolds based on MWCNT developed till date and their associated aspects in regard to artificial bone grafting [22–26].

The performance of pure MWNTs and N-doped MWNTs was investigated in vivo systems of mice. It was observed that the direct administration of extremely high concentrations of N-doped MWNT into the mice's trachea responded with only mild intensity of granulomatous inflammation. It is worth mentioning that all other passages of installation of N-doped MWNT did not exhibit any signs of distress or tissue transmutation in any operated mouse. The results also revealed that the functionalizations of N-doped MWNT nanotubes have an added advantage of being less detrimental than that of undoped MWNTs or SWNTs leading to their dynamic applications in biotechnology and allied areas [27]. Functionalization of CNTs is significantly effectual to attain the optimal properties of CNTs in a composite as functionalization brings about the enhancement in the overall properties of the composite by improving both the wettability and dispersibility of CNTs in the composite matrix [28–31].

Generally, the most common and easiest method to carry out the functionalization of CNTs is chemical oxidation that augments both the reactivity and assists their efficient dispersion in both aqueous and organic media. Chemical oxidation not only binds functional groups [hydroxyl (OH), carbonyl (C=O), sulfonyl ($-SO_3$) groups, etc.] to the CNT tips but also fills its sidewall defects by attaching the functional groups to them that imparts further functionality to CNT. This method also plays an important role in purifying the nanotubes, enabling them to get rid of unwanted nanoparticles or any amorphous carbon. So, the load transfer efficiency of CNT in the composite network can be upgraded by functionalization although it is also accompanied with the probability of little bit reduction in the strength of CNTs, leading to a trade-off in properties.

Another merit of functionalized CNTs is that they exhibit remarkable downfall in their toxicity level enabling them to be used for various biomedical applications such as gene delivery, drug delivery, bone regeneration, neural regeneration, etc. This fact has been proved by the observation of decrease in the toxicity of CNT functionalized with poly-L-lysine intended to apply it in cell adhesion [32–34].

5.4 Multiwalled CNT composite biomaterials for artificial bone grafting

5.4.1 *Multiwalled CNT-polymer nanocomposite*

The multidimensional properties especially electrical and mechanical properties of MWCNTs are the key features for realizing their potential in bone tissue engineering. The mechanical properties of MWCNT provide a platform to reinforce the strength of the bone. The mechanical properties of MWCNT are likely to imitate the function of the bone while the electrical properties are chiefly responsible for the stimulation of MSCs for causing osteogenic differentiation [35,36]. Tremendous work has been done in the domain of nanocomposites composed of polymers and MWCNT and the field of tissue engineering is also being influenced by these MWCNT-polymer nanocomposites due to the need for almost ideal bone replacements [37]. The multifunctional nature of polymers assignable to their eco-compatibility, nontoxicity, biodegradability including their fairly good mechanical properties shapes them to serve as an excellent matrix for the dispersion of MWCNTs and to be successfully employed in artificial bone grafting. A report showed the positive response of MWCNT-polymer nanocomposites when subjected to human lung epithelial cells, primary osteoblast cells, and osteoblast-like cells, in terms of cell adhesion, proliferation, and differentiation and acquiring direct communication with body fluids in vivo conditions [38]. As per the statistical analysis of the recent reports, the emerging class of polymers for the application in artificial bone grafting, surgery, dentistry, and drug delivery contains a group of chitosan, polylactic acid, poly(ε-caprolactone), polyglycolic acid, and collagen.

5.4.2 *CNT coating on the polymeric surface*

Various coatings of MWCNTs on the surface of polymers have been developed and studied so far. In a case study, a comparison was drawn between bare TiO_2 nanotubes and MWCNT-coated TiO_2 nanotubes in order to examine the osteoblast cell behavior toward the different surface chemistries of both the nanotubes. The results obtained made it apparent that surface chemistry of the nanotubes had no adverse effect on the morphology, proliferation, or cell adhesion but played a crucial role in modulating the alkaline phosphatase (ALP) activity of osteoblast cells, in comparison to the surface chemistry of the nonmodified TiO_2 nanotubes bearing higher ALP levels. Although, both the TiO_2 nanotubes came out to be beneficial for osteogenic differentiation in MSCs but the MWCNT-coated TiO_2 nanotubes led to the dramatic improvement in cell differentiation process [39]. Another example of MWCNTs coating was found on the surface of three-dimensional (3D) collagen scaffold. Posttreatment ALP activity along with the osteopontin and calcium contents and cell differentiation of rat primary osteoblasts (ROBs) cultured on the MWCNT-coated collagen sponges substantially increased with respect to those of uncoated sponges.

A MWCNT nanofiber layer composite with agarose displayed nontoxicity along with high conductivity that was used for cell adhesion and response both in vivo and in vitro systems [40].

A detailed investigation of the effects of MWCNT composite coatings versus pure titanium coating (layer-by-layer configuration) on the cell proliferation was performed by Bhattacharya et al. [41] wherein they observed similar results for both the materials. However, on the implantation of both the modified osteoblasts in critical-sized rat calvarial defect, the MWCNT composite stimulated bone formation and its recovery showing no signs of inflammation or failure.

5.4.3 Multiwalled CNT-collagen nanocomposite

The efforts for creating improved scaffolds leaning toward the application of biomedical tissue implants has prompted research to discover a variety of new approaches employing natural polymers. Collagen is one such natural polymer that can be a promising biomaterial for artificial bone grafting. In the class of natural polymers, collagen is being extensively investigated in biomedicine for its interesting physiological and biochemical services in the extracellular matrix of many living systems [42,43]. A wide range of properties of the bone tissue involving surface morphology, topography, cell proliferation, and differentiation can be improved by the reinforcement of collagen with MWCNT. The composite scaffolds of MWCNT-reinforced collagen fibers also play an important role in increasing the DNA content on the MWNT-coated collagen after the span of 1 week. However, the determined ALP activity on the coated collagen did not make much difference in comparison to the uncoated sponge scaffolds [44].

Another study done on ROB cell demonstrated the quick and significant response of the MWNT-coated collagen scaffolds in terms of remarkable proliferation, quick responsive differentiation, and higher ALP activity with higher osteopontin contents versus uncoated scaffolds. In vivo studies also proved that MWNT-coated collagen sponges can result into substantial bone formation than the uncoated sponges [44,45]. Thus, it is concluded that MWNT-coated collagen scaffolds have the potential to be used as a matrix for regenerative medicine [46].

5.4.4 Multiwalled CNT-polylactic acid nanocomposite

Polyhydroxy acids are the biopolymers that have covered the domain of therapeutic devices on a large scale. Both colloidal and electrospinning techniques have been employed to produce MWCNT-based poly(lactic-*co*-glycolic acid) composite materials. It has been proved that the fibroblast adhesion, propagation, and proliferation considerably improve by the reinforcement of poly(lactic-*co*-glycolic acid) (PLGA) scaffolds with MWCNTs when assessed against uncoated PLGA fibrous mats and macroporous PLGA membranes [47]. Various preparation techniques such as solvent casting method have been used for the synthesis of PLGA-MWCNTs biocompatible and biodegradable composites. The results of mechanical properties of the PLGA-MWCNT nanocomposite exhibited a great improvement due to the incorporation of MWCNTs. An in vitro study was done to check the hydrolytic degradation of PLGA wherein significantly enhanced degradation was observed by the addition of carboxylate-functionalized MWCNTs (c-MWCNT). The performance of MSCs cultured onto PLGA-based c-MWCNT scaffolds for 21 days, was found superior in regard to better cell viability, attachment, and higher ALP activity as compared with control groups [48]. The proliferation and elongation of the tissues cultured on the electrospun conductive nanofibers of polylactic acid reinforced by MWCNT were found to be influenced by the electrical stimulation while the topography had a feeble impact on the bioactivity of the cells [49].

5.4.5 Multiwalled CNT-chitosan nanocomposite

In recent times, chitosan scaffolds have emerged as potentially active biopolymers which have exclusively been implanted in various living cells to treat bone defects and failures. Chitosan and HA are those two biomaterials which possess excellent biodegradability, biological activity, and osteoconductivity together and it is the cumulative effect of these three characteristics that makes them fall in the category of the best bioactive biomaterials being looked for artificial bone grafting [50]. Chitosan besides being extraordinarily biocompatible, also carries inherent antibacterial properties including the ability of cell propagation and pore formation [51,52]. However, chitosan-originated biocomposites suffer from the issues of inadequate interconnection between the pores of the composites as well as optimum mechanical strength that requires further improvement for producing better results. CNTs can play a major role in surpassing the issue of mechanical strength by incorporating CNTs into the biocomposites of chitosan. Recently, significant work has been done in the field of chitosan-CNTs composite biomaterials for bone tissue engineering. Several comparative studies of both SWCNT and MWCNT-based chitosan composites have been published. Detailed studies about the preparation of chitosan-CNTs composite scaffolds, their mechanical properties, types of interactions, and in vitro investigations of bioactivity have been performed by several research units [12]. A report was published on the in vitro examination of different chitosan-SWCNT composites wherein SWCNT functionalized by glucosamine, acid-functionalized SWCNT, and pristine SWCNT were taken in the concentration range of 0.001–1.0% wt/vol. The performance of the osteoblast cell was found to decline with the increase in concentration of SWCNT, this implies that there is an inverse relation between the cell viability and SWCNT concentration. The chitosan composite scaffold grafted (g) with functionalized MWCNT (f-MWCNT) and f-MWCNT-g-chitosan/HA composite scaffold were prepared by Venkatesan et al. [50] using freeze-drying technique. The results exhibited that the cell proliferation was twice in comparison to that of nonmodified chitosan. Thus, this biomaterial can hopefully bring a revolution in artificial bone grafting. Another effort intended to develop reliable MWCNT-chitosan composites for treating bone defects was made by Abarrategi et al. [53]. They found proper adsorption of rhBMP-2 in MWNT-chitosan scaffolds which were successfully implanted in muscle tissue. It was also discovered that the bone tissues dramatically formed along with the favorable in vivo results. Hence, it is apparent that MWCNTs are capable for boosting the growth and differentiation of osteoblast while upgrading the mechanical properties of the composite, simultaneously.

5.4.6 Multiwalled CNT-polycaprolactone nanocomposites

Polycaprolactone is a biodegradable polymer that undergoes hydrolytic degradation under physiological conditions. PC has become a subject of continuous research in regard to the development of artificial bone implants because of its fairly good dissolution tendency, excellent biocompatibility, and low MP (59–64°C). The exceptional viscoelasticity and rheological properties of PC make it a suitable candidate for various biomedical engineering applications. Several tissue scaffolds have been synthesized using polycaprolactone [54–58]. A tissue scaffold composed of a hydrogel, PC, and MWCNT oriented toward skeletal muscle implant was fabricated by coaxial electrospinning. The addition of MWCNT not only improved the mechanical strength of the muscle implant but also greatly increased its conductivity as compared with the pig and rat skeletal tissue. Moreover, the PC-MWCNT composite scaffolds demonstrated their potential to be used as an artificial nanoactuator for the skeletal tissues [59]. The PC-MWCNT composite scaffolds prepared by the solution evaporation method have been reported. The mechanical properties and phase composition along with the biomimetic activity of the scaffolds were examined on rat bone marrow-derived stroma cells (BMSCs). It has been suggested that the small loading of MWCNT (0.5 wt%) into the polymer matrix can give rise to better scaffolds whose cellular growth and differentiation would be higher than that of higher MWCNT-loaded scaffolds. This indicates that MWCNT incorporated polycaprolactone can be used to develop potentially active composite scaffolds for biomedical bone tissue implants under the condition of using relatively low concentration of MWCNTs [37].

5.4.7 CNT-HA nanocomposite

Over the last few decades, HA has been extensively studied concerning the application in artificial bone engineering [60]. HA is basically a bioceramic material whose chemical composition exhibits similarity to the inorganic phase of natural bone and teeth [61,62]. It is considered to be an incredible material for bone implants in clinical bone grafting in various form of composites and coatings. However, the major limitation associated with this biomaterial is its low mechanical strength as compared with that of the natural bone that retrains it for heavy load-bearing applications. The high aspect ratio and remarkable mechanical properties of MWCNTs help strengthening and toughening HA without deteriorating its bioactivity [63]. Composite scaffolds coated with HA have been prepared from PLGA incorporated with MWCNTs for biomimetic mineralization which was carried out by a solution of sodium dodecyl sulfate (SDS) with calcium-phosphate [64,65]. Poly-L-lactic acid-HA (PLLA/HA) composite scaffolds reinforced by both SWCNT and MWCNT have been developed. The bone morphogenetic protein-2 (BMP-2) loaded PLLA/HA/SWCNT 3D scaffolds amazingly responded to the cell adhesion, proliferation, and differentiation leading to bone regeneration [66]. The PLLA/MWCNT/HA membrane scaffolds were prepared by electrospinning wherein dramatic improvement was observed in biomimetic activity of the periodontal ligament cells (PDLCs) by about 30%, accompanied with the 30% inhibition of gingival epithelial cells response with reference to the limiting group [67]. Synthesis of collagen/MWCNT composites and self-assembled nHA/MWCNT has also been worked upon in some literatures. The resultant spindle-shaped units separated from the MWNT template maintained the parallel alignment of the nHA polycrystalline fibrils [68]. The solution-blended HA/MWCNT composites led to the tremendous improvement in the flexural strength and fracture toughness by 28% and 50% respectively for 7% volume of MWCNTs [69]. Recently, a nanocomposite of HA based on PMMA and MWCNTs has come into existence which seems to be a superior material targeting the domain of artificial bone grafts. The reason for selecting PMMA was its superior compatibility with HA including its wide usage as bone cement. Moreover, PMMA causes functionalization of MWCNTs and crosslinking of the composite when integrated with HA.

5.4.8 CNT-bioglass nanocomposite

It is a well-known fact that reasonably high concentration of Ca-P can promote the formation of apatite in bone; bioglass is one such biomaterial consisting high amount of calcium and phosphorus with other elements as specific ratios of Na_2O, SiO_2, P_2O_5, and CaO. Thus, the addition of bioglass into MWCNTs will emerge as highly advanced functional materials for the application of artificial bone grafting. One such composite scaffolds based on highly porous 45S5 Bioglass has been synthesized that was coated with MWCNTs using electrophoretic deposition. These composite scaffolds had enhanced electricity conductivity owing to the layer of MWCNTs on their surface that might be beneficial to improve the cell differentiation and thus its feasibility when cultured under electrical simulation [70]. A composite composed of poly(3-hydroxybutyrate), bioglass, and MWCNTs was also prepared. It was found that the composite bearing (2–7 wt%) of MWCNTs produced nonuniform and coarse morphology and produced insignificant results while the sample containing less amount of MWCNTs resulted into substantial increment in biomimetic activity of MG-63 osteoblast-like cells [71]. The general mechanism of gene expression regulation by bioactive glass [72] is as shown in Fig. 5.5.

5.5 Challenges and future directions

Although MWCNT is considered to be wonderful biomaterial and has number of applications in different fields of science and technology ranging from tissue engineering, medicine, electronics, sensors, actuators, and so on but there are still plenty of issues left behind which are required to be addressed; some of them are listed below:

1. The toxicity and biocompatibility being the critical parameters for biomedical applications, these should be considered by discovering and inventing new techniques to determine the exact toxicity of CNT and MWCNT which poses a serious challenge to the ongoing research. The available tools and techniques are not competent enough to deal with the toxicity of MWCNTs with accuracy and precision since the toxicity is a sensitive factor that varies with the size, concentration, functionalization, presence of toxic metals, production methods, etc.

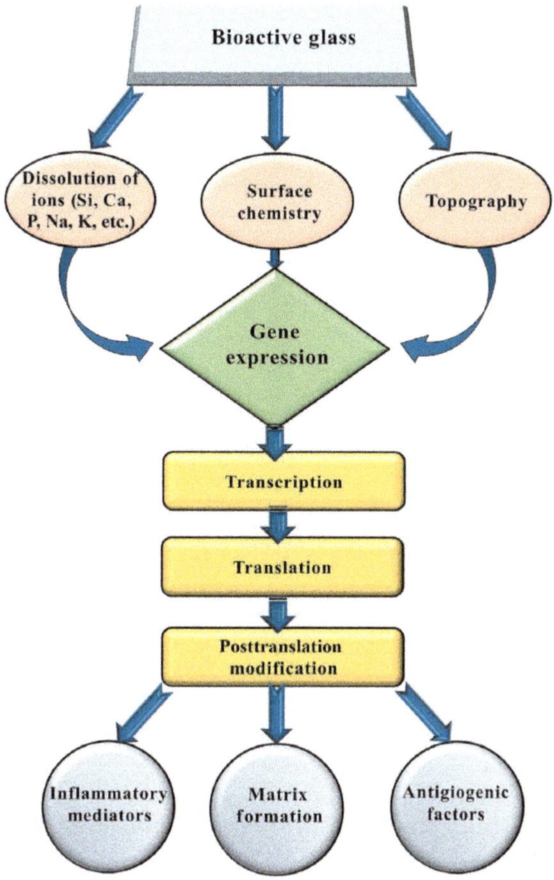

Fig. 5.5 Mechanism of gene expression regulation by bioactive glasses.

2. The direct contact and compatibility of the tissues in proximity to the tissue operated with MWCNT composites must be taken care of.
3. Since proliferation and differentiation of osteoblasts are also governed by the extent of MWCNTs loading in the polymeric material, appropriate balance of the composition is required to obtain the desired results.
4. The field of MWCNTs composites for bone tissue grafting is a relatively fledgling area that requires extensive research to draw any conclusion, but there is definitely a scope to envisage its applicability in tissue engineering if challenges associated with it are tackled rationally.
5. The publications till date provide insufficient picture of the whole scenario, a proper comparison of the studies must be carried out with reference to a standard. For example, different sintering parameters have been used by various groups of researchers but only few of them have shown their comparison. So, a detailed study with its clear cut comparison should be done so as to realize the applications of MWCNTs composites in artificial bone grafting.

5.6 Conclusions

Taking into consideration the foundational stage of MWCNTs for its application toward regenerative medicine, it will be injustice to conclude anything at this point of time. Likewise everything carries both pros and cons along with it at its initial stage, MWCNTs also have some limitations inspite of being an excellent material for enormous applications. As we have seen, a large number of MWCNTs composites have been synthesized as artificial bone grafts, for example, composite scaffolds formed by the combination of polymer with ceramics and MWCNTs pictured themselves to be promising materials for regeneration of the damaged bones. In some experiments, it has exhibited outstanding results as discussed above in the manuscript, yet there are areas which demands improvement and better performance than the current situation, say, its toxicity, due to which MWCNTs can't be used alone in bone implants but it can be suppressed by functionalization of MWCNTs with suitable functional groups. Here, the scope of improvement lies in the method of functionalization that should be considered and worked up for making advancements in this particular field. Functionalization can be an effective tool to circumvent various drawbacks of MWCNTs by which it can be used as a dynamic material for future research.

Acknowledgments

The authors are grateful to the Elsevier for giving us a great opportunity to be a part of this valuable work and we would also like to express our heartfelt gratitude to the researchers and expertise whose contribution to the field of artificial bone grafting made it possible for us to review this discipline for future perspectives. They are also thankful to the chairman, Department of Applied Chemistry for the necessary provisions.

References

[1] J.J. Li, D.L. Kaplan, H. Zreiqat, Scaffold-based regeneration of skeletal tissues to meet clinical challenges, J. Mater. Chem. B 2 (42) (2014) 7272–7306.
[2] M. Orciani, M. Fini, R. Di Primio, M. Mattioli-Belmonte, Biofabrication and bone tissue regeneration: cell source, approaches, and challenges, Front. Bioeng. Biotechnol. 5 (2017).
[3] M. Fan, Y. Ma, J. Mao, Z. Zhang, H. Tan, Cytocompatible in situ forming chitosan/hyaluronan hydrogels via a metal-free click chemistry for soft tissue engineering, Acta Biomater. 20 (2015) 60–68.
[4] P.S. Thayer, A.S. Goldstein, Bio-instructive scaffolds for tendon/ligament regeneration, in: Bio-instructive Scaffolds for Musculoskeletal Tissue Engineering and Regenerative Medicine, Academic Press, 2017, pp. 87–112. (chapter 4).
[5] A.P. Pêgo, et al., Preparation of degradable porous structures based on 1,3-trimethylene carbonate and D,L-lactide (co)polymers for heart tissue engineering, Tissue Eng. 9 (5) (2003) 981–994.

[6] C.L. Cummings, D. Gawlitta, R.M. Nerem, J.P. Stegemann, Properties of engineered vascular constructs made from collagen, fibrin, and collagen–fibrin mixtures, Biomaterials 25 (17) (2004) 3699–3706.
[7] G. Chen, et al., The use of a novel PLGA fiber/collagen composite web as a scaffold for engineering of articular cartilage tissue with adjustable thickness, J. Biomed. Mater. Res. 67A (4) (2003) 1170–1180.
[8] D. Tang, R.S. Tare, L.-Y. Yang, D.F. Williams, K.-L. Ou, R.O.C. Oreffo, Biofabrication of bone tissue: approaches, challenges and translation for bone regeneration, Biomaterials 83 (2016) 363–382.
[9] U. Simon, P. Augat, M. Utz, L. Claes, A numerical model of the fracture healing process that describes tissue development and revascularisation, Comput. Methods Biomech. Biomed. Eng. 14 (1) (2011) 79–93.
[10] P.V. Giannoudis, H. Dinopoulos, E. Tsiridis, Bone substitutes: an update, Injury 36 (3) (2005) S20–S27.
[11] E. Gentleman, A.N. Lay, D.A. Dickerson, E.A. Nauman, G.A. Livesay, K.C. Dee, Mechanical characterization of collagen fibers and scaffolds for tissue engineering, Biomaterials 24 (21) (2003) 3805–3813.
[12] I. Olivas-Armendariz, S.A. Martel-Estrada, M.E. Mendoza-Duarte, F. Jiménez-Vega, P. García-Casillas, C.A. Martínez-Pérez, Biodegradable chitosan/multiwalled carbon nanotube composite for bone tissue engineering, J. Biomater. Nanobiotechnol. 04 (02) (2013) 204–211.
[13] L. Wang, R.M. Shelton, P.R. Cooper, M. Lawson, J.T. Triffitt, J.E. Barralet, Evaluation of sodium alginate for bone marrow cell tissue engineering, Biomaterials 24 (20) (2003) 3475–3481.
[14] P.R. Supronowicz, P.M. Ajayan, K.R. Ullmann, B.P. Arulanandam, D.W. Metzger, R. Bizios, Novel current-conducting composite substrates for exposing osteoblasts to alternating current stimulation, J. Biomed. Mater. Res. 59 (3) (2002) 499–506.
[15] T.J. Webster, M.C. Waid, J.L. McKenzie, R.L. Price, J.U. Ejiofor, Nano-biotechnology: carbon nanofibres as improved neural and orthopaedic implants, Nanotechnology 15 (1) (2004) 48–54.
[16] H. Haniu, et al., Basic potential of carbon nanotubes in tissue engineering applications, J. Nanomater. 2012 (2012) 1–10.
[17] A. Fraczek, E. Menaszek, C. Paluszkiewicz, M. Blazewicz, Comparative in vivo biocompatibility study of single- and multi-wall carbon nanotubes, Acta Biomater. 4 (6) (2008) 1593–1602.
[18] W. Wang, et al., Preparation and characteristics of a binderless carbon nanotube monolith and its biocompatibility, Mater. Sci. Eng. C 28 (7) (2008) 1082–1086.
[19] S. Namgung, K.Y. Baik, J. Park, S. Hong, Controlling the growth and differentiation of human mesenchymal stem cells by the arrangement of individual carbon nanotubes, ACS Nano 5 (9) (2011) 7383–7390.
[20] Y. Usui, et al., Carbon nanotubes with high bone-tissue compatibility and bone-formation acceleration effects, Small 4 (2) (2008) 240–246.
[21] D. Liu, C. Yi, D. Zhang, J. Zhang, M. Yang, Inhibition of proliferation and differentiation of mesenchymal stem cells by carboxylated carbon nanotubes, ACS Nano 4 (4) (2010) 2185–2195.
[22] N. Saito, et al., Carbon nanotubes: biomaterial applications, Chem. Soc. Rev. 38 (7) (2009) 1897.
[23] Y. Zhang, Y. Bai, B. Yan, Functionalized carbon nanotubes for potential medicinal applications, Drug Discov. Today 15 (11–12) (2010) 428–435.

[24] K. Sahithi, M. Swetha, K. Ramasamy, N. Srinivasan, N. Selvamurugan, Polymeric composites containing carbon nanotubes for bone tissue engineering, Int. J. Biol. Macromol. 46 (3) (2010) 281–283.

[25] X. Li, Y. Fan, F. Watari, Current investigations into carbon nanotubes for biomedical application, Biomed. Mater. 5 (2) (2010) 022001.

[26] X. Li, X. Liu, J. Huang, Y. Fan, F. Cui, Biomedical investigation of CNT based coatings, Surf. Coat. Technol. 206 (4) (2011) 759–766.

[27] J.C. Carrero-Sánchez, et al., Biocompatibility and toxicological studies of carbon nanotubes doped with nitrogen, Nano Lett. 6 (8) (2006) 1609–1616.

[28] A. Hirsch, Functionalization of single-walled carbon nanotubes, Angew. Chem. Int. Ed. 41 (11) (2002) 1853.

[29] T. Lin, V. Bajpai, T. Ji, L. Dai, Chemistry of carbon nanotubes, Aust. J. Chem. 56 (7) (2003) 635.

[30] S. Niyogi, et al., Chemistry of single-walled carbon nanotubes, Acc. Chem. Res. 35 (12) (2002) 1105–1113.

[31] D. Tasis, N. Tagmatarchis, A. Bianco, M. Prato, Chemistry of carbon nanotubes, Chem. Rev. 106 (3) (2006) 1105–1136.

[32] P.A. Tran, L. Zhang, T.J. Webster, Carbon nanofibers and carbon nanotubes in regenerative medicine, Adv. Drug Deliv. Rev. 61 (12) (2009) 1097–1114.

[33] D.W. Lin, C.J. Bettinger, J.P. Ferreira, C.L. Wang, Z. Bao, A cell-compatible conductive film from a carbon nanotube network adsorbed on poly-l-lysine, ACS Nano 5 (12) (2011) 10026–10032.

[34] K. Balasubramanian, M. Burghard, Chemically functionalized carbon nanotubes, Small 1 (2) (2005) 180–192.

[35] J. Venkatesan, S.K. Kim, Carbon nanotube for bone repair, in: K.K. Kar, J.K. Pandey, S. Rana (Eds.), Handbook of Polymer Nanocomposites: Processing, Performance and Application, Springer Verlag, Berlin, Heidelberg, 2015.

[36] I. Firkowska, M. Olek, N. Pazos-Peréz, J. Rojas-Chapana, M. Giersig, Highly ordered MWNT-based matrixes: topography at the nanoscale conceived for tissue engineering, Langmuir 22 (12) (2006) 5427–5434.

[37] L.S. Sundar, R. Hawaldar, E. Titus, J. Gracio, M.K. Singh, Integrated Biomimemic Carbon Nanotube Composites for Biomedical Applications, IntechOpen, 2012.

[38] R. Ravichandran, S. Sundarrajan, J.R. Venugopal, S. Mukherjee, S. Ramakrishna, Advances in polymeric systems for tissue engineering and biomedical applications, Macromol. Biosci. 12 (3) (2012) 286–311.

[39] K.S. Brammer, C. Choi, C.J. Frandsen, S. Oh, G. Johnston, S. Jin, Comparative cell behavior on carbon-coated TiO_2 nanotube surfaces for osteoblasts vs. osteo-progenitor cells, Acta Biomater. 7 (6) (2011) 2697–2703.

[40] D.Y. Lewitus, et al., Biohybrid carbon nanotube/agarose fibers for neural tissue engineering, Adv. Funct. Mater. 21 (14) (2011) 2624–2632.

[41] M. Bhattacharya, et al., Bone formation on carbon nanotube composite, J. Biomed. Mater. Res. Part A 96A (1) (2011) 75–82.

[42] E. Sayin, R.H. Rashid, J.C. Rodríguez-Cabello, A. Elsheikh, E.T. Baran, V. Hasirci, Human adipose derived stem cells are superior to human osteoblasts (HOB) in bone tissue engineering on a collagen-fibroin-ELR blend, Bioact. Mater. 2 (2017) 71–81.

[43] S. Kuttappan, D. Mathew, M.B. Nair, Biomimetic composite scaffolds containing bioceramics and collagen/gelatin for bone tissue engineering—a mini review, Int. J. Biol. Macromol. 93 (2016) 1390–1401.

[44] E. Hirata, M. Uo, H. Takita, T. Akasaka, F. Watari, A. Yokoyama, Development of a 3D collagen scaffold coated with multiwalled carbon nanotubes, J. Biomed. Mater. Res. Part B Appl. Biomater. 90B (2) (2009) 629–634.
[45] E. Hirata, M. Uo, H. Takita, T. Akasaka, F. Watari, A. Yokoyama, Multiwalled carbon nanotube-coating of 3D collagen scaffolds for bone tissue engineering, Carbon N. Y. 49 (10) (2011) 3284–3291.
[46] R.A. MacDonald, B.F. Laurenzi, G. Viswanathan, P.M. Ajayan, J.P. Stegemann, Collagen-carbon nanotube composite materials as scaffolds in tissue engineering, J. Biomed. Mater. Res. Part A 74A (3) (2005) 489–496.
[47] F. Liu, et al., Effect of the porous microstructures of poly(lactic-co-glycolic acid)/carbon nanotube composites on the growth of fibroblast cells, Soft Mater. 8 (3) (2010) 239–253.
[48] C. Lin, et al., Incorporation of carboxylation multiwalled carbon nanotubes into biodegradable poly(lactic-co-glycolic acid) for bone tissue engineering, Colloids Surf. B: Biointerfaces 83 (2) (2011) 367–375.
[49] S. Shao, et al., Osteoblast function on electrically conductive electrospun PLA/MWCNTs nanofibers, Biomaterials 32 (11) (2011) 2821–2833.
[50] J. Venkatesan, S.-K. Kim, Chitosan composites for bone tissue engineering—an overview, Mar. Drugs 8 (8) (2010) 2252–2266.
[51] A. Di Martino, M. Sittinger, M.V. Risbud, Chitosan: a versatile biopolymer for orthopaedic tissue-engineering, Biomaterials 26 (30) (2005) 5983–5990.
[52] M. Peter, N.S. Binulal, S.V. Nair, N. Selvamurugan, H. Tamura, R. Jayakumar, Novel biodegradable chitosan–gelatin/nano-bioactive glass ceramic composite scaffolds for alveolar bone tissue engineering, Chem. Eng. J. 158 (2) (2010) 353–361.
[53] A. Abarrategi, et al., Multiwall carbon nanotube scaffolds for tissue engineering purposes, Biomaterials 29 (1) (2008) 94–102.
[54] C.M. Agrawal, R.B. Ray, Biodegradable polymeric scaffolds for musculoskeletal tissue engineering, J. Biomed. Mater. Res. 55 (2) (2001) 141–150.
[55] E. Sachlos, J. Czernuszka, Making tissue engineering scaffolds work. Review: The application of solid freeform fabrication technology to the production of tissue engineering scaffolds, Eur. Cells Mater. 5 (2003) 29–40.
[56] P.S.P. Poh, D.W. Hutmacher, B.M. Holzapfel, A.K. Solanki, M.M. Stevens, M.A. Woodruff, In vitro and in vivo bone formation potential of surface calcium phosphate-coated polycaprolactone and polycaprolactone/bioactive glass composite scaffolds, Acta Biomater. 30 (2016) 319–333.
[57] M. Yazdimamaghani, M. Razavi, D. Vashaee, V.R. Pothineni, J. Rajadas, L. Tayebi, Significant degradability enhancement in multilayer coating of polycaprolactone-bioactive glass/gelatin-bioactive glass on magnesium scaffold for tissue engineering applications, Appl. Surf. Sci. 338 (2015) 137–145.
[58] L. Roseti, et al., Scaffolds for bone tissue engineering: state of the art and new perspectives, Mater. Sci. Eng. C 78 (2017) 1246–1262.
[59] K.D. McKeon-Fischer, D.H. Flagg, J.W. Freeman, Coaxial electrospun poly(ε-caprolactone), multiwalled carbon nanotubes, and polyacrylic acid/polyvinyl alcohol scaffold for skeletal muscle tissue engineering, J. Biomed. Mater. Res. Part A 99A (3) (2011) 493–499.
[60] G. Wei, P.X. Ma, Structure and properties of nano-hydroxyapatite/polymer composite scaffolds for bone tissue engineering, Biomaterials 25 (19) (2004) 4749–4757.
[61] I. Sopyan, M. Mel, S. Ramesh, K.A. Khalid, Porous hydroxyapatite for artificial bone applications, Sci. Technol. Adv. Mater. 8 (1–2) (2007) 116–123.

[62] R. Pallela, J. Venkatesan, S.K. Kim, Polymer assisted isolation of hydroxyapatite from Thunnus obesus bone, Ceram. Int. 37 (8) (2011) 3489–3497.
[63] A.A. White, S.M. Best, I.A. Kinloch, Hydroxyapatite? Carbon nanotube composites for biomedical applications: a review, Int. J. Appl. Ceram. Technol. 4 (1) (2007) 1–13.
[64] H. Zhang, J. Liu, Z. Yao, J. Yang, L. Pan, Z. Chen, Biomimetic mineralization of electrospun poly(lactic-co-glycolic acid)/multi-walled carbon nanotubes composite scaffolds in vitro, Mater. Lett. 63 (27) (2009) 2313–2316.
[65] Q. Tan, K. Zhang, S. Gu, J. Ren, Mineralization of surfactant functionalized multi-walled carbon nanotubes (MWNTs) to prepare hydroxyapatite/MWNTs nanohybrid, Appl. Surf. Sci. 255 (15) (2009) 7036–7039.
[66] G. Ciapetti, et al., Enhancing osteoconduction of PLLA-based nanocomposite scaffolds for bone regeneration using different biomimetic signals to MSCs, Int. J. Mol. Sci. 13 (12) (2012) 2439–2458.
[67] F. Mei, et al., Improved biological characteristics of poly(L-lactic acid) electrospun membrane by incorporation of multiwalled carbon nanotubes/hydroxyapatite nanoparticles, Biomacromolecules 8 (12) (2007) 3729–3735.
[68] S. Liao, et al., Self-assembly of nano-hydroxyapatite on multi-walled carbon nanotubes, Acta Biomater. 3 (5) (2007) 669–675.
[69] Y.H. Meng, C.Y. Tang, C.P. Tsui, D.Z. Chen, Fabrication and characterization of needle-like nano-HA and HA/MWNT composites, J. Mater. Sci. Mater. Med. 19 (1) (2008) 75–81.
[70] D. Meng, J. Ioannou, A.R. Boccaccini, Bioglass®-based scaffolds with carbon nanotube coating for bone tissue engineering, J. Mater. Sci. Mater. Med. 20 (10) (2009) 2139–2144.
[71] S.K. Misra, et al., Characterization of carbon nanotube (MWCNT) containing P(3HB)/bioactive glass composites for tissue engineering applications, Acta Biomater. 6 (3) (2010) 735–742.
[72] I.D. Xynos, A.J. Edgar, L.D.K. Buttery, L.L. Hench, J.M. Polak, Gene-expression profiling of human osteoblasts following treatment with the ionic products of bioglass 45S5 dissolution, J. Biomed. Mater. Res. 55 (2) (2001) 151–157.

Nanocomposite materials for prosthetic devices

6

M. Saquib Hasnain*, Syed Anees Ahmad†, Mohammad Akram Minhaj‡, Tahseen Jahan Ara§, Amit Kumar Nayak¶
*Department of Pharmacy, Shri Venkateshwara University, Gajraula, India, †Department of Pathology, King George's Medical University, Lucknow, India, ‡Department of Pharmacology, Maulana Azad Medical College and Hospital, New Delhi, India, §Department of Chemistry, L.N.M University, Dharbhanga, India, ¶Department of Pharmaceutics, Seemanta Institute of Pharmaceutical Sciences, Mayurbhanj, India

6.1 Introduction

Nanotechnology is identified as one of the prosperous areas of current researches and developments in more or less all biomedical applications [1–8]. The "nano" word is derived from Greek, which means "dwarf." A nanometer is a spatial unit of measurement (1.0×10^{-9} m). As the particle size is decreased from macro- or micrometer size to nanometer, it exhibits completely different physical and chemical properties of the same large-sized particles. In insulators, silicon converts into conductors. Gold converts into liquid and red in color at the nanoscale. In current years, numerous nanocrystalline materials and nano-skill have been significantly developed [9]. At present, synthesis and characterization of nanoparticles is the rising and speedily growing field in various biomedical applications, like orthopedic, dentistry, drug delivery, etc. Nanostructured materials are generally nanoparticles (ceramic, fullerenes, nanotubes, nanofibers, nanowires, dendrimers, etc.), nanointermediates (dispersions, high expanse materials, nanostructured films, supramolecular assemblies, etc.), and nanocomposites [10]. Nanostructured materials are also classified in Fig. 6.1.

Nanocomposites are the nanoscopic composites. A composite could be a mixture of one or more different materials combined together. The engineering of nanocomposites for various applications has been generally tackled within the preceding decade, as verified within the literature; "the potential of nanocomposites lies in their multifunctionality, the likelihood of comprehending distinctive mixtures of properties unrealizable with ancient materials" [11]. Nanocomposites are mainly multiple nanomaterials or nanomaterials incorporated into another bulk material. At present, a variety of nanocomposites are explored and exploited in almost all fields of materials, metals, plastics, ceramics, biomaterial, electronic materials, etc. [12–15]. Numerous nanocomposites are being designed with industrial viewpoints for the fields of biomedical, biotechnology, pharmaceutical, coating and packaging, electronics, aerospace, and materials industries [16,17]. There are numerous prospective biomedical nanocomposites for the most important biomedical uses like orthopedics, dentistry, tissue engineering, antimicrobial properties, wound dressings, drug delivery, cancer

Applications of Nanocomposite Materials in Orthopedics. https://doi.org/10.1016/B978-0-12-813740-6.00007-7

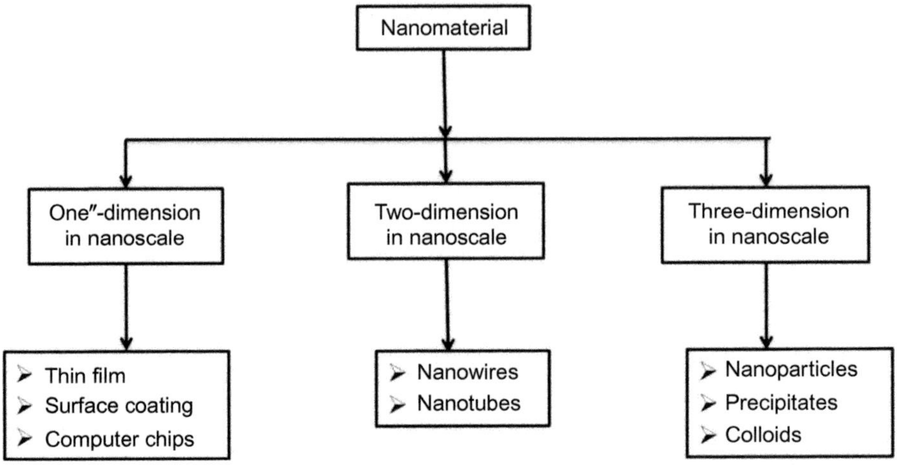

Fig. 6.1 Classification of nanomaterials.

therapy, stem cell therapy, biosensors, etc. [16–19]. This chapter describes various nanocomposites for the use in prosthetic devices. In addition, some basics of nanocomposites like preparation, classifications, and applications are also described.

6.2 Preparation of nanocomposites

A number of approaches are employed to prepare different types of nanocomposites like physically mixing, solution mixing, in situ preparation, dip coating, film casting, coprecipitation, ionotropic gelation, covalent coupling, polymerization, electrospinning, colloidal assembly, layer-by-layer assembly, etc. [20]. Most of these approaches for the preparation of nanocomposites facilitate advantages of hydrogen bonding, ionic interactions, electrostatic interactions, hydrophobic effects, coulombic interactions, etc. The challenges in preparation of nanocomposites comprise controlling procedures of fabrication or synthesis; assurance of compatibility for the different material components; and accomplishing desirable and distinctive material properties [20]. Polymerization is a procedure of reacting monomeric molecules together to form polymeric-chains or 3-dimensional net-works by means of chemical reaction(s). In polyethylene/clay nanocomposites, the catalyst between the clay layers and chemical process takes place within the gap [21]. Therefore, as the polymerization process progresses, the spacing between the clay's layers will increase bit by bit and therefore, the dispersion state of the clays changes from intercalated to exfoliated (delamination with demolition of the clay sheet array). The benefits of this technique are as: (1) the solitary pace synthesis of the metallocene polymer nanocomposites; (2) enhanced compatibility of the clay as well as the polymer milieu; and (3) amplified clay dispersity. Nanocomposites also can be prepared by dispersing five carbon nanofibers into a swarm polymer, usually at 5 wt% levels. This method is also termed "exfoliation." Once nanoclay is well distributed, it is same to be exfoliated. Exfoliation is expedited by the surface compatibilization chemistry that expands the nanoclay platelets to the purpose wherever individual platelets may be separated from another by mechanical shear or heat of polymerization. In some cases, the ultimate nanocomposites are prepared within the reactor throughout the polymerization stage. For alternative polymer systems, processes are being developed to include nanomers/nanoclays into a hot-melt combining operation [22]. The initial step within the preparation of the nanoclay involves purifying just about 99% of the montmorillonite. Montmorillonite is a clay material and comparatively contrary with most hydrophobic polymers. Therefore, montmorillonite can be treated with the chemicals to form its surface additional receptive to the dispersion. The succeeding step involves surface treatment of the clay, montmorillonite. Once the clays are treated with chemicals, these are distributed within the compound [23]. The dispersion method needs a custom answer for every polymer utilized to prepared nanocomposites. Therefore, the preparation of polymer nanocomposites is based on the capital intensive analysis and the effective development planning [24].

6.3 Classification of nanocomposites

On the basis of composition of materials, nanocomposites can be classified into two classes (Fig. 6.2):

(1) nonpolymer-based nanocomposites, and
(2) polymer-based nanocomposites.

6.3.1 Nonpolymer-based nanocomposites

These types of nanocomposites do not contain polymers or polymer-based materials in the composition formula. These kinds of nanocomposites further classified into three subcategories: metal-metal nanocomposites, metal-ceramic nanocomposites, and ceramic-ceramic nanocomposites.

6.3.1.1 Metal-metal nanocomposites

Metal matrix nanocomposites (MMNCs) are the substances comprises of malleable metal or alloy matrixes, in which a little nanosized reinforcement material are entrenched. MMNCs are appropriate for the development of materials with excessive energy in shear/compression approaches and excessive provider of temperature abilities.

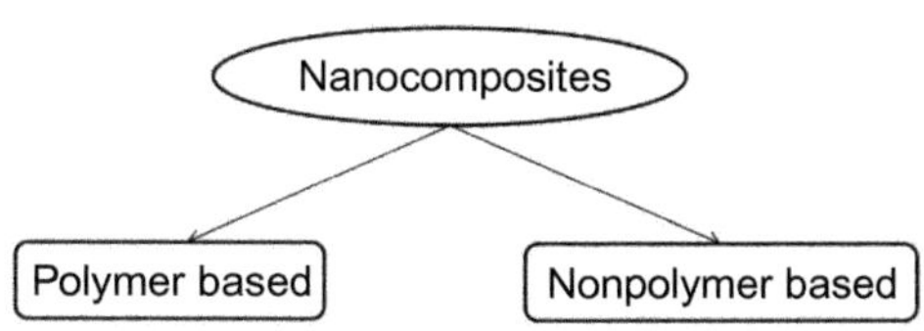

Fig. 6.2 Classification of nanocomposites on the basis of composition materials.

These show a rare capability of application in several regions, like nanocomposite-enabled part in the automobile industries and improvement of structural substances [25], for examples, ZnO-CuO nanocomposite [26] and Ag/TiO_2 nanocomposite [27].

6.3.1.2 Metal-ceramic nanocomposites

Within metal-ceramic nanocomposites, metallic and ceramic constituents are evenly distributed in one another so as to get materials with preferred nanoscopic properties. Traditional metallurgical methods of powder and solution chemical processes, such as sol-gel and methods of co-precipitation have been used to produce composite powders of ceramic-metal nanocomposites such as aluminum oxide (Al_2O_3) with tungsten (W), molybdenum (Mo), nickel (Ni), copper, cobalt (Co), or iron (Fe); zirconium dioxide (ZrO_2) with nickel or molybdenum and magnesium oxide (MgO) with iron, cobalt, or nickel [28,29].

6.3.1.3 Ceramic-ceramic nanocomposites

Two or more ceramic materials are employed to prepare ceramic-ceramic nanocomposites. These nanocomposites exhibit enhanced hardness, strength, toughness, and creep resistance in comparison to the traditional ceramic matrix composites. The joining of high-quality nanofibers into earthenware grids has permitted the arrangement of cutting-edge nanocomposites with high strength and better disappointment qualities looked at than the sudden disappointments of earthenware materials [30].

6.3.2 Polymer-based nanocomposites

Polymer-based composites [31–38] and nanocomposites are the mostly understood and have been extensively researched for an extensive variety of utilizations including numerous biomedical applications. These nanocomposites show a better execution looked at than their associate composites and this makes open doors for various modern applications. Polymer-based nanocomposites speak to a contrasting option to customary filled polymers or polymer mixes—a most reasonable material for the present day plastics industry. Polymer-based nanocomposites contain generously less filler (1–5 vol%), and this empowers more prominent maintenance of the inalienable processability and sturdiness of the flawless resin [39]. Polymer-based nanocomposites can be categorized into seven classes:

- Polymer-polymer nanocomposites
- Polymer-ceramic nanocomposites
- Inorganic-organic polymer-based nanocomposites
- Inorganic-organic half and half nanocomposites
- Polymer-layered silicate nanocomposites

Nanocomposites can be classified into two broad categories on the basis of matrix type: microcrystal matrix (micro-nano type) and nanocrystal matrix. The classification of nanocomposites is presented in Fig. 6.3.

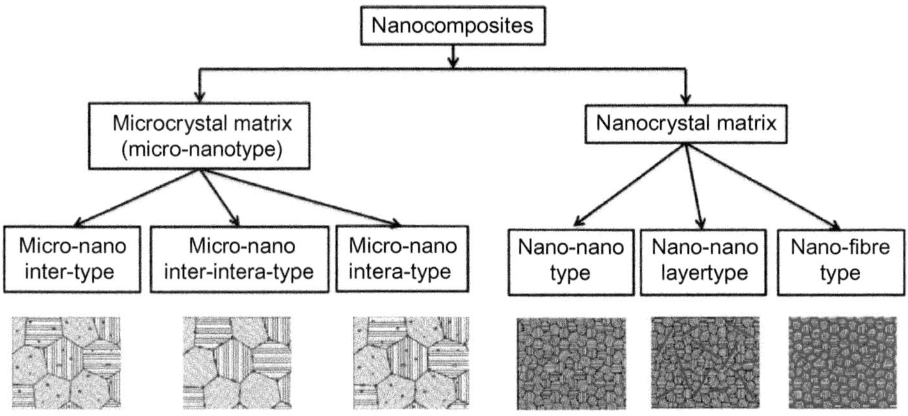

Fig. 6.3 Classification of nanocomposites on the basis of matrix type.

6.4 Application of nanocomposites

Nanotechnology is changing the universe of materials in the materials science and material engineering areas. It has a high effect in building up another era of nanocomposites with improved usefulness and an extensive variety of applications. The information on handling, characterization, and applications makes difference in comprehension by the scientist and researcher communities using the extraordinary production and material standards hidden these forefront polymer-based nanocomposites. Despite the fact that nanocomposites are acknowledging many key applications in the mechanical fields, various key specialized and financial obstructions exist to allow broad commercialization. Recent years, nanocomposites are being employed in broad range of applications [19,40–50]:

1. Applications in prosthetic devices
2. Applications in dentistry
3. Applications in orthopedic tissue engineering and long bone construction devices and appliances
4. Applications in wound dressings
5. Applications in drug delivery
6. Applications in sensors
   - Chemoresistive sensors
   - Optical sensors
   - Biosensors
7. Applications in optical devices
8. Applications in magnetic devices
9. Applications in electrical devices
10. Applications in food packaging
11. Automobile sectors
12. Battery applications
13. Supercapacitors

14. Solar cells
15. Fire retardation
16. Photocatalytic and antimicrobial nanocomposites for the fictionalization of textile fabrics
17. Fuel tanks
18. Lubricants
19. Applications in defense
20. Corrosion inhibitors
21. Nanocomposite-based inkjet for digitally printed electronics
22. Coatings
23. In water purification

6.5 Prosthetics

In biomedical science, prosthesis (Greek *prósthesis*, "expansion, application, attachment") is a device, which is used in place of a missing or lost body part, which may be either lost through damage, affliction, or intrinsic conditions [51,52]. Prosthetic amputee rebuilding is chiefly sorted out by a prosthetist, surgeon, and an interdisciplinary group of specialists, like physiotherapist and occupational therapist.

6.5.1 *Types of prosthetics*

Prosthetics ought to be outlined and gathered as indicated by the patient's appearance and utilitarian needs [52]. For example, a patient may require transradial prosthesis and need to choose between aesthetic functional devices, myoelectric, body-controlled or actively movable devices. The future objectives of patient and practical abilities might enable them to select at least one device. Intraoral and additional oral prostheses are being included in the craniofacial prostheses.

The extraoral prostheses are classified as follows:

(a) Hemifacial prostheses
(b) Auricular (ear) prostheses
(c) Nasal prostheses
(d) Orbital prostheses
(e) Ocular prostheses

The intraoral prostheses include various dental prostheses such as:

(a) Dentures
(b) Obturators
(c) Dental implants

In the neck region the prostheses include:

(a) Larynx substitutes
(b) Trachea
(c) Upper oesophageal replacements

6.5.2 Limb prosthetics

Limb prostheses incorporate both upper-and lower-furthest point prostheses [53]. Upper-extremity prostheses are employed when there is loss of limb by trauma or medial illness:

(a) Shoulder disarticulation
(b) Transhumeral prosthesis
(c) Elbow disarticulation
(d) Transradial prosthesis
(e) Wrist disarticulation
(f) Full hand
(g) Partial hand
(h) Partial finger

Lower-extremity prostheses are used for the replacement of different lower body part. These include:

(a) Hip disarticulation
(b) Transfemoral prosthesis
(c) Knee disarticulation
(d) Trans-tibial prosthesis
(e) Syme's amputation

Transfemoral prosthesis is a manufactured extremity, which replaces a missing or lost leg over the knee [54]. Transfemoral amputees can have an exceptional difficulty in recouping common advancement. In general, a transfemoral amputee must exploit around 80% extra vitality to walk around a man with two whole legs [55]. This is a direct result of the complexities being developed related with the knee. In addition, progressive and extra upgraded designs, hydraulics, mechanical linkages, carbon fibers, PC microchips, and imaginative blends of these developments are used to provide more control to the patient of prosthetic devices. In the prosthetic business, a transfemoral prosthetic leg is as often as possible suggested as "AK" or an over the knee prosthesis [56]. Transtibial prosthesis is a counterfeit extremity that replaces a missing/lost leg underneath the knee. While a transtibial amputee is prepared to recoup typical development more swiftly than someone with a transfemoral removal, due in extensive part to hold the knee, which takes into account less demanding development. Lower extremity prosthetics renders falsely supplanted attachments situated at the hip or lower level. In prosthetic industry, a transtibial prosthetic leg is frequently alluded to as a "BK" or beneath the knee prosthesis [56]. In addition, the majority of limb prostheses (upper- and lower-limb) are prepared nowadays from composites or nanocomposites with basic polymer matrix. These polymer matrix-based composites or nanocomposites are favorable because of their excellent strength to the weight characteristics with good biocompatibility and biodegradability [57,58].

6.5.3 Patient course of action

Prosthesis is practically a replacement for a severe or inherently twisted or missing appendage [52]. Prosthetists are in-charge of the prescription, designing of the device, and management of a prosthetic device. Mostly, the prosthetist starts by taking a plaster cast of the affected limb of patient. The hard materials (known as cutting edge) like carbon fibers, titanium (Ti), etc., provide quality and solidness while preparation of the new prosthesis lighter. More complex prostheses are furnished with cutting edge hardware, giving extra stability and control.

6.5.4 Current innovation and assembling

Throughout the years, there have been extensive progressions in the simulated appendages. New plastics and other different materials, i.e., carbon fibers, have allowed fake appendages to be extra grounded and lighter, restricting the measure of additional vitality significant to work the appendage. This is particularly critical for transfemoral amputees [55]. Additional materials have enabled counterfeit appendages to appear a great deal which is more reasonable, essential to trans-outspread and transhumeral amputees since these will probably have the fake appendage exposed [59]. In spite of new materials, the utilization of hardware has turned out to be exceptionally normal in fake appendages. Myoelectric appendages, which control the appendages by changing over muscle developments to electrical signs, have turned out to be significantly more typical than link worked appendages. Myoelectric signs are gotten by terminals, the flag gets incorporated and once it surpasses the specific edge, control flag of the prosthetic appendage is activated, which is the reason intrinsically, all myoelectric controls slack. On the other hand, link control is quick and physical, which offers a specific level of direct compel input that myoelectric control doesn't have. Computers are additionally utilized broadly in the assembling of appendages. Computer-aided design (CAD) and computer-aided manufacturing are regularly employed to aid the outline and avoid counterfeit limbs production [59,60]. Present day's manufactured appendages are connected to the stump of the amputee through belts and sleeves or by means of suction. The stump either specifically fits into an attachment on the prosthetic or a liner is utilized which then settled to the attachment either by using vacuum (suction attachments) or a stick bolt. Liners are delicate and by that, these can make a much better suction fit than hard attachments. Silicone liners can be produced in standard sizes for the most part with a roundabout cross segment, yet for some other stump shape and custom liners can be produced. Attachment is uniquely designed to fit the lingering appendage and to circulate the strengths of the fake appendage over the zone of the stump (as opposed to only one little spot), which lessens wear on the stump. Custom attachment is made using a plaster cast of the stump or, all the more regularly today, of the liner put on over the stump, and after that making a form from the plaster cast. Fresher techniques incorporate laser-guided measuring that can be reported and analyzed specifically using a computerized system into consideration of a more advanced outline. One issue with these stump and attachment connections are that an awful fit can decreases the contact region in-between stump and attachment or liner, and increment pockets between stump skin and attachment or liner. Weight is then higher, which can be unbearable. Air pockets can enable sweat to aggregate that can diminish the skin.

At last, this is a regular reason for irritated skin rashes. After some time, this may promptly cause skin breakdown. Counterfeit appendages are normally fabricated employing the associated steps [59]:

1. Estimation of the stump
2. Estimation of the body to decide the size required for the simulated appendage
3. Fitting of a silicone liner
4. Production of a model of the liner worn over the stump
5. Arrangement of thermoplastic sheet around the model (then used to test the able-bodied of the prosthetic)
6. Development of perpetual attachment
7. Development of plastic parts of the manufactured appendage (different strategies are utilized, including vacuum framing and infusion shaping)
8. Formation of metal parts of the manufactured appendage utilizing die casting
9. Get together of whole appendage

6.5.5 Body-controlled arms

Current innovation permits body controlled arms to weigh just about one-half to 33% of that a myoelectric arm performs.

Sockets: Modern body-controlled arms have attachments that are worked from hard epoxy or carbon fiber. These attachments or "interfaces" can be prepared additional agreeable by means of fixing these with a gentler, compressible froth material which gives cushioning to the bone prominences. A configuration of self-suspending or supra-condylar attachment is helpful for those having lower to medium range beneath elbow nonappearance. Longer appendages may entail the utilization of a locking move on sort inward liner or more unpredictable saddling to help enlarge suspension.

Wrists: These units are either screw-on connectors including the UNF 1/2-20 string (United States) or brisk discharge connector available in diverse models.

Deliberate opening and deliberate closing: Two kinds of body controlled frameworks exist, i.e., willful opening "draw to open" and intentional shutting "draws to close." For all intents and purposes all "split snare" prostheses work with a deliberate opening sort framework. More present day's "prehensors" (also called as GRIPS) use deliberate shutting frameworks. The differences are huge. Clients of intentional opening frameworks depend on flexible groups or springs for holding power, while clients of deliberate shutting frameworks depend on all alone body power and vitality to make grasping power. Intentional shutting clients can produce prehension powers comparable to the typical hand, aloft to or surpassing 100 pounds. Intentional shutting GRIPS needs steady strain to hold, similar to a human hand, and in that characteristic, these came nearer to coordinating human hand execution.

Terminal devices: These types of gadgets have the scope of snares, prehensors, hands, or different gadgets.

Snares: Voluntary opening split snare frameworks are straightforward, advantageous, light, hearty, adaptable, and generally reasonable. A snare does not coordinate with a typical human hand for manifestation or general adaptability, yet its material resistances may surpass and outperform the ordinary hand of human being for mechanical anxiety (person can even utilize a snare to cut open boxes or like a sledge while the identical is unrealistic with a typical hand), for warm solidness (person can utilize a snare to grasp things from bubbling water, to turn meat on a flame broil, to grasp a match until it has torched totally) and for concoction perils (as a metal snare resists acids or lye, and doesn't respond to solvents similar to human skin or prosthetic glove).

6.5.6 Lower-extremity prosthetics

6.5.6.1 Hands, hips, and knees

These hands are accessible in both desirable opening and deliberate shutting renditions and in view of their more mind boggling mechanics and corrective glove covering require a moderately vast initiation compel, which, contingent upon the sort of saddle utilized, might be uncomfortable. The improvement of mechanical prosthetic hands has been dismissed in midst of the previous decades. The squeeze constrains level of the majority modern mechanical hands are measured too low for down to earth application. The best-tried hand was a prosthetic hand created around 1945 [61,62]. Bring down limit prosthetics depicts falsely supplanted appendages situated at the hip or lower level. Pertaining to all ages Ephraim and their team reported an overall gauge of all-cause bring down furthest point removals of 2.0–5.9 for each 10,000 tenants [63]. For birth pervasiveness rates of inborn appendage inadequacy, these found a gauge in-between 3.5 and 7.1 cases for every 10,000 births [63]. Two principle subcategories of lower furthest point prosthetic gadgets are transtibial (any confiscation transecting the tibia bone or an inborn oddity bringing on a tibial lack) and transfemoral (any confiscation transecting the femur bone or an inherent oddity bringing on a femoral inadequacy). In business of prosthetic devices, a transtibial prosthetic leg is known as a "BK" or beneath the knee prosthesis; whereas, transfemoral prosthetic leg is known as "AK" or over the knee prosthesis [56]. In addition, the mechanical performance and fixation method plays the vital role in full replacement of hip. Design and development of nanocomposite materials for the application in hip-joint fixation nowadays concentrates typically on the cementing approaches and bone in growth [58]. Other, less pervasive lower furthest point cases incorporate the taking after:

(1) Hip disarticulations: This more often than not alludes to when an amputee or intrinsically tested patient has either a removal or abnormality at or in nearness to the hip joint.
(2) Knee disarticulations: Knee disarticulation for the most part alludes to a removal by means of the knee disarticulating the femur from the tibia.
(3) Symes: This is a lower leg displacement while saving the heel cushion.

6.5.6.2 Socket

This vital portion fills in as the interface in-between residuum and prosthesis, permitting agreeable weight bearing, development control and proprioception. Its fitting is a standout among the most difficult parts of whole prosthesis. Challenges went with the attachment are that it requires a flawless fit, with total surface bearing to anticipate excruciating weight spots. It should be adaptable, yet solid, to permit typical walk development yet not twist underweight [64].

6.5.6.3 Shank and connectors

Makes separation and support in-between knee joint and foot (if there should be an occurrence of upper-leg prosthesis) or among the attachment and foot. The sort of connectors which are utilized between shank and knee or foot decides if prosthesis is secluded or not. Particular implies that the point and uprooting of the foot with regard to the attachment can be altered in the wake of fitting. While considering kids measured quality of point, stature is essential due to their normal development of 1.9 cm, annually [65].

6.5.6.4 Foot

Giving contact to the surface or ground, the foot gives stun ingestion and strength amid stance. Furthermore, it impacts the biomechanics of walk by its shape and firmness. This is on account of the direction of focal point of weight and edge of the ground or surface response strengths is controlled by the shape as well as solidness of the foot and needs to coordinate the subject's form so as to create an ordinary walk pattern. J. Andrysek (2010) discovered 16 unique sorts of feet with enormously differing outcomes concerning sturdiness and biomechanics [52]. The primary issue found with current feet is sturdiness, continuance going from 16 to 32 months. These outcomes are for grown-ups and will most likely be more regrettable for kids because of higher action levels and scale impacts [52,66,67].

6.5.6.5 Knee joint

If there should arise an occurrence of a transfemoral removal there additionally is a requirement for a mind boggling connector giving verbalization, permitting flexion amid swing-stage however not amid position.

6.5.6.6 Microprocessor control

To mirror the knee's usefulness amid walk, knee joints controlled by microchip have been created which control the flexion of the knee. Otto Bock's C-leg (1997), Power Knee by Ossur (2006), Ossur's Rheo Knee discharged (2005), Plié Knee, Self Learning Knee (SLK) are some of the examples. The thought was initially created by a Canadian architect Kelly James, at the University of Alberta [59,68]. A microchip is utilized to decipher and investigate signals from knee-edge sensors and minute sensors. This chip gets signals from their sensors to decide the kind of movement being utilized by the amputee. Maximum numbers of chip controlled knee joints are fuelled by a battery present in the interior of the prosthesis. The tactile signs registered by the chip are utilized to control the resistance produced by pressure driven barrels in the knee joint. Little valves control the measure of water driven liquid, which may go into and out of chamber, therefore directing the expansion and pressure of the cylinder associated with the upper area of the knee [69]. The primary preferred standpoint of microchip controlled prosthesis is nearer estimation to an amputee's common stride. Several enable amputees to turn close strolling velocity or run. Varieties in speed are likewise conceivable and are considered by sensors and conveyed to the microchip, which acclimates to these progressions as needs be. It additionally empowers the amputees to stroll down stairs by means of a stage overstride approach, as opposed to the slowly and carefully approach utilized with mechanical knees. However, some have some noteworthy downsides that hinder its utilization. These can be defenseless to water harm and consequently awesome care must be taken to guarantee that the prosthesis stays dry [70].

6.5.7 Myoelectric prosthetics

A myoelectric prosthesis utilizes electromyography signs or possibilities from intentionally contracted muscles inside a man's remaining appendage on the exterior of the skin to manage the developments of prosthesis, for examples, elbow flexion/augmentation, wrist supination/pronation (turn), or opening/shutting of the fingers. Prosthesis of this type uses the lingering neuromuscular arrangement of human body to control the elements of an electric controlled prosthetic wrist, elbow, hand, or foot. Again, this is not quite the same as an electric switch prosthesis, which requires straps as well as links activated by body developments to impel or work switches that control the developments of the prosthesis. The USSR was the first to build up a myoelectric arm in 1958, [71], while the main myoelectric arm ended up plainly business in 1964 by the CPRI, USSR (Central Prosthetic Research Institute, USSR), and disseminated by the Hangar Limb Factory of the United Kingdom [51,72]. Specialists at the Rehabilitation Institute of Chicago reported in 2013 that they have built up a mechanical leg that interprets neural driving forces from the client's thigh muscles into development, which is the main prosthetic leg to do as such. Nowadays, it is in testing [73].

6.5.8 Orthopedic prosthetics

Composite materials are being widely used in numerous orthopedic applications including orthopedic prosthetics [58]. In bone obsession plates, hip joint substitution, bone concrete, and bone unions. In absolute hip substitution, basic materials for the femoral-stem segment, for example, 316L stainless steel, hydroxyapatite combinations, and Ti-6Al-4V titanium composite have high solidness contrasted and these are replaced [58,74]. For fracture fixation, a completely resorbable bone plate is desirable to keep away from the requirement for a moment operation to evacuate the embedded in the mending of fractured part. This as a custom made low solidness composite additionally evades the issue of stress protecting problem of stress shielding. Examples of composite bone plates include laminated continuous carbon fiber in a polylactide (PLA) matrix, which is partially absorbable, and calcium phosphate glass fibers also in PLA, which is fully resorbable. At present, numerous fiber reinforced polymeric nanocomposites are widely employed as multiphase materials in orthopedic prosthetic devices as these are capable of accomplishing lower elastic modulus and higher strength in an effectual manner [58].

6.5.9 Robotic prostheses

Robots can be utilized to produce target measures of patient's disability and treatment result, aid determination, redo treatments in light of patient's engine capacities, also guarantee consistence with treatment regimens and keep up patient's records. It is appeared in many reviews that there is a critical change in upper appendage engine work after stroke utilizing applies autonomy for upper appendage rehabilitation [75]. All together for a mechanical prosthetic appendage to work, it must have a few parts to coordinate it into the body's capacity: biosensors identify signals from the client's apprehensive or strong frameworks. It then transfers this data to a controller situated inside the gadget, and procedures criticism from the appendage and actuator (e.g., position, constrain) and sends it to the controller. Cases incorporate surface cathodes that identify electrical movement on the skin, needle terminals embedded in muscle, or strong state anode exhibits with nerves developing through them. One type of these biosensors are utilized in myoelectric prostheses.

A gadget known as the controller is associated with the client's nerve and strong frameworks and the gadget itself. It sends aim summons from the client to the actuators of the gadget, and translates criticism from the mechanical and biosensors to the client. The controller is likewise in charge of the checking and control of the developments of the gadget. An actuator imitates the activities of a muscle in creating power and development. Illustrations incorporate an engine that guides or replaces unique muscle tissue. Directed muscle reinnervations is a strategy in which engine nerves, which formerly controlled muscles on a severed appendage are surgically rerouted with the end goal that these reinnervate a little district of an expansive in place muscle, for example, the pectoralis major. Therefore, when a patient considers moving the thumb of his missing hand, a little range of muscle on his trunk will contract. By setting sensors over the reinnervated muscle, these constrictions can be made to control development of a proper some portion of the automated prosthesis [76,77]. A variation of this system is called focused on tactile reinnervation. This methodology is like directed muscle reinnervations, with the exception of those tactile nerves are surgically rerouted to skin on the trunk, as opposed to engine nerves rerouted to muscle. As of late, automated appendages have enhanced in their capacity to take signals from the human mind and make an interpretation of those signs into movement in the manufactured appendage.

In prosthetic applications, nanocomposites are being used in various orthopedic prosthetics, dental prosthetics, maxillofacial prosthetics, cardiac prosthetics, etc. [58,78–80]. Various biomedical nanocomposites used in prosthetic devices are presented in Table 6.1.

6.6 Conclusion

During past few decades, considerable research efforts have been directed in the nanotechnological research to design and develop newer nanomaterials for a range of industrial applications including biomedical purposes. Thus, nanocomposites are the

Table 6.1 Biomedical nanocomposites used in prosthetic devices

Biomedical uses in prosthetics	Nanocomposites
Orthopedic prosthetics	Calcium phosphate-alumina Hydroxyapatite-polyolefins Hydroxyapatite-alumina Hydroxyapatite-zirconia Hydroxyapatite-collagen Hydroxyapatite-stainless steel Bioactive glass-collagen Poly (vinyl alcohol)-TiO_2 Bioactive glass-hydroxyapatite Hydroxyapatite-poly(lactic acid), Hydroxyapatite-poly(lactic-*co*-glycolic acid)
Dental prosthetics	Hydroxyapatit-bioactive glass Hydroxyapatite-poly(lactic acid), Hydroxyapatite-poly(lactic-*co*-glycolic acid)
Maxillofacial prosthetics	Calcium phosphate-alumina Hydroxyapatite-zirconia Hydroxyapatite-collagen Hydroxyapatite-stainless steel
Cardiac prosthetics	Modified polyhedral oligomeric silsesquioxane PVA-bacterial cellulose

newer nanomaterials, which are comprised of multiple nanoscale materials or a nanoscale material incorporated into a bulk material. The advancement in nanocomposite development is noteworthy and includes numerous ventures. The investigation of nanocomposites is an interdisciplinary region, including physics, chemistry, biology, materials science, and engineering. A variety of nanocomposites have already been explored and exploited in the fields of materials, metals, plastics, ceramics, biomaterial, electronic materials, etc. In biomedical applications, these nanocomposites are being used in orthopedics, dentistry, tissue engineering, antimicrobial properties, wound dressings, drug delivery, cancer therapy, stem cell therapy, biosensors, etc. At present, numerous nanocomposites are being utilized as improved biomaterials in various kinds of prosthetic devices. In prosthetic applications, nanocomposites are used in orthopedic prosthetics, dental prosthetics, maxillofacial prosthetics, cardiac prosthetics, etc. Even, most of these nanocomposites already have shown good prospects for diverse prosthetic applications.

References

[1] D. Pal, A.K. Nayak, Nanotechnology for targeted delivery in cancer therapeutics, Int. J. Pharm. Sci. Rev. Res. 1 (2010) 1–7.

[2] A.K. Nayak, A.K. Dhara, Nanotechnology in drug delivery applications: a review, Arch. Appl. Sci. Res. 2 (2010) 284–293.

[3] S. Jana, N. Maji, A.K. Nayak, K.K. Sen, S.K. Basu, Development of chitosan-based nanoparticles through inter-polymeric complexation for oral drug delivery, Carbohydr. Polym. 98 (2013) 870–876.
[4] S. Jana, A. Gangopadhaya, B.B. Bhowmik, A.K. Nayak, A. Mukherjee, Pharmacokinetic evaluation of testosterone-loaded nanocapsules in rats, Int. J. Biol. Macromol. 72 (2015) 28–30.
[5] S. Jana, S. Manna, A.K. Nayak, K.K. Sen, S.K. Basu, Carbopol gel containing chitosan-egg albumin nanoparticles for transdermal aceclofenac delivery, Colloids Surf. B: Biointerf. 114 (2014) 36–44.
[6] J. Malakar, A. Ghosh, A. Basu, A.K. Nayak, Nanotechnology: A promising carrier for intracellular drug delivery system, Int. Res. J. Pharm. 3 (2012) 36–40.
[7] J. Malakar, S.O. Sen, A.K. Nayak, K.K. Sen, Formulation, optimization and evaluation of transferosomal gel for transdermal insulin delivery, Saudi Pharm. J. 20 (2012) 355–363.
[8] B. Das, S.O. Sen, R. Maji, A.K. Nayak, K.K. Sen, Transferosomal gel for transdermal delivery of risperidone: Formulation optimization and *ex vivo* permeation, J. Drug Deliv. Sci. Technol. 38 (2017) 59–71.
[9] R. Abraham, S. Thomas, S. Kuryan, J. Isac, K. Varughese, S. Thomas, Mechanical properties of ceramic-polymer nanocomposites, Express Polym. Lett. 3 (2009) 177–189.
[10] R. Roy, R.A. Roy, D.M. Roy, Alternative perspectives on "quasi-crystallinity": non-uniformity and nanocomposites, Mater. Lett. 4 (1986) 323–328.
[11] P.M. Ajayan, L.S. Schadler, P.V. Braun, Nanocomposite Science and Technology, Wiley-VCH Verlag GmbH, Weinheim, 2003.
[12] E.P. Giannelis, R. Krishnamoorti, E. Manias, Polymer-silicate nanocomposites: model systems for confined polymers and polymer brushes, in: Polymers in Confined Environments, Springer, Berlin, 1999.
[13] D.E. Hill, Y. Lin, L.F. Allard, Y.P. Sun, Solubilization of carbon nanotubes via polymer attachment, Int. J. Nanosci. 1 (2002) 213–221.
[14] S. Beg, M. Rizwan, A.M. Sheikh, M.S. Hasnain, K. Anwer, K. Kohli, Advancement in carbon nanotubes: basics, biomedical applications and toxicity, J. Pharm. Pharmacol. 63 (2011) 141–163.
[15] S. Abdulalmohsin, S. Al-mutoki, Z. Li, Al/PANI-MWNT/au-plastic Schottky diode solar cells, J. Arkansas Acad. Sci. 66 (2012) 36–40.
[16] D.R. Paul, L.M. Robeson, Polymer nanotechnology: Nanocomposites, Polymer 49 (2008) 3187–3204.
[17] R. Zafar, K.M. Zia, S. Tabasum, F. Jabeen, A. Noreen, M. Zuber, Polysaccharide based bionanocomposites, properties and applications: a review, Int. J. Biol. Macromol. 92 (2016) 1012–1024.
[18] C.J. Grande, F.G. Torres, C.M. Gomez, M.C. Bañó, Nanocomposites of bacterial cellulose/hydroxyapatite for biomedical applications, Acta Biomater. 5 (2009) 1605–1615.
[19] H. Ullah, F. Wahid, H.A. Santos, T. Khan, Advances in biomedical and pharmaceutical applications of functional bacterial cellulose-based nanocomposites, Carbohydr. Polym. 150 (2016) 330–352.
[20] Y. Zheng, J. Monty, R.J. Linhardt, Polysaccharide-based nanocomposites and their applications, Carbohydr. Res. 405 (2015) 23–32.
[21] P. Panupakorn, E. Chaichana, P. Praserthdam, B. Jongsomjit, Polyethylene/clay nanocomposites produced by *in situ* polymerization with zirconocene/MAO catalyst, J. Nanomater. 2013 (2013) 1–9. Article ID 154874.
[22] P. Zapata, R. Quijada, J. Retuer, E. Moncada, Preparation of nanocomposites by *in situ* polimerization, J. Chil. Chem. Soc. 53 (2008) 1359–1360.

[23] C. Capanescu, I. Capanescu, C. Cincu, Nano-Clays in Polyester Gelcoats, 18, Business News Publishing Co., Paint & Coatings Industry, Troy, Michigan, 200254–58.
[24] C.C. Okpala, Nanocomposites–an overview, Int. J. Eng. Res. Dev. 8 (2013) 17–23.
[25] S. Tjong, G. Wang, High-cycle fatigue properties of Al-based composites reinforced with in situ TiB_2 and Al_2O_3 particulates, Mater. Sci. Eng. A 386 (2004) 48–53.
[26] B. Li, Y. Wang, Facile synthesis and photocatalytic activity of ZnO-CuO nanocomposite, Superlattices Microstruct. 47 (2010) 615–623.
[27] H. Zhang, G. Chen, Potent antibacterial activities of Ag/TiO2 nanocomposite powders synthesized by a one-pot sol-gel method, Environ. Sci. Technol. 43 (2009) 2905–2910.
[28] K. Niihara, Development of high performance ceramic base nanocomposites, Met. Powder. Rep. 3 (1999) 37.
[29] W.P. Lim, C.T. Wong, S.L. Ang, H.Y. Low, W.S. Chin, Phase-selective synthesis of copper sulfide nanocrystals, Chem. Mater. 18 (2006) 6170–6177.
[30] J. She, T. Inoue, M. Suzuki, S. Sodeoka, K. Ueno, Mechanical properties and fracture behavior of fibrous Al_2O_3/Sic ceramics, J. Eur. Ceram. Soc. 20 (2000) 1877–1881.
[31] A.K. Nayak, S. Kalia, M.S. Hasnain, Optimization of aceclofenac-loaded pectinate-poly (vinyl pyrrolidone) beads by response surface methodology, Int. J. Biol. Macromol. 62 (2013) 194–202.
[32] A.K. Nayak, S. Khatua, M.S. Hasnain, K.K. Sen, Development of diclofenac sodium-loaded alginate-PVP K 30 microbeads using central composite design, DARU J. Pharm. Sci. 19 (2011) 356–366.
[33] A.K. Nayak, J. Malakar, D. Pal, M.S. Hasnain, S. Beg, Soluble starch-blended Ca^{2+}-Zn^{2+}-alginate composites-based microparticles of aceclofenac: formulation development and *in vitro* characterization. Future J. Pharm.. Sci. (2017) https://doi.org/10.1016/j.fjps.2017.10.001.
[34] M.S. Hasnain, P. Rishishwar, S. Ali, Floating-bioadhesive matrix tablets of hydralazine HCL made of cashew gum and HPMC K4M, Int. J. Pharm. Pharmaceut. Sci. 9 (2017) 124–129.
[35] M.S. Hasnain, P. Rishishwar, S. Ali, Use of cashew bark exudate gum in the preparation of 4% lidocaine HCL topical gels, Int. J. Pharm. Pharmaceut. Sci. 9 (2017) 146–150.
[36] P.R. Guru, H. Bera, M.P. Das, M.S. Hasnain, A.K. Nayak, Aceclofenac-loaded *Plantago ovata* F. Husk mucilage-Zn^{+2}-pectinate controlled-release matrices, Starch—Stärke 70 (2018) 1700136.
[37] A.K. Nayak, D. Pal, J. Pradhan, M.S. Hasnain, Fenugreek seed mucilage-alginate mucoadhesive of metformin HCl: design, optimization and evaluation, Int. J. Biol. Macromol. 54 (2013) 144–154.
[38] A.K. Nayak, D. Pal, M.S. Hasnain, Development, optimization and *in vitro-in vivo* evaluation of pioglitazone-loaded jackfruit seed starch-alginate beads, Curr. Drug Deliv. 10 (2013) 608–619.
[39] H.D. Wagner, R.A. Vaia, Nanocomposites: issues at the interface, Mater. Today 7 (2004) 38–42.
[40] P.H.C. Camargo, K.G. Satyanarayana, F. Wypych, Nanocomposites: synthesis, structure, properties and new application opportunities, Mater. Res. 12 (2009) 1–39.
[41] A.K. Nayak, K.K. Sen, Hydroxyapatite-ciprofloxacin implantable minipellets for bone delivery: preparation, characterization, *in vitro* drug adsorption and dissolution studies, Int. J. Drug Dev. Res. 1 (2009) 47–59.
[42] M.S. Hasnain, A.K. Nayak, R. Singh, F. Ahmad, Emerging trends of natural-based polymeric systems for drug delivery in tissue engineering applications, Sci. J. UBU 1 (2010) 1–13.

[43] M. Baghayeri, E.N. Zare, M.M. Lakouraj, A simple hydrogen peroxide biosensor based on a novel electro-magnetic poly (p-phenylenediamine)@Fe_3O_4 nanocomposite, Biosens. Bioelectron. 55 (2014) 259–265.
[44] A.K. Nayak, A. Bhattacharya, K.K. Sen, Hydroxyapatite-antibiotic implantable minipellets for bacterial bone infections using precipitation technique: Preparation, characterization and *in-vitro* antibiotic release studies, J. Pharm. Res. 3 (2010) 86–88.
[45] A.K. Nayak, A. Bhattacharya, K.K. Sen, *In vivo* ciprofloxacin release from hydroxyapatite-based bone implants in rabbit tibia: a preliminary study, ISRN Orthop. 2011 (2011) 1–4. Article ID 420549.
[46] A.K. Nayak, B. Laha, K.K. Sen, Development of hydroxyapatite-ciprofloxacin bone-implants using "Quality by Design", Acta Pharma. 6 (2011) 25–36.
[47] A.K. Nayak, M.S. Hasnain, J. Malakar, Development and optimization of hydroxyapatite-ofloxacin implants for possible bone-implantable delivery in osteomyelitis treatment, Curr. Drug Deliv. 10 (2013) 241–250.
[48] S. Radhakrishnan, C. Sumathi, A. Umar, S.J. Kim, J. Wilson, V. Dharuman, Polypyrrole-poly (3, 4-ethylenedioxythiophene)-Ag (PPy-PEDOT-Ag) nanocomposite films for label-free electrochemical DNA sensing, Biosens. Bioelectron. 47 (2013) 133–140.
[49] M.S. Hasnain, A.K. Nayak, M. Singh, M. Tabish, M.T. Ansari, T.J. Ara, Alginate-based bipolymeric-nanobioceramic composite matrices for sustained drug release, Int. J. Biol. Macromol. 83 (2016) 71–77.
[50] N.B. Singh, S. Agarwal, Nanocomposites: an overview, Emerg. Mater. Res. 5 (2016) 5–43.
[51] A. Muzumdar, Powered Upper Limb Prostheses: Control, Implementation and Clinical Application; 11 Tables, 2004.
[52] J. Andrysek, Lower-limb prosthetic technologies in the developing world: a review of literature from 1994–2010, Prosthet. Orthot. Int. 34 (2010) 378–398.
[53] K.A. Raichle, M.A. Hanley, I.M. Molton, N.J. Kadel, K. Campbell, E. Phelps, D. Ehde, D.G. Smith, Prosthesis use in persons with lower- and upper-limb amputation, J. Rehabil. Res. Dev. 45 (2008) 961–972.
[54] M.J. Highsmith, J.T. Kahle, R.M. Miro, L.J. Mengelkoch, Bioenergetic differences during walking and running in transfemoral amputee runners using articulating and non-articulating knee prostheses, Technol Innov 18 (2016) 159–165.
[55] V.J. Eberly, S.J. Mulroy, J.K. Gronley, J. Perry, W.J. Yule, J.M. Burnfield, Impact of a stance phase microprocessor-controlled knee prosthesis on level walking in lower functioning individuals with a transfemoral amputation, Prosthet. Orthot. Int. 38 (2014) 447–455.
[56] E. Hagberg, O.K. Berlin, P. Renström, Function after through-knee compared with below-knee and above-knee amputation, Prosthet. Orthot. Int. 16 (1992) 168–173.
[57] L.J. Marks, J.W. Michael, Artificial limbs, Br. Med. J. 323 (2001) 732–735.
[58] M.S. Scholz, J.P. Blanchfield, L.D. Bloom, B.H. Coburn, M. Elkington, J.D. Fuller, M.E. Gilbert, S.A. Muflahi, M.F. Pernice, S.I. Rae, J.A. Trevarthen, S.C. White, P.M. Weaver, I.P. Bond, The use of composite materials in modern orthopaedic medicine and prosthetic devices: a review, Compos. Sci. Technol. 71 (2011) 1791–1803.
[59] N.E. Kyriazi, AI and Prosthetics, Universitat Politècnica de Catalunya, 2016.
[60] A. Mamalis, J.J. Ramsden, A. Grabchenko, L. Lytvynov, V. Filipenko, S. Lavrynenko, A novel concept for the manufacture of individual sapphire-metallic hip joint endoprostheses, J. Biol. Phys. Chem. 6 (2006) 113–117.
[61] G. Smit, D.H. Plettenburg, Efficiency of voluntary closing hand and hook prostheses, Prosthet. Orthot. Int. 34 (2010) 411–427.
[62] G. Smit, R.M. Bongers, C.K. van der Sluis, D.H. Plettenburg, Efficiency of voluntary opening hand and hook prosthetic devices, 24 years of development? J. Rehabil. Res. Dev. 49 (2012) 523–534.

[63] P.L. Ephraim, T.R. Dillingham, M. Sector, L.E. Pezzin, E.J. Mackenzie, Epidemiology of limb loss and congenital limb deficiency: a review of the literature, Arch. Phys. Med. Rehabil. 84 (2003) 747–761.
[64] A.F. Mak, M. Zhang, D.A. Boone, State-of-the-art research in lower-limb prosthetic biomechanics-socket interface: a review, J. Rehabil. Res. Dev. 38 (2001) 161–174.
[65] E. Strait, Prosthetics in developing countries, Prosthet. Resident. (2006) 1–40.
[66] Y. Jian, D.A. Winter, M.G. Ishac, L. Gilchrist, Trajectory of the body COG and COP during initiation and termination of gait, Gait Posture 1 (1993) 9–22.
[67] G. Stark, Perspectives on how and why feet are prescribed, J. Prosthet. Orthot. 17 (2005) S18–S22.
[68] M. Marriott, Titanium and sensor replace Ahab's peg leg, The New York Times (2005).
[69] A. Pike, The new high-tech prostheses, Motion 9 (1999) 17–20.
[70] C.W. Martin, Otto Bock C-Leg: A Review of its Effectiveness, WCB Evidence Based Group, 2003.
[71] R.W. Wirta, D.R. Taylor, F.R. Finley, Pattern-recognition arm prosthesis: a historical perspective—a final report, Bull. Prosthet. Res. 10 (1978) 8–35.
[72] E.D. Sherman, A Russian bioelectric-controlled prosthesis: report of a research team from the rehabilitation institute of Montreal, Can. Med. Assoc. J. 91 (1964) 1268.
[73] G. Bekey, R. Ambrose, V. Kumar, D. Lavery, A. Sanderson, B. Wilcox, J. Yuh, Y. Zheng, Robotics: State of the Art and Future Challenges, Imperial College Press, London, 2008.
[74] A.K. Nayak, Hydroxyapatite synthesis methodologies: an overview, Int. J. ChemTech Res. 2 (2010) 903–907.
[75] D.J. Reinkensmeyer, Robotic assistance for upper extremity training after stroke, Stud. Health Technol. Inform. 145 (2009) 25–39.
[76] T.A. Kuiken, L.A. Miller, R.D. Lipschutz, B.A. Lock, K. Stubblefield, P.D. Marasco, P. Zhou, G.A. Dumanian, Targeted reinnervation for enhanced prosthetic arm function in a woman with a proximal amputation: a case study, Lancet 369 (2007) 371–380.
[77] G.L. Andriole, E.D. Crawford, R.L. Grubb III, S.S. Buys, D. Chia, T.R. Church, M.N. Fouad, E.P. Gelmann, P.A. Kvale, D.J. Reding, Mortality results from a randomized prostate-cancer screening trial, N. Engl. J. Med. 360 (2009) 1310–1319.
[78] Z.M. Huang, S. Ramakrishna, Development of knitted fabric reinforced composite material for prosthetic application, Adv. Compos. Lett. 8 (1999) 289–294.
[79] G.A.X. Acasigua, G.M. Olyveira, L.M.M. Costa, D.I. Braghirolli, A. Fossati, A.C.M. Carlos Guastaldi, et al., Novel chemically modified bacterial cellulose nanocomposite as potential biomaterial for stem cell therapy applications, Curr. Stem. Cell. Res. Ther. 9 (2014) 117–123.
[80] H. Ghanbari, D. Radenkovic, S.M. Marashi, S. Parsno, N. Roohpour, G. Burriesci, A.M. Seifalian, Novel heart valve prosthesis with self-endothelialization potential made of modified polyhedral oligomeric silsesquioxane-nanocomposite material, Biointerphases 11 (2016) 029801.

Nanocomposites for improved orthopedic and bone tissue engineering applications

7

M. Saquib Hasnain, Amit Kumar Nayak*[†]

*Department of Pharmacy, Shri Venkateshwara University, Gajraula, India, [†]Department of Pharmaceutics, Seemanta Institute of Pharmaceutical Sciences, Mayurbhanj, India

7.1 Introduction

Nanotechnological research is basically recognized as one of the booming areas of current researches and developments in almost all technological disciplines [1–6]. The current advancements in the material sciences and technologies covers the areas including biomaterials, nanoparticle drug delivery, scaffolds, conductive polymers, hydrogels, polymer blends, composites and nanocomposites, nanoceramics, nanofibers, membranes, nanofilms, nanotubes, and nanorods [7–16].

Nanocomposites are mainly multiple nanomaterials or nanomaterials incorporated into another bulk material [17]. These belong to a broad class of materials, with the structures modulated in zero to three dimensions on length scales <100 nm [18]. Several approaches are available to fabricate or synthesize different kinds of nanocomposites. Important and widely employed approaches of fabrication or synthesis of nanocomposites comprise physically mixing, in situ preparation, film casting, dip coating, ionotropic gelation, coprecipitation, covalent coupling, electrospinning, colloidal assembly, and layer-by-layer assembly [19]. Almost all these approaches facilitate benefits of hydrogen bonding, hydrophobic effects, ionic interactions, electrostatic interactions, and coulombic interactions. Challenges in fabrication or synthesis of nanocomposites comprise controlling of fabrication/synthesis procedures; guaranteeing compatibility of diverse material constituents; and attaining desirable and unique material characteristics [19,20]. Nanocomposites differ from the conventional composite materials because of the remarkably higher surface-to-volume ratio of the reinforcing-phase and/or their extremely higher characteristic ratio, increased ductility with no decrease of strength and scratching resistance [21]. Reinforcing materials generally improve the physical as well as mechanical characteristics of the matrix. Reinforcing materials can be composed of inorganic materials (e.g., metals, metal oxides, other inorganic components), fibers (e.g., electrospun fibers, carbon nanotubes, etc.), or sheets (e.g., exfoliated clay stacks) [22–24]. Thus, numerous nanocomposites containing biomolecules, multiple polysaccharides, functional polymers, metals, metal oxides, other inorganic components, structured carbons, etc., have been investigated and developed [19,20,25]. The uses of nanocomposites

Applications of Nanocomposite Materials in Orthopedics. https://doi.org/10.1016/B978-0-12-813740-6.00008-9

guide newer technologies along with industrial viewpoints for the numerous fields of biomedical, pharmaceutical, biotechnology, packaging, coating, electronics, and aerospace industries [8,26,27]. This chapter presents a comprehensive review on the nanocomposites for improved orthopedic drug delivery and bone tissue engineering applications.

7.2 Biomedical nanocomposites

Nanocomposites designed for various biomedical purposes, which are often called "biomedical nanocomposites." There are numerous prospective biomedical nanocomposites, which can be classified into various overlapping categories. The most important biomedical uses of these biomedical nanocomposites comprise drug delivery, antimicrobial properties, tissue engineering, wound dressings, stem cell therapy, cancer therapy, cardiac prosthesis, artificial blood vessels, biosensors, and enzyme immobilization [28,29]. Various applications of biomedical nanocomposites are summarized in Table 7.1. Four important advanced applications of biomedical nanocomposites are illustrated in Fig. 7.1.

Table 7.1 **Various applications of biomedical nanocomposites**

Applications	Biomedical nanocomposites/nanocomposite-based systems	References
Drug delivery	Nanocomposites of cassava starch acetate-PEG/gelatin	[30]
	Chitosan-starch nanocomposite particles	[31]
	Pectin microspheres-$CaPO_4$ composite bone cement	[32]
	MMT-alginate nanocomposite	[33]
	Alginate-PVP-nano-HAp composite matrices	[34]
	Chitosan-sodium alginate nanocomposites	[35]
	Melt-blended halloysite nanotubes/wheat starch nanocomposites	[36]
	Chitosan-clay nanocomposite microparticles	[37]
	Sodium alginate/nano-HAp nanocomposite beads	[38]
	Modified K-carrageenan nanocomposite hydrogels	[39]
	Nano-HAp-antibiotic composites	[40–44]
Stem cell therapy	Chemically modified bacterial cellulose nanocomposite	[45]
Cardiac prosthesis	Modified polyhedral oligomeric silsesquioxane-nanocomposite	[46]
	PVA-bacterial cellulose nanocomposite	[47]
Artificial blood vessels	Nanocrystalline cellulose-fibrin nanocomposites	[48]

Table 7.1 Continued

Applications	Biomedical nanocomposites/nanocomposite-based systems	References
Tissue engineering	Bacterial cellulose/HAp nanocomposite scaffolds	[49]
	a-chitin/nanobioactive glass ceramic composite	[50]
	Bacterial cellulose/heparin hybrid nanofiber composites	[51]
	First otoliths/collagen/bacterial cellulose nanocomposites	[52]
	Porous K-carrageenan/$CaPO_4$ nanocomposite scaffolds	[53]
	Chitosan-PVP-TiO_2 nanocomposite	[54]
Wound dressings	Chitosan-pectin-TiO_2 nanocomposite film	[55]
	Banana peel powder/chitosan nanocomposites	[56]
	Nano-Ag/ b-chitin composite scaffolds	[57]
	MMT-chitosan-silver sulfadiazine nanocomposites	[58]
	Alginate/Ag/nicotinamide nanocomposites	[59]
	Bacterial cellulose-ZnO nanocomposites	[60]
	Sodium alginate/PVA/nano-ZnO composite nanofibers	[61]
Antimicrobial properties	Chitosan-Ag/PVP nanocomposite films	[62]
	HAp/Ti nanocomposites	[63]
	Zn-mineralized alginate nanocomposites	[64]
	Grafted sugarcane bagasse/Ag nanocomposites	[65]
	Fe_3O_4-HAp nanocomposites	[66]
Cancer therapy	Polypyrrole-poly (3, 4 ethylene dioxy thiophene)-Ag nanocomposite films	[67]
Biosensors	Polypyrrole polyaniline-Au nanocomposite films	[68]
	Electromagnetic poly(p-phenylene diamine) @Fe_3O_4 nanocomposite	[69]
	Graphene/PVP/polyaniline nanocomposite	[70]
	Graphene-polyaniline nanocomposite	[71]
	Polyaniline-bismuth oxide nanocomposite	[72]

7.3 Nanocomposites in orthopedic drug delivery applications

Drug delivery application field entails an ability of delivering drugs or other therapeutic agents to the body [73]. Currently, it is recognized as one of the most important research areas, where the delivery of drugs or other therapeutic agents are required to attain the optimum therapeutic results to treat or to protect from various kinds of diseases. Similar to other organs of the body, bones are susceptible to different orthopedic diseases, such as arthrosis, osteomyelitis, osteoporosis, osteosarcoma, bone metastasis, etc. [74,75]. During last few years, an extraordinary improvement has been noticed in the designing and testing orthopedic drug delivery strategies as well as

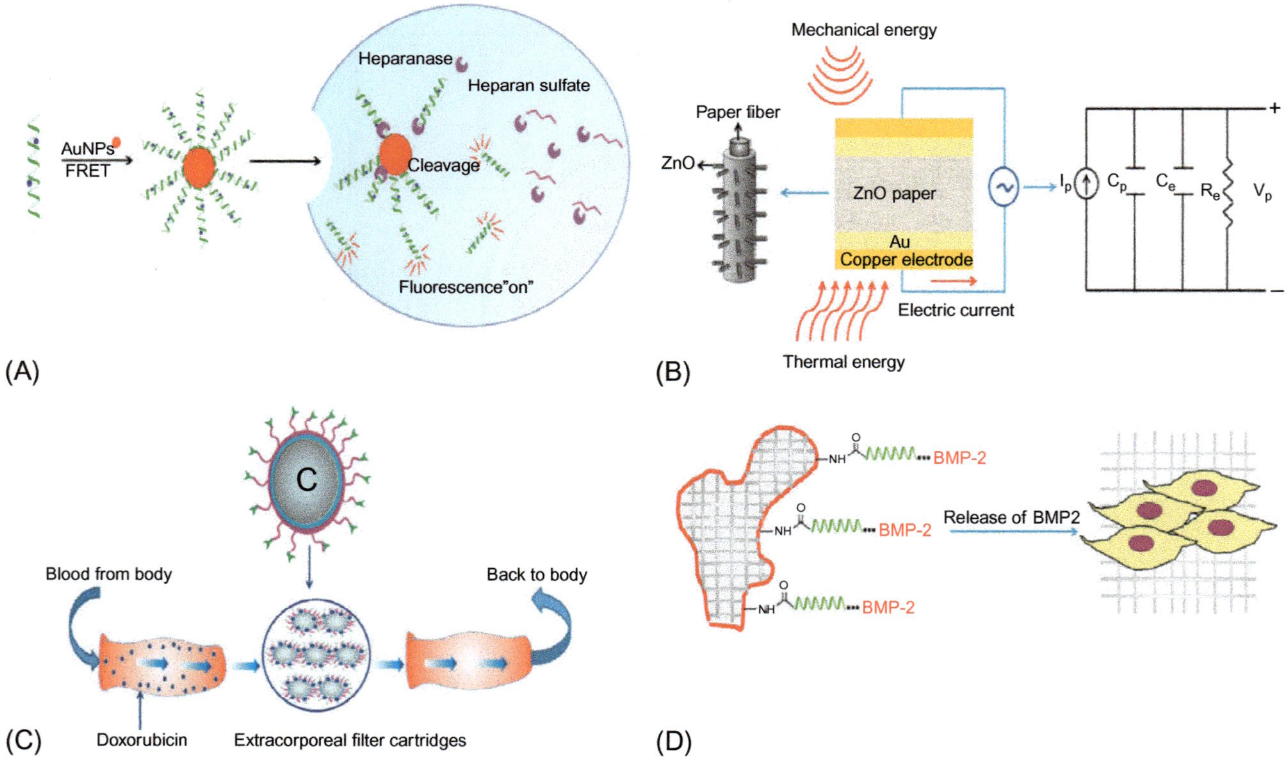

Fig. 7.1 Four important applications of biomedical nanocomposites: (A). quenching of heparin-labeled with fluorescent when bound to gold nanoparticles and releasing via heparanase quenching to decrease the allowing of heparinase generating cancer cells' recognition; (B) zinc-oxide nanoparticles developed on the cellulose fibers transducing mechanical as well as thermal energies to electrical energy; (C) doxorubicin (an anticancer drug) circulating via blood vessels passed through an extracorporeal cartridge containing activated carbon particles coated with a nanoporous PMMA/chitosan/heparin nanocomposite film to release doxorubicin; (D) heparin coupled with gelatin-lining matrix containing $CaPO_4$ employing to release BMP-2 to support the cell differentiation. From Y. Zheng, J. Monty, R. J. Linhardt, Polysaccharide-based nanocomposites and their applications, Carbohydr. Res. 405 (2015) 23–32; Copyright © 2014 Elsevier Ltd.

drug delivery devices to treat these different orthopedic diseases [1,42,76,77]. The orthopedic deliveries of drugs are dependent on the extravasation by the vessels in or near the bone [77]. During past few years, various kinds of nanocomposites are being researched as orthopedic delivery carrier matrices for a variety of drugs, proteins, hormones, and cells to treat various orthopedic diseases [78–80].

Inorganic-inorganic nanocomposites have been developed by the orthopedic researchers for the delivery of drug as well as other bioactive molecules to treat various orthopedic diseases [81]. These inorganic-inorganic nanocomposites are helpful to meet the mechanical strength as orthopedic implantable materials. In most of the cases, various bioresorbable inorganics like calcium phosphate ($CaPO_4$) and hydroxyapatite (HAp) have been widely employed to prepare these inorganic-inorganic nanocomposites for orthopedic drug delivery applications [82,83]; nanocomposites of calcium sulfate ($CaSO_4$; 50 wt%) and nanocrystalline apatite (50 wt%) have been designed for the orthopedic drug delivery carrier. Within these prepared $CaSO_4$-nanocrystalline apatite nanocomposites, a nonsteroidal antiinflammatory drug (NSAID), indomethacin was incorporated and in vitro indomethacin release from this inorganic-inorganic nanocomposite was examined. The simulated body fluid (pH 7.25) was utilized as the medium for drug release since the chemical composition of it was comparable to the human plasma composition. The prepared $CaSO_4$-nanocrystalline apatite nanocomposite demonstrated a slow and sustained drug-releasing performance, where 32% and 80% of the loaded indomethacin were released after 24 h and 144 h, respectively. On the other hand, a faster release of indomethacin was detected from the pure $CaSO_4$ cement where approximately 97% of indomethacin was measured to be released over 48 h. In elution testing, the in vitro degradations of nanocomposite samples and pure $CaSO_4$ were determined by calculating the concentration of calcium ions present in the eluting solutions. The calcium-ion concentration in the simulated body fluid (pH 7.25) augmented constantly for both the samples (i.e., nanocomposite samples and pure $CaSO_4$) with the progress of time. The calcium-ion concentration in the simulated body fluid containing pure $CaSO_4$ was found to be higher as compared to that of $CaSO_4$-nanocrystalline apatite nanocomposite. This authenticates the higher degrees of aqueous solubility of pure $CaSO_4$ as compared to $CaSO_4$-nanocrystalline apatite nanocomposite. The injectability as well as setting time of these nanocomposite cement samples was found comparatively greater than those of the pure $CaSO_4$ cement samples, whereas a reverse effect was evidenced for the compressive strength results. In addition, the loading of indomethacin into the $CaSO_4$-nanocrystalline apatite nanocomposite improved injectability as well as setting time to some extent, which did not alter the compressive strength of the nanocomposite cement sample. In vitro biocompatibility tests showed no cytotoxic effect for calcium sulfate-nanocrystalline apatite nanocomposites cement and significant difference ($P < .05$) was not found between the number of viable mouse fibroblast L929 cell line seeded in contact with the samples (i.e., $CaSO_4$-nanocrystalline apatite nanocomposite samples and pure $CaSO_4$) and the culture plate as control (Fig. 7.2). The morphology of fibroblast L929 cell line affixed onto the plate-floor of the $CaSO_4$-nanocrystalline apatite nanocomposite samples was observed comparatively more slender with polygonal morphological form of the control. Fibroblast L929 cell lines with polygonal morphological form

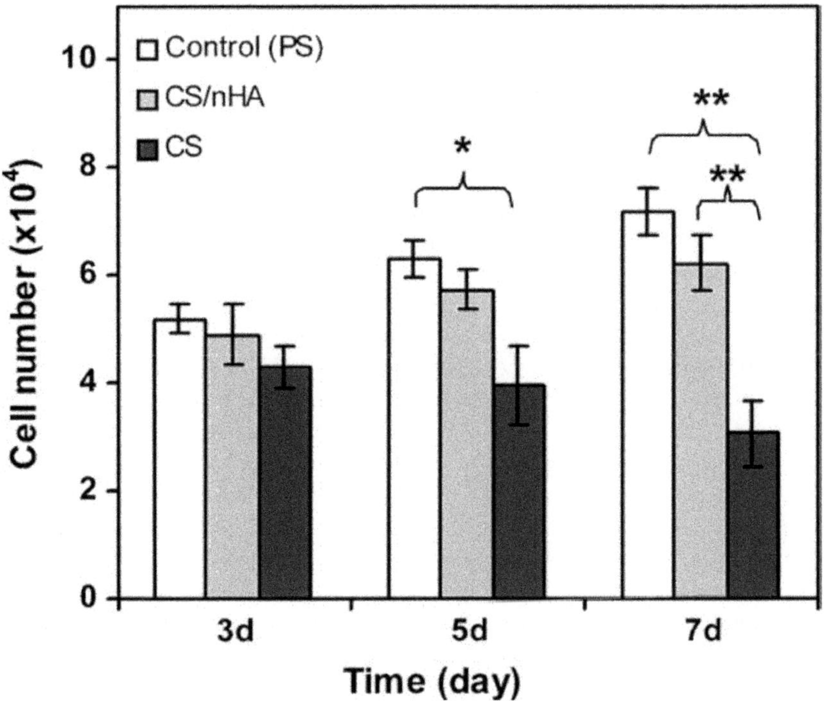

Fig. 7.2 Proliferation of mouse fibroblast L929 cell line cultured in the presence of pure $CaSO_4$ and $CaSO_4$-nanocrystalline apatite nanocomposites cements for various time intervals ($*P<.05$ and $**P<.005$).
From S. Hesaraki, F. Moztarzadeh, R. Nemati, N. Nezafati, Preparation and characterization of calcium sulfate-biomimetic apatite nanocomposites for controlled release of antibiotics, J. Biomed. Mater. Res. B. Appl. Biomater. 91 (2009) 651–661; Copyright © 2009 IPEM.

and the evidence of extended cytoplasmic membrane indicated the cytocompatibility of $CaSO_4$-nanocrystalline apatite nanocomposite with the fibroblast L929 cell line. In contrast, a partial cytotoxicity of the pure $CaSO_4$ cement samples was observed because of the decreased numbers in cell contacts with the samples tested and spherical morphological form. As a result, the $CaSO_4$-nanocrystalline apatite nanocomposite can be employed as a more suitable orthopedic carrier matrix than pure $CaSO_4$ facilitating controlled releasing of indomethacin over a longer period for the management of orthopedic defects.

Rauschmann et al. have studied biocompatibility and antibiotic-releasing capability of a bioresorbable composite, PerOssal composed of $CaSO_4$ and nanocrystalline HAp [81]. In this study, to prepare PerOssal pellets, $CaSO_4$ α-hemihydrate powders were thoroughly mixed up with the nanocrystalline HAp paste (synthesized, 35% of HAp content) in 48.50%–51.50% of ratio (solid phase). The resultant paste of $CaSO_4$ α-hemihydrate powder-nanocrystalline HAp was successfully injected into a mold and then, kept until hardened. The size of the PerOssal pellets was 5.90 mm as height and 5.90 mm as diameter. The prepared PerOssal pellets were then transferred into the glass vials and sterilized through the γ-irradiation power of 25 Gy. These sterilized PerOssal pellets were loaded with vancomycin and gentamicin. About 96.30% elution of vancomycin was measured for PerOssal within 10 days, whereas gentamicin release was found as 96.30% within the same time period. PerOssal exhibited higher content of initial and lower content of vancomycin and gentamicin releases after a period of about 5 days as compared to that of $CaSO_4$. The cytocompatibility evaluation of PerOssal was tested on fibroblasts L929 and human osteoblast cell lines, in vitro. No significant ($P < .05$) cytotoxicity variation was observed. PerOssal displays excellent characteristics concerning resorption capability, excellent biocompatibility, and suitable local release of antibiotics, which suggested its use as local drug-releasing carrier in the management of various kinds of orthopedic infections.

Aghaei et al. have synthesized in situ mesoporous silicate MCM-48/HAp nanocomposite and evaluated the orthopedic drug-releasing prospective, in vitro bioactivity and cytocompatibility of this nanocomposite [84]. Ibuprofen (a NSAID) was loaded within these formulated nanocomposites. The in vitro ibuprofen releasing in simulated body fluid from 1 g of mesoporous silicate MCM-48/HAp nanocomposite powders loaded with ibuprofen was studied at 37°C. The results displayed in vitro release of ibuprofen was slightly more rapid from the nanocomposite powders loaded at 25 mg/mL ibuprofen concentration. However, at both the loadings of ibuprofen concentration (15 and 25 mg/mL of ibuprofen), the in vitro release from MCM-48/HAp nanocomposite powders was more rapid than from MCM-48 alone (Fig. 7.3). This occurrence could probably be associated to the smaller pore sizes and surface area of the tested nanocomposites, which would militate against the retention of ibuprofen, especially on the surface of the adsorbent. The bioactivity of this MCM-48/HAp nanocomposite was determined through measuring the declining of calcium-ion concentrations in simulated body fluid after soaking the nanocomposite. The results

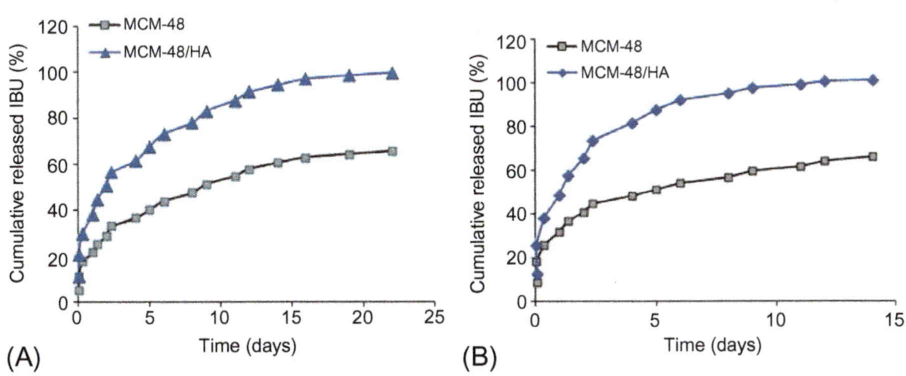

Fig. 7.3 Ibuprofen release in simulated body fluid from MCM-48 and mesoporous silicate MCM-48/HAp nanocomposite after uptake from (A) a solution containing 15 mg/mL, (B) a solution containing 25 mg/mL of ibuprofen.

From H. Aghaei, A.A. Nourbakhsh, S. Karbasi, R. Kalbasid, M. Rafienia, N. Nourbakhsh, S. Bonakdar, K.J.D. Mackenzie, Investigation on bioactivity and cytotoxicity of mesoporous nano-composite MCM-48/hydroxyapatite for ibuprofen drug delivery, Ceram. Int. 40 (2014) 7355–7362; © 2013 Elsevier Ltd. and Techna Group S.r.l.

of the in vitro bioactivity evaluation of mesoporous silicate MCM-48/HAp nanocomposite powders exhibited that the nanocomposite powder-surface was entirely covered up by the crystalline aggregate of spherical particles over 28 days of observation after soaking in the simulated body fluid. Measured calcium-ion concentrations in simulated body fluid continued to decrease as the nanocomposite powders were allowed to soak over 20 days of incubation. This occurrence advocated the apatite crystals' formation. Bioactive HAp layer formation on the surface of MCM-48/HAp nanocomposite powders after just 28 days of soaking in the simulated body fluid suggested that the bioactivity of the new nanocomposite was superior to that of the host mesoporous silicate MCM-48, which was already reported to require soaking for 60 days in the simulated body fluid to form bioactive HAp [85]. Thus, bioactivity results suggested a prospect of these nanocomposites as a bioactive material for biomedical use like tissue engineering applications. MTT assay was employed to evaluate the in vitro cytocompatibility of the mesoporous silicate MCM-48/HAp nanocomposite sample on the proliferation of tested osteosarcoma cell line. It was found that the nanocomposite powders loaded with ibuprofen exhibited >40% toxicity (Fig. 7.4), confirming safe as orthopedic biomaterial for bone drug delivery application. The results indicated the potential biomedical applications of the nanocomposite as a bioactive ceramic and as a drug delivery biomaterial with a higher rate of drug-releasing capacity as compared to mesoporous silicate MCM-48.

Recently, various organic-inorganic composites have been reported by numerous research groups for the use as orthopedic delivery matrices to treat various orthopedic diseases [78]. Even, these organic-inorganic composites were investigated as carriers for the delivery of proteins, hormones, and cells to promote bone repair

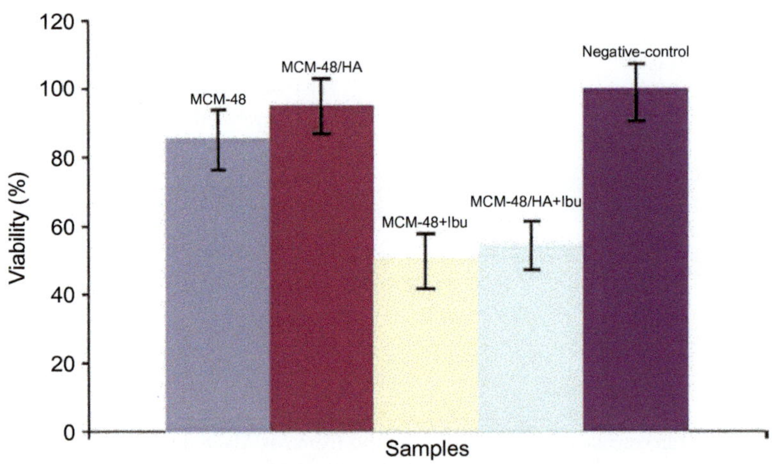

Fig. 7.4 Viability (%) of MCM-48, MCM-48/HAp nanocomposite, MCM-48 containing Ibuprofen, MCM-48/HAp nanocomposite containing Ibuprofen and negative control. From H. Aghaei, A.A. Nourbakhsh, S. Karbasi, R. Kalbasid, M. Rafienia, N. Nourbakhsh, S. Bonakdar, K.J.D. Mackenzie, Investigation on bioactivity and cytotoxicity of mesoporous nano-composite MCM-48/hydroxyapatite for ibuprofen drug delivery, Ceram. Int. 40 (2014) 7355–7362; © 2013 Elsevier Ltd. and Techna Group S.r.l.

[79,80]. Itoh et al. have developed nano-HAp-collagen nanocomposite medical device using a collagen scaffold with nano-HAp and rh-bone morphogenic protein-2 (rh-BMP-2) [86,87]. They have studied biocompatibility and osteoconductive potential of nano-HAp-collagen nanocomposite [87]. They also have performed in vivo studies using the implant model in dogs [86] and found positive results suggesting its use as bone delivery carrier for rh-BMP-2.

Otsuka and Hirano have developed an orthopedic drug delivery carrier with biodegradable apatite-collagen nanocomposite cements [88]. They evaluated bone cell activity responsive releasing of menatetrenone (vitamin K2). The in vitro drug-releasing capabilities from these biodegradable apatite-collagen nanocomposite cements were investigated for the nanoscale biointerface among bone cells and bone matrices. In vitro drug-releasing testing was performed under the osteoclast and osteoblast-like conditions. The apatite cement bulk powders comprised of an equimolar combination of dicalcium phosphate dehydrate and tetracalcium phosphate was obtained through grinding operation at 90 strokes/s for 10 min using a vibration mixer mill. Bulk powder of apatite-collagen nanocomposite cement having 2.50% of menatetrenone and 20.00% of bovine collagen was attained through grinding operation, sequentially kneaded with the H_3PO_4, and then, poured into the molds to produce fixed blocks of 0–60 perforated macropores. These nanocomposite cement samples were characterized through Fourier transformed and infrared (FT-IR) spectroscopy and X-ray powder diffraction (XRD) analyses. The results of FT-IR spectroscopy and XRD analyses clearly suggested that these biodegradable apatite-collagen nanocomposite cements were found comparable to the physicochemical attributes of the natural bone. In vitro menatetrenone-releasing experiments were carried out under the osteoblast-like conditions in simulated body fluid of pH 7.8 and after that, under the osteoclast-like conditions in acetate buffer of pH 4.5 at the $37.0°C \pm 0.1°C$. The experimental procedure was replicated two times. The biodegradable apatite-collagen nanocomposite cement device exhibited approximately no menatetrenone in the osteoblast-like conditions in simulated body fluid of pH 7.8. However, a significant amount of the menatetrenone was found to be released in vitro in the osteoclast-like conditions in the acetate buffer solution of pH 4.5 from these nanocomposite cement device. The menatetrenone release from the biodegradable apatite-collagen nanocomposite cement in the osteoclast-like conditions in acetate buffer of pH 4.5 was found to be hastening with the increasing porosity because of the dissolution of HAp matrices (Fig. 7.5). In the osteoblast-like condition the biodegradable apatite-collagen nanocomposite cement was found covered up by the little quantities of fine HAp slowering down the menatetrenone release. In addition, the collagen matrices speeded up the apatite crystal formation in the holes of surface, indicating a dissolution medium responsive-releasing potential from these biodegradable apatite-collagen nanocomposite cements to reveal geometrical structure of the micropores in-between collagen fibers and apatite crystals. The in vitro menatetrenone releasing in the osteoclast-like conditions was not always proportional to the macropore numbers in the first experiment; but, it was proportional to the macropore numbers in the second experiment. Thus, these biodegradable apatite-collagen nanocomposite cement devices demonstrated a dissolution medium responsive releasing of drugs in controlled manner and could be employed as scaffolds of bone cells because of their biodegradable properties.

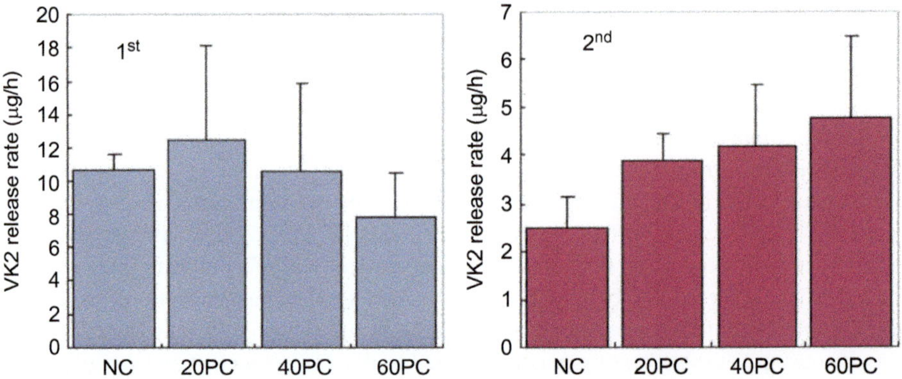

Fig. 7.5 Influence of macropore numbers on the maximum in vitro menatetrenone-release rates of different biodegradable apatite-collagen nanocomposite cements.
From M. Otsuka, R. Hirano, Bone cell activity responsive drug release from biodegradable apatite/collagen nano-composite cements—in vitro dissolution medium responsive vitamin K2 release, Colloids Surf. B: Biointerf. 85 (2011) 338–342; © 2011 Elsevier B.V.

Nanocomposite of nano-HAp-alginate using wet chemical system was reported by Venkatasubbu et al., to evaluate it as orthopedic drug delivery carrier matrix [89]. Ciprofloxacin (a fluoroquinolone derivative drug) was loaded within these nano-HAp-alginate nanocomposites. In this study, it was found that the nano-HAp crystallites were well intact with the alginate molecules. In vitro ciprofloxacin releasing from these nano-HAp-alginate nanocomposites was assessed and sustained release of ciprofloxacin from these nanocomposites was noticed. This work reported that the nano-HAp-alginate nanocomposites synthesized under the optimum control extended the ciprofloxacin-releasing pattern compared with the ciprofloxacin-loaded HAp. In another research, Nithya et al., developed ciprofloxacin-loaded HAp-polycaprolactone (PCL) nanocomposite-based film for the use as sustained drug-releasing orthopedic drug delivery bone implantable carrier for the management of osteomyelitis (an orthopedic diseases caused by bacteria) treatment [90]. The nanocrystalline HAp (used to prepare these nanocomposite film-based bone implants) was synthesized via precipitation process by means of the precursor derived from eggshell. The HAp-PCL nanocomposite film loaded with ciprofloxacin was formulated by means of solvent evaporation technique. In vitro drug releasing as well as in vitro biodegradation analyses were performed through immersing the ciprofloxacin-loaded nanocomposite film in the saline phosphate buffer. The nanocomposite film loaded with ciprofloxacin displayed a heavyweight reduction because of the faster ciprofloxacin releasing. The nanocomposite film loaded with ciprofloxacin also exhibited an elevated water absorbing property compared to PCL film. The increment of the Hap concentrations increased the water absorbing property of the nanocomposite film. In vitro cytotoxicity evaluation was carried out using fibroblast NIH-3T3 cell line and osteoblast MG-63 cell line. The in vitro cytotoxicity evaluation results of HAp-PCL nanocomposite film loaded with ciprofloxacin displayed an excellent biocompatibility as well as bioactivity. Therefore, HAp-PCL nanocomposite film can be an effective orthopedic drug delivery carrier to produce sustained drug-releasing capability.

Jia et al. prepared novel porous graphene oxide (GO)/chitosan-HAp nanocomposite films via combining layer-by-layer technique and biomimetic mineralization methodology [91]. In general, the layer-by-layer technique is a versatile method utilized to control the film thickness and to stimulate biomimetic mineralization of biocompatible HAp, which is extensively utilized in the bone tissue engineering area [92]. The preparation procedure of porous GO/chitosan-HAp nanocomposite film is illustrated in Fig. 7.6. The pore sizes of GO/chitosan-HAp nanocomposite film were measured as approximately 300nm. Thus, the porous architecture of the film made it efficient to load drug in higher concentration. A NSAID candidate, aspirin was loaded within these porous GO/chitosan-HAp nanocomposite films and the aspirin loadings in these nanocomposite-based films were within the range between 57.70% and 77.80%. The aspirin contents of these films were measured within the range between 1.364 and 3.505mg/mg. It was apparent that the increments in aspirin loading and aspirin content in the nanocomposite films were found to be increased with the increment of the number of the bilayers present in the films. The increased aspirin loading and amount in GO/chitosan-HAp nanocomposite films could result from the highly porous and layer-by-layer architecture of the film. The higher aspirin loading and aspirin content might hold enhanced antiinflammatory outcome. The in vitro release of loaded aspirin exhibited a sustained-releasing pattern. This sustained-releasing pattern of aspirin from the porous GO/chitosan-HAp nanocomposite films could be due to slower dissolution process (Fig. 7.7). The preliminary burst releasing of aspirin was seen within first 30min of the drug-releasing study for the reason that the aspirin particles present in the outermost porous HAp-bilayer of GO/chitosan-HAp nanocomposite films released

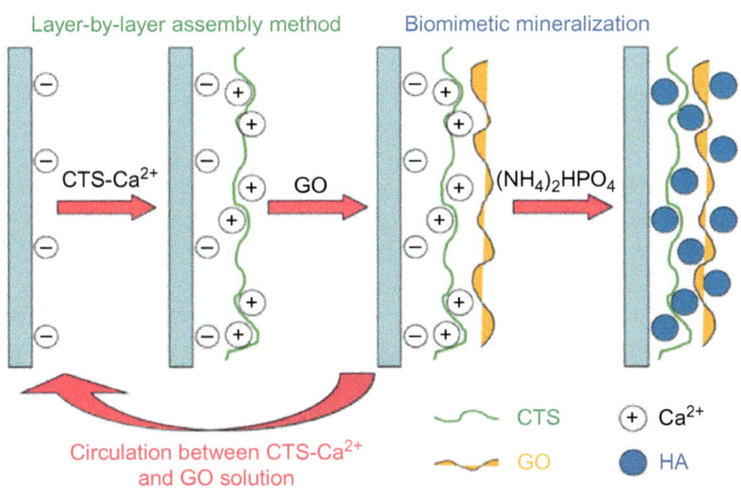

Fig. 7.6 Schematic illustration of the porous GO/chitosan-HAp nanocomposite films structured on the substrate through combining layer-by-layer technique with biomimetic mineralization.
From M. Jia, H. Li, H. Guo, A. Xie, S. Wang, F. Huang, S. Lia, Y. Shen, J. He, A novel porous aspirin-loaded (GO/CTS-HA)n nanocomposite films: Synthesis and multifunction for bone tissue engineering. Carbohydr. Polym. 153 (2016) 124–132; © 2016 Elsevier Ltd.

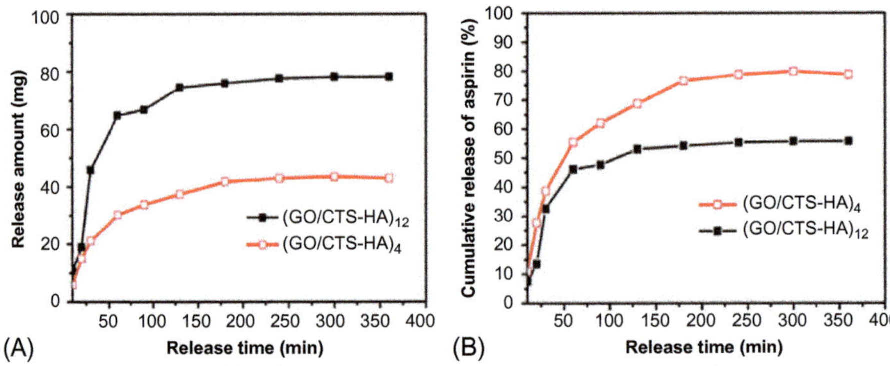

Fig. 7.7 (A) In vitro aspirin-releasing contents and (B) cumulative in vitro aspirin releasing from porous GO/chitosan-HAp nanocomposite films.
From M. Jia, H. Li, H. Guo, A. Xie, S. Wang, F. Huang, S. Lia, Y. Shen, J. He, A novel porous aspirin-loaded (GO/CTS-HA)n nanocomposite films: Synthesis and multifunction for bone tissue engineering. Carbohydr. Polym. 153 (2016) 124–132; © 2016 Elsevier Ltd.

very easily. After that, the drug-releasing rate progressively reduced until 240 min of the study for the reason that the entrapped drug released bit by bit. GO/chitosan-HAp nanocomposite films might able to release the entrapped aspirin to the adjacent tissues and thereby, the inflammation region would be decreased. Furthermore, the accumulated aspirin loading dosage might be regulated via film thickness and sustained releasing of entrapped aspirin could make sure antiinflammatory outcome, effectively. In vitro cytocompatibility evaluation of these nanocomposite films was done using mouse mesenchymal stem cells. The porous GO/chitosan-HAp nanocomposite films with 12 bilayers exhibited excellent in vitro cytocompatibility as compared to that of nanocomposite films with 4 bilayers. The percentage cell viability of this nanocomposite films was measured 71.40%. These multifunctional properties of aspirin-loaded new porous GO/chitosan-HAp nanocomposite films showed the potential application as antiinflammatory biocoating for the use in orthopedic applications.

Recently, Kolanthai et al., have evaluated agarose encapsulated mesoporous carbonated HAp nanocomposite powder for the use as orthopedic drug delivery carrier matrix and filling bone voids [93]. In this study, these nanocomposite powders were synthesized by the solvothermal process and afterward, calcined for studying the effects of thermal exposure on the physicochemical alterations. The phase of synthesized nanocomposite powder was carbonated HAp-agarose; while calcinated samples were nonstoichiometric HAp. In addition, the calcinations generated sphere-shaped nanoparticles as well as nanorods. The dimensions of mesoporous carbonated HAp-agarose nanorods were of 10–80 nm length and 40–190 nm width. Calcination route generated carbonated HAp-agarose spheres (10–50 nm) and carbonated HAp-agarose nanorods with reduced size (40–120 nm length and 20–30 nm width). In simulated body fluid, in vitro bioactivity evaluation of mesoporous carbonated HAp-agarose nanocomposites was carried out. The nanocomposites were found partially dissolved in simulated body fluid and afterward, displayed improved in vitro bioactivity compared to the nonstoichiometric HAp authenticated by the gravimetric process.

The formation of apatite was indicated through calculating differences among the weights of nanocomposites, before and after immersion in the simulated body fluid. All these nanocomposite samples demonstrated a gradual increment in weight up to the immersion of 4 weeks. Even after the immersion, the surfaces of these nanocomposite samples were observed to contain spherical apatite particles (3–5 μm), which were distributed randomly and an apatite layer on the all nanocomposite samples. The results of in vitro dissolution of mesoporous carbonated HAp-agarose nanocomposites specified that the crystallinity was found to be reduced devoid of varying crystal structures caused by discharge of calcium ions, phosphate ions, and agarose molecule. The mesoporous carbonated HAp-agarose nanocomposites were loaded with an antibiotic, amoxicillin, and an anticancer drug, 5-fluorouracil. The in vitro drug releases from these drugs-loaded nanocomposites were carried out in saline phosphate buffer. The amoxicillin-releasing pattern demonstrated a faster preliminary release of amoxicillin followed by slower releasing from nanocomposite. This faster amoxicillin-releasing pattern may be due to physical adsorption of amoxicillin molecules on to the surface of HAp-agarose nanocomposites. The complete release of the amoxicillin from the mesoporous carbonated HAp-agarose nanocomposites prepared at 120°C was found to be extended by another 60 h, which indicated the enhanced sustained release of amoxicillin from these nanocomposites. All these nanocomposite samples demonstrated faster 5-fluorouracil-releasing pattern initially and afterward, it slowed down. The carbonated HAp-agarose nanocomposite powders exhibited a sustained-releasing pattern 5-fluorouracil indicating an excellent promise for the use in the cancer chemotherapeutics. The carbonated HAp-agarose nanorods prepared at 120°C showed prolonged drug release of drugs (amoxicillin and 5-fluorouracil) compared to that of prepared at 150°C and calcined samples (Fig. 7.8). These nanocomposites also displayed a better amoxicillin and 5-fluorouracil-releasing pattern than that from the calcined samples. These mesoporous carbonated HAp-agarose nanocomposites demonstrated a strong antimicrobial action against different Gram positive as well as Gram negative bacteria such as *Escherichia coli* (MTCC 2939), *Staphylococcus aureus* (MTCC 3381), and *Staphylococcus epidermidis* (MTCC 3382). The carbonated HAp-agarose nanorods prepared at 120°C demonstrated excellent antimicrobial effectiveness compared to that of prepared at 150°C and calcined samples. The hemocompatibility evaluation of these nanocomposites showed percentage hemolysis of <1% confirming high degree of hemocompatibility. Biocompatibility of these nanocomposites was evaluated using human osteoblast MG-63 cells derived from human osteosarcoma. Hemocompatibility and biocompatibility were found unaltered by the occurrence of carbonate or agarose in nanocomposite formula. Biocompatibility of the orthopedic biomaterial is very essential for the biomedical uses mainly in orthopedic drug delivery biomaterials and bone repair implants. The procedure for the preparation of nanocomposites was found simple. The results of this investigation indicate the potential of mesoporous carbonated HAp-agarose nanocomposites as improved biomaterial for orthopedic local drug-releasing carriers to prevent orthopedic infections or inflammatory reactions and filling of voids in the damaged bone tissues. Besides these nanocomposites, some other nanocomposites investigated as carrier for orthopedic drug delivery applications are presented in Table 7.2.

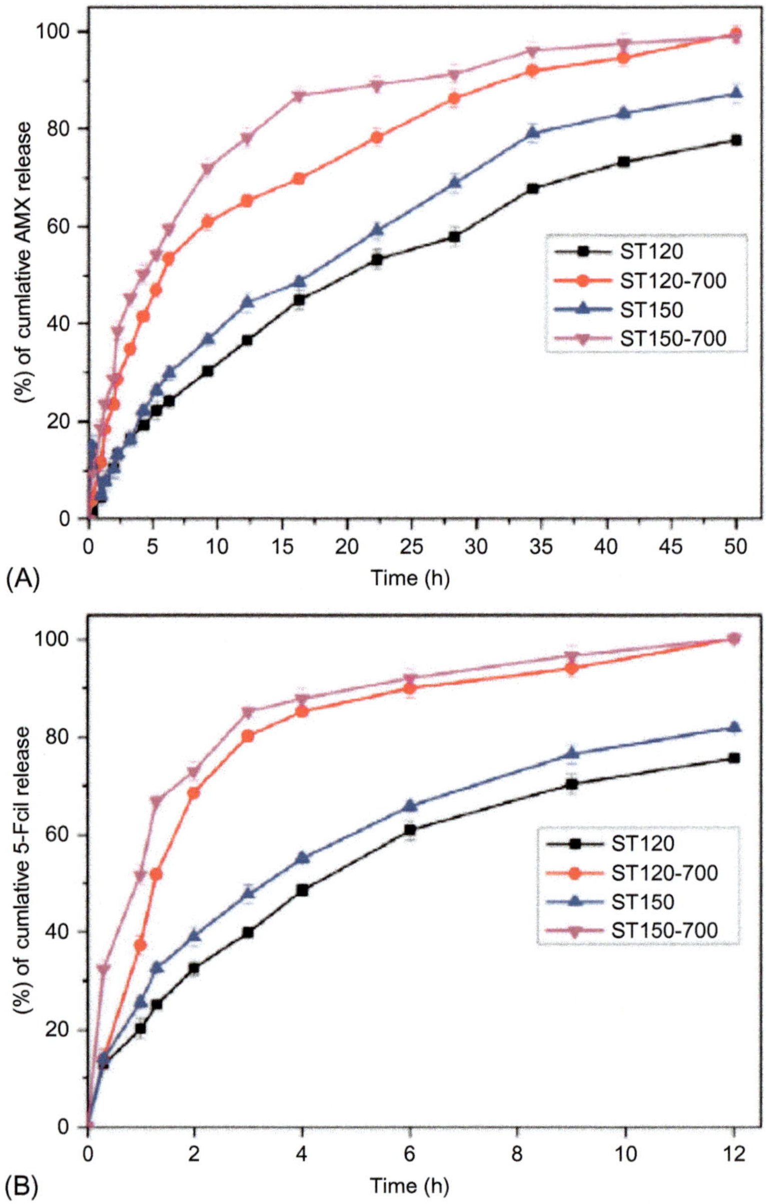

Fig. 7.8 Release of (A) amoxicillin and (B) 5-fluorouracil from mesoporous carbonated HAp-agarose nanocomposites powder.
From E. Kolanthai, P.A. Sindu, K.T. Arul, V.S. Chandra, E. Manikandan, S.N. Kalkura, Agarose encapsulated mesoporous carbonated hydroxyapatite nanocomposites powder for drug delivery, J. Photochem. Photobiol. B: Biol. 166 (2017) 220–231; © 2016 Elsevier B.V.

Table 7.2 Some examples of nanocomposites investigated as carrier for orthopedic drug delivery applications

Nanocomposites in orthopedic drug delivery	Drugs	References
Nano-HAp-anionic collagen composites	Antibiotic	[94]
Nano-HAp-anionic collagen composite	Gentamicin and norfloxacin/ ciprofloxacin	[95]
Nano-HAp-antibiotic composites in bone implants	Ciprofloxacin HCl, gatifloxacin HCl, ofloxacin, amoxicillin trihydrate	[40–44]
HAp-PCL nanocomposite film	Ciprofloxacin	[90]
PMMA-HAp bone cement	Gentamicin, amikacin, ceftiofur	[96]
PLA/HAp nanocomposite thin films	Antibiotic	[97]
PCL-forsterite nanocomposite fibrous membranes	Dexamethasone	[98]
Glass-reinforced nano-HAp composites	Sodium ampicillin	[82]
$CaSO_4$-biomimetic apatite nanocomposites	Gentamicin sulfate	[99]
Scaffolds containing $CaPO_4$/ chitosan composite	Antibiotic	[100]
Chitosan-*graft*-poly (acrylic acid-*co*-acrylamide)/HAp nanocomposite scaffold	Celecoxib	[101]
Nanocomposite scaffolds made of nanobioactive glass and alginate crosslinked with therapeutic ions such as calcium and copper	Alendronate	[102]
HAp-magnetite-MWCNT nanocomposite	Clodronate	[103]
Silica-$CaPO_4$ nanocomposite	Vancomycin and rh-BMP2	[104]
Nano-HAp/collagen composite	BMP-2	[105]
HAp/collagen-alginate nanocomposite bone filler	rh-BMP-2	[106]
Nanocomposite containing alendronate-grafted hyaluronate	BMP-2	[107]
Nanocomposite scaffold containing chemically cross-linked c-PCL and HAp nanoparticles	BMP-2	[108]
Bio-hybrid silk fibroin/$CaPO_4$/ PLGA nanocomposite scaffold	VEGF	[109]
Silk/ $CaPO_4$/PLGA nanocomposite scaffold	PDGF	[110]

7.4 Nanocomposites in bone tissue engineering applications

The regenerative medicine and tissue engineering have been recognized among various important most biomedical research fields aiming to replace and/or repair the damaged or diseased tissues and/or organs of the body [80,111]. In the year of 1988, "tissue engineering" was officially defined in a workshop by National Science Foundation (NSF), United States as "the application of principles and methods of engineering and life sciences toward fundamental understanding of structure-function relationships in normal and mammalian tissues and the development of biological substitutes to restore, maintain or improve tissue function" [79,80,112]. Such strategies allow the development of scaffolds exploiting a variety of biomaterials and these scaffolds are being implanted in the patients' body for the induction of tissue regenerations in the damaged and/or diseased areas. In general, these scaffolds are the fundamental constituents of a variety of tissue engineering approaches for the reason that these offer an architectural milieu in which extracellular matrix, cell-cell interactions, and cell-growth factor interactions combined to facilitate the regenerative niche [79,80,113]. Thus, the tissue engineering approaches mimic the cellular matrix and also make available necessary information such as cell attachment signaling, cell proliferations, and cell differentiations to meet the requisite of dynamic reciprocity for the tissue engineering applications [79].

Inorganic-inorganic nanocomposites have been designed and evaluated for their bone tissue regeneration applications as inorganic materials are able to add sufficient mechanical strength to the scaffold for load-bearing capability [114,115]. Several inorganic–inorganic bone scaffolds and implants have already reported in the literature, which have shown their prospects in the bone tissue repairing along with excellent load-bearing characteristics [63,116]. Nath et al. have also designed nano-HAp-mullite composites for orthopedic application and evaluated for their in vivo responses up to 12 weeks of implantation using rabbit model [63]. Bone growth was noticed over the 12 weeks of implantation period. Thus, the in vivo study of nano-HAp-mullite composites clearly suggested their use in bone repair application.

Very recently, Swain et al. have reported dense bioresorbable β-tricalcium phosphate-(iron-silver) nanocomposites, which were fabricated by means of attrition milling technique subsequently high-pressure processing consolidation (cold sintering at 2.50 GPa) [115]. They have incorporated β-tricalcium phosphate within the nanocomposite structure as it is chemically comparable to the composition of bone minerals. Therefore, β-tricalcium phosphate has been regarded as a suitable bioinorganic material for designing of various bioresorbable bone repairing scaffolds. Yet, intrinsic brittleness as well as low bending strength makes β-tricalcium phosphate as inappropriate material for the use in load-bearing bone sites. The mechanical properties, before and after immersion, were tested in compression and bending. All the compositions of these bioresorbable nanocomposites exhibited high mechanical strength (the strength in bending). Partial substitution of iron with silver led to an increase in both strength and ductility. The degradation behavior of the developed β-tricalcium phosphate-(iron-silver) nanocomposites was studied by immersion in Ringer's solutions and saline solutions for up to 1 month. After 1 month immersion, the composites retained about 50% of their initial bending strength. The in vitro cytotoxicity evaluation of these developed bioresorbable β-tricalcium phosphate-(iron-silver) nanocomposites on the primary human osteoblast cell lines was carried out. In in vitro cell culture experiments, these nanocomposites exhibited no signs of cytotoxicity toward human osteoblast cell lines indicating that these can be used as an implant material.

Reinforcement of inorganic nanomaterials inside the organic polymeric matrices is always proposed to enhance the cell adhesion as well as proliferation [117]. These are also essential for the bone tissue engineering purposes. During past few decades', numerous researchers have designed nanocomposites through incorporating nano-HAp along with other biocompatible polymers for the use in bone tissue engineering applications [118–123]. HAp is a bioceramic extensively utilized in a variety of orthopedics as well as dentistry applications by reason of possessing the chemical composition very close to the inorganic mineral phase of the natural bone and teeth [76]. It possesses excellent biocompatibility and osteoconductivity properties [124]. Du et al. designed HAp-collagen nanocomposite-based implantable systems and evaluated the tissue responses of the implants in the bone marrow cavity [118]. In another research, Du et al. have designed three-dimensional nano-HAp-collagen matrix which was loaded with osteogenic cells to study osteogenic responses [119]. Kikuchi et al. developed bone-like HAp-collagen nanocomposite and examined in vivo biological responses were measured [120]. The results of these investigations suggested the prospective applications of HAp-collagen nanocomposites in the bone tissue repairing applications. Like HAp-collagen nanocomposites, HAp-gelatin nanocomposites were also reported by various research groups. Kim et al. have developed biomimetic HAp-gelatin nanocomposites [121,122]. These biomimetic HAp-gelatin nanocomposites exhibited significant osteoblast responses indicating their prospective use as bone regeneration substance in the bone tissue repairing. In another study, Ishikawa et al. have prepared HAp-gelatin nanocomposite foams for the use as a hard tissue engineering scaffold [125]. These HAp-gelatin nanocomposite foams were prepared using freeze drying along with cross-linking technique.

Mohanapriya et al. fabricated and characterized nanocomposite films containing PVA-titanium dioxide (TiO_2) for the use in orthopedic purposes [126]. In this research, PVA was reinforced by TiO_2 nanoparticles with the purpose of enhancing thermomechanical stabilities, surface properties, and the adhesion of osteoblastic cells. PVA-TiO_2 nanocomposite films with desired mechanical, thermal as well as biocompatible properties were prepared via solution casting technique followed by dehydrothermal cross-linking action. The content of TiO_2 nanoparticles in the film composition was standardized to obtain mechanically stable films on the basis of tensile strength. The optimal content of TiO_2 was 3% by weight. The UV-VIS spectral analysis results of PVA and PVA-TiO_2 nanocomposite films demonstrated successful interaction between PVA and TiO_2. On the basis of absorption spectral studies, Mohanapriya et al., suggested that the TiO_2 got imbibed into PVA-matrix and interacted powerfully with the hydroxyl groups of PVA by means of hydrogen bonding [126]. FT-IR spectroscopy results indicated incorporation of TiO_2 into the PVA-matrix, cross-linking of PVA by the thermal dehydration action and interaction between PVA and TiO_2. Scanning electron microscopy (SEM) results demonstrated dense and increased surface roughness structural morphological characteristics of PVA-TiO_2 nanocomposite films with homogenously distributed TiO_2 nanoparticles, which facilitated to increase the thermomechanical performances (Fig. 7.9). The elemental mapping analysis confirmed the inclusion of TiO_2 nanoparticles within PVA-matrix. With the increase of the quantity of TiO_2 nanoparticles within PVA-matrix, the tensile strength was found to be increased to an utmost level and afterward, it began to decrease. The tensile strength results spotlighted on the fact that the survival of an interfacial interaction in-between the organic–inorganic phases powerfully controls the mechanical characteristics of PVA-TiO_2 nanocomposite films. Therefore, PVA-TiO_2 represented a promising bionanocomposite with benefits of both flexibility of a biopolymer as well as hardness of a metal, which would be more valuable for the use as orthopedic implantable material. The antibacterial actions of the PVA-TiO_2 nanocomposite films were examined against different Gram positive as well as Gram negative bacteria such as *S. aureus*, *Pseudomonas aeruginosa, Klebsiella pneumonia, Bacillus subtilis*, and *E. coli*, as model bacteria. The viable cell count process was carried out against these bacteria to

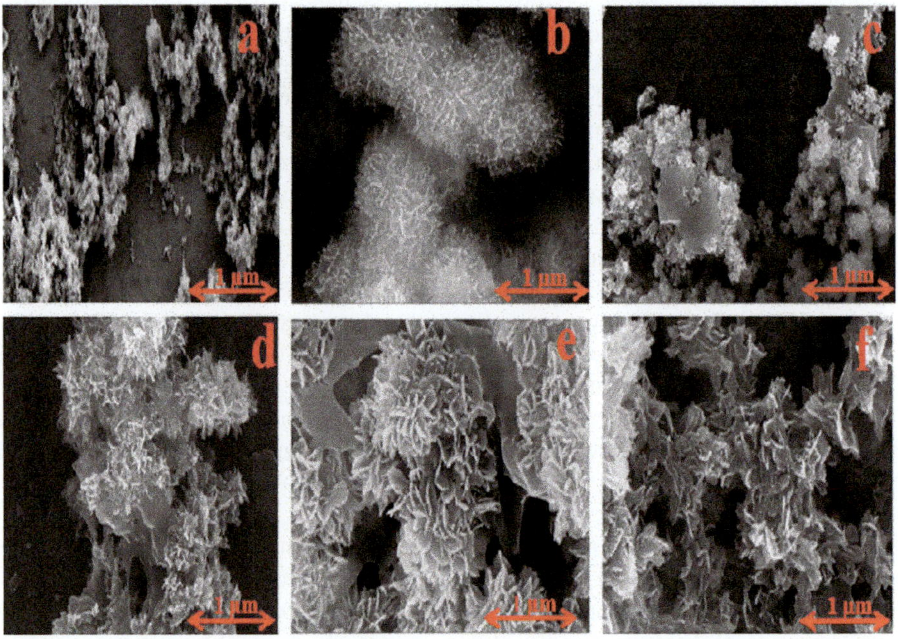

Fig. 7.9 SEM images of apatite formed on the samples dipped in simulated body fluid for 7 days: (A) PVA-TiO_2 nanocomposite films-I (B) PVA-TiO_2 nanocomposite films-III, (C) PVA-TiO_2 nanocomposite films-V nanocomposite films. After 14 days (D) PVA-TiO_2 nanocomposite films-I (E) PVA-TiO_2 nanocomposite films-III, (F) PVA-TiO_2 nanocomposite films-V.
From S. Mohanapriya, M. Mumjithaa, K. PurnaSai, V. Raja, Fabrication and characterization of poly (vinyl alcohol)-TiO2 nanocomposite films for orthopedic applications, J. Mech. Behav. Med. Mater. 63 (2016) 141–156; © 2016 Elsevier Ltd.

measure bacterial growth. From the antibacterial action measurement results, it was detected that PVA-TiO_2 nanocomposite films exhibited higher antibacterial action in comparison with that of the PVA film. Actually, TiO_2 contributed to the antibacterial action and the mechanism of the antibacterial action of TiO_2 nanoparticles over bacteria is still not clearly understood. Previous reports proposed that TiO_2 nanoparticles with positively charged surface might attract the negatively charged microorganisms by the electromagnetic attraction, which may lead to killing of microorganisms by means of oxidation. Nanoparticles could deactivate DNA as well as various cellular enzymes via coordinating with the electron donating groups, which may cause raptures and holes in the bacterial cell walls causing higher range of permeability and finally, death of the bacterial cells [127]. The dimensional stabilities of the nanocomposite films containing PVA-TiO_2 were assessed by means of soaking films in the simulated body fluid for a period of 24h suggested that TiO_2 nanoparticles influenced the controlling of swelling. It was also apparent that the incorporation of TiO_2 nanoparticles could reduce the dimensional alters in PVA polymeric matrix. TiO_2 nanoparticles could perform as a filler nanomaterial within PVA polymeric chains reducing the swelling degree and accordingly strengthening the polymeric-network. In the cell adhesion and cytotoxicity evaluations, the proliferative potential of MG-63 osteoblast cell lines on various PVA-TiO_2 nanocomposite films were analyzed by MTT assay. The results of in vitro bioactivity and cell adhesion experiments also indicated that the occurrence of titanium dioxide was beneficial to improve the growth of apatite and to encourage the cell-substrate interactions. The TiO_2 nanoparticles formed the cell adhesive hydrophilic nanodomains, which performed as the prospective cell adhesion sites and also, promoted the osteointegration. Biocompatibility evaluation showed that the thermally cross-linked PVA was found nontoxic with respect to glutaraldehyde cross-linked PVA. These results also indicated that the incorporation of TiO_2 nanoparticles into PVA polymeric-matrix influenced improved cell response as compared to that of PVA film. In vitro cytotoxicity of these PVA-TiO_2 nanocomposite films was assessed via the MMT assay for the measurement of percentage cell viability by the crystal violet staining demonstrated that PVA-TiO_2 nanocomposite contain TiO_2 of 3% by weight. These cytotoxicity results also clarified the role of TiO_2 nanoparticles toward supporting the cell viability by the improved surface chemistry and roughness along with surface nanotopography. These PVA-TiO_2 dioxide nanocomposite films could perform as an outstanding biomedical nanocomposite for orthopedic implantable applications.

Recently, Bhowmick et al. have developed bone-like organic–inorganic hybrid porous nanocomposites of chitosan, PEG, and nano-HAp, where zirconium oxide nanoparticles were incorporated [128]. The ratio of chitosan, PEG, and nano-HAp-zirconium oxide in three formulations of zirconium oxide nanoparticles-doped chitosan-PEG-nano-HAp was 55:40:5, 55:35:10 and 55:30:15, respectively. The surface morphologies as well as pore dimensions of these nanocomposites were investigated by SEM (Fig. 7.10). SEM pictures of the surface morphologies of these nanocomposites exhibited porous morphology. Numerous pores of different sizes were observed on the surface of nanocomposites. In addition, numerous macrovoids were seen on these bone-like nanocomposites. These nanocomposites have porosities of 65.71%–73.02%. Therefore, zirconium oxide nanoparticles-doped chitosan-PEG-nano-HAp was highly

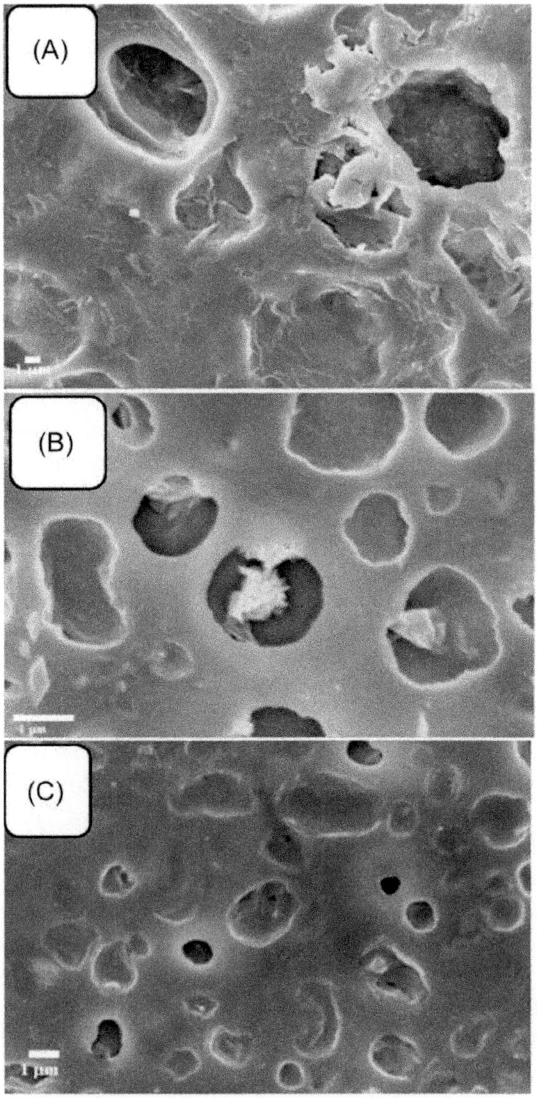

Fig. 7.10 Scanning electron microscopic images of (A) zirconium oxide nanoparticles-doped chitosan-PEG-nano-HAp nanocomposites I, (B) II, and (C) III.
From A. Bhowmick, N. Pramanik, P. Jana, T. Mitra, A. Gnanamani, M. Das, P.P. Kundu, Development of bone-like zirconium oxide nanoceramic modified chitosan based porous nanocomposites for biomedical application, Int. J. Biol. Macromol. 95 (2017) 348–356; © 2016 Elsevier B.V.

porous in nature. The porosity of these nanocomposites was almost similar to that of the human cancellous bone, and thus, helpful for permitting cell migration, exchanges of nutrients, etc. Various mechanical characteristics, such as tensile strength and stiffness of these bone-like organic-inorganic hybrid porous nanocomposites, were obtained in tensile mode. Tensile strength and stiffness of these nanocomposites were found to be improved gradually from 21.69 to 24.34 and from 17,829.93 to 55,304.40 MPa, respectively, with the increment of inorganic materials in the nanocomposite formula (nano-HAp-zirconium oxide) (Fig. 7.11). The antimicrobial activities of zirconium oxide nanoparticles-doped chitosan-PEG-nano-HAp were investigated against Gram-positive bacteria, such as *Bacillus cereus* and *Lysinibacillus fusiformis* and a gram-negative bacterium, such as *E. coli* XL1 B. The antimicrobial zone of inhibition enlarged insignificantly with the increased incorporation of inorganic materials in nanocomposite formulation. The hemocompatibility of these bone-like nanocomposites was investigated by in vitro hemolytic assay and the hemolysis percentage was calculated as below 5%, only, indicating hemocompatibility of these nanocomposites. These zirconium oxide nanoparticles-doped chitosan-PEG-nano-HAp nanocomposites were subjected to in vitro cytotoxic studies (cell proliferation analysis and MTT assay) with human osteoblastic MG-63 cell lines. The in vitro cell proliferation analysis and MTT assay results suggested nontoxic nature of these bone-like nanocomposites indicating excellent cytocompatibility with the human osteoblastic MG-63 cell lines (Fig. 7.12). Good mechanical strength, porous nature, stronger antimicrobial actions, and higher proliferation of osteoblastic MG-63 cell lines with appropriate hemocompatibility make these zirconium oxide nanoparticles-doped chitosan-PEG-nano-HAp nanocomposites as an appropriate bone tissue engineering material.

Shi et al. [129] manufactured bacterial cellulose-based calcium-deficient carbonate-containing HAp nanocomposites by means of a biomimetic mineralization procedure. In this study, HAp was utilized for the manufacturing of three-dimensional network-based nanocomposites with bacterial cellulose using a biomimetic methodology since it is an osteoconductive bioceramic material. The calcium-deficient carbonate containing HAp in the nanocomposites contained carbonate partly and this had a resemblance to the natural bone apatite. The investigation results also revealed that a three-dimensional network of bacterial cellulose presented an excellent template for well-ordered deposition of bacterial cellulose-based calcium-deficient carbonate-containing HAp particles that were made of nano-sized squama-shaped apatite crystals. Therefore, bacterial cellulose-calcium-deficient carbonate containing HAp nanocomposites have shown a great prospective as an appropriate bone tissue engineering material.

Bacterial cellulose-HAp nanocomposites scaffolds were fabricated by means of biomimetic procedure and also, examined the cell differentiations and proliferations of human bone marrow-originated stromal cells on these scaffolds [49]. The human bone marrow-derived stromal cells seeded on the bacterial cellulose-HAp nanocomposites demonstrated a better cell attachment, cell differentiations, and proliferations than the bacterial cellulose. Real-time reverse-transcription polymerase chain reaction (PCR) results confirmed that the expression of molecular players for the bone formation as well as bone regeneration was enhanced by these nanocomposites compared to that by bacterial cellulose. Alkaline phosphatase activity, expression of alkaline phosphatase, osteopontin osteocalcin, and bone-sialoprotein was found favorable for up to 7 days with human bone marrow-derived stromal cells cultured on the nanocomposites compared to that of bacterial cellulose in the absence as well as presence of osteogenic supplements, such as dexamethasone, glycerophosphate, and L-ascorbic acid. Thus, bacterial cellulose-HAp nanocomposites scaffold

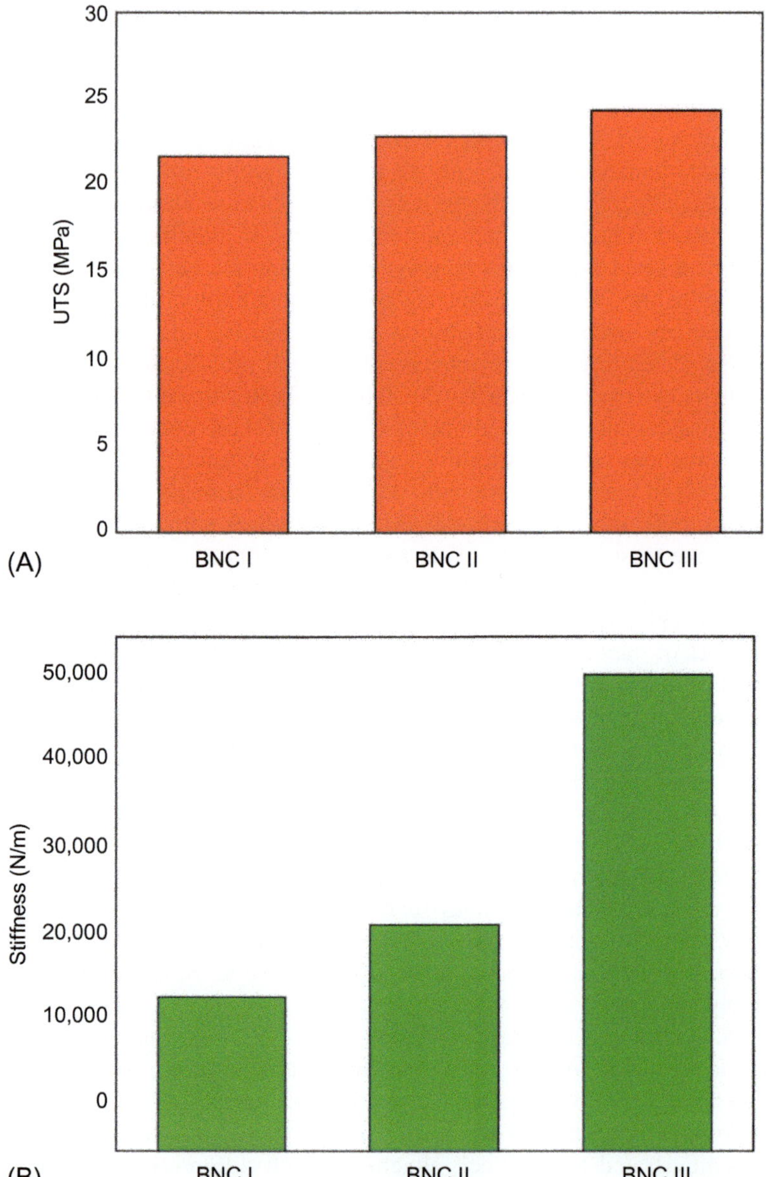

Fig. 7.11 Mechanical properties (A) ultimate tensile strength (UTS) and (B) stiffness of zirconium oxide nanoparticles-doped chitosan-PEG-nano-HAp nanocomposites I–III. From A. Bhowmick, N. Pramanik, P. Jana, T. Mitra, A. Gnanamani, M. Das, P.P. Kundu, Development of bone-like zirconium oxide nanoceramic modified chitosan based porous nanocomposites for biomedical application, Int. J. Biol. Macromol. 95 (2017) 348–356; © 2016 Elsevier B.V.

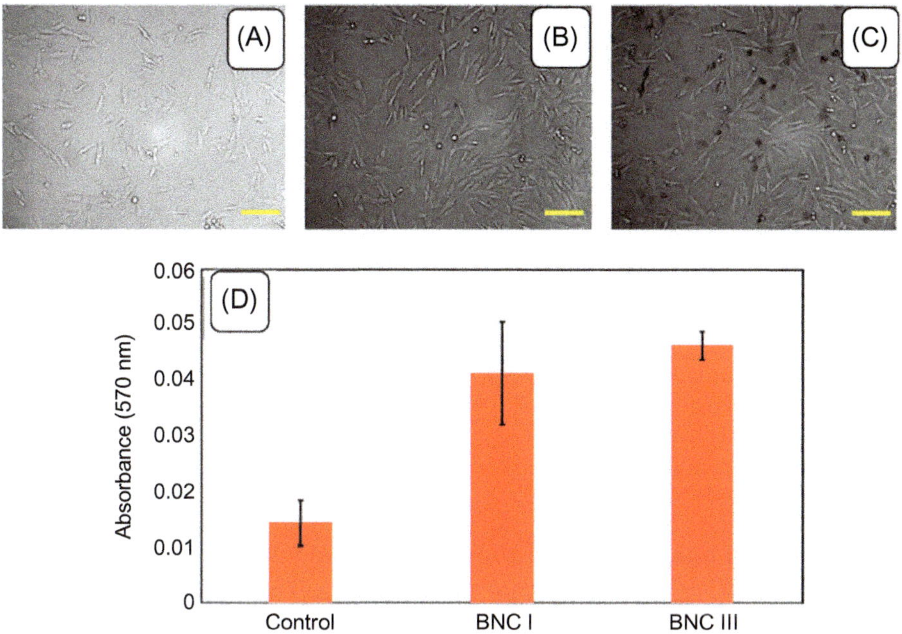

Fig. 7.12 Inverted phase contrast micrographs showing human osteoblastic MG-63 cell lines grown over control (A), zirconium oxide nanoparticles-doped chitosan-PEG-nano-HAp nanocomposites I (B), zirconium oxide nanoparticles-doped chitosan-PEG-nano-HAp nanocomposites III (C), and MTT assay with these nanocomposites after 24 h (D).
From A. Bhowmick, N. Pramanik, P. Jana, T. Mitra, A. Gnanamani, M. Das, P.P. Kundu, Development of bone-like zirconium oxide nanoceramic modified chitosan based porous nanocomposites for biomedical application, Int. J. Biol. Macromol. 95 (2017) 348–356; © 2016 Elsevier B.V.

demonstrated in vitro biocompatibility and suggested a promise to be employed in the bone tissue engineering.

Recently, starch-based organic-inorganic nanocomposites are being researched, which present numerous advantages, such as environmental friendliness and biocompatibility for biomedical and pharmaceutical uses in various large-scale productions [130]. In another investigation, Fama et al. [117] developed multiwalled carbon nanotube-starch-based nanocomposites. In this investigation, the multiwalled carbon nanotubes were incorporated within the starch-matrices as the carbon nanotubes are recognized from the medical perspective as a suitable reinforcement. Besides their use as reinforcement, carbon nanotubes are able to stimulate bone formation [131]. These multiwalled carbon nanotube-starch-based nanocomposites have shown good prospect for the use as scaffolds in the bone regenerating therapy [117]. Besides these nanocomposites, some other nanocomposites investigated in bone tissue engineering applications are presented in Table 7.3.

Table 7.3 Some examples of nanocomposites investigated in bone tissue engineering applications

Nanocomposites in bone tissue engineering	References
HAp/Ti bionanocomposites	[63,132]
HAp-TiO_2-based nanocomposites	[133]
Zirconia-alumina nanocomposite	[114]
β-tricalcium phoshate-PCL nanocomposites	[134]
Graphene-HAp nanocomposite hydrogels	[135]
Bacterial cellulose-collagen nanocomposite	[136]
Bacterial cellulose-HAp nanocomposites	[137]
Chitosan/HAp nanocomposites	[138]
Nanocomposites of HAp with aspartic acid and glutamic acid	[139]
HAp-collagen-chondroitin sulfate nanocomposite	[140]
Nanocomposite scaffolds of bioactive glass ceramic nanoparticles disseminated chitosan matrix	[141]
Scaffold containing first otoliths/collagen/bacterial cellulose nanocomposites	[52]
Nanocomposite scaffold containing Porous K-carrageenan/$CaPO_4$	[53]
Bone-like composite of self-assembled collagen fibers/HAp nanocrystals	[142]
Bone-like zirconium oxide nanoceramic modified chitosan based porous nanocomposites	[128]
Bioactive glass nanofiber-collagen nanocomposite	[143]
Nanocomposites of bioactive glass nanofiber filled PLA	[144]
Nanocomposites of HAp with aspartic acid and glutamic acid	[139]
Ferrimagnetic glass-ceramic nanocomposites	[145]

7.5 Conclusion

Nanocomposites are mainly multiple nanomaterials or nanomaterials incorporated into another bulk material. Nanocomposites vary from the conventional composite materials because of the remarkably higher surface to volume ratio of the reinforcing phase and/or extremely higher characteristic ratio, increased ductility with no decrease of strength and scratching resistance. Reinforcing materials generally improve the physical as well as mechanical characteristics of the matrix. Recent years, various kinds of nanocomposites such as inorganic-inorganic, organic-inorganic, and organic-organic nanocomposites have been synthesized and evaluated for variety of industrial applications in numerous areas, such as biotechnology, biomedical, pharmaceutical, packaging, coating, electronics, and aerospace industries. Biomedical nanocomposites have also designed for the use in various biomedical applications such as drug delivery, antimicrobial properties, tissue engineering, wound dressings, stem cell therapy, cancer therapy, cardiac prosthesis, artificial blood vessels, biosensors, and enzyme immobilization. Even, various biomedical nanocomposites have shown prospects for various improved orthopedic applications, like orthopedic drug delivery and bone tissue engineering applications. Various nanocomposites are also being researched as orthopedic drug delivery carriers for a variety of drugs, proteins, hormones, and cells, to treat different orthopedic diseases such as arthrosis, osteomyelitis, osteoporosis, osteosarcoma, and bone metastasis. In addition, a variety of nanocomposites are being developed as improved nanobiomaterials for the use in bone tissue engineering applications, which possess sufficient mechanical strength, excellent bioactivity, biocompatibility, cell attachment capability, and osteoconductivity to meet the requirement of dynamic reciprocity for tissue engineering. The current chapter should be helpful to get various reported information on the nanocomposites for improved orthopedic drug delivery and bone tissue engineering applications.

References

[1] A.K. Nayak, A.K. Dhara, Nanotechnology in drug delivery applications: a review, Arch. Appl. Sci. Res. 2 (2010) 284–293.

[2] D. Pal, A.K. Nayak, Nanotechnology for targeted delivery in cancer therapeutics, Int. J. Pharm. Sci. Rev. Res. 1 (2010) 1–7.

[3] J. Malakar, A. Ghosh, A. Basu, A.K. Nayak, Nanotechnology: a promising carrier for intracellular drug delivery system, Int. Res. J. Pharm. 3 (2012) 36–40.

[4] S. Jana, N. Maji, A.K. Nayak, K.K. Sen, S.K. Basu, Development of chitosan-based nanoparticles through inter-polymeric complexation for oral drug delivery, Carbohydr. Polym. 98 (2013) 870–876.

[5] S. Jana, A. Gangopadhaya, B.B. Bhowmik, A.K. Nayak, A. Mukhrjee, Pharmacokinetic evaluation of testosterone-loaded nanocapsules in rats, Int. J. Biol. Macromol. 72 (2015) 28–30.

[6] S. Jana, S. Manna, A.K. Nayak, K.K. Sen, S.K. Basu, Carbopol gel containing chitosan-egg albumin nanoparticles for transdermal aceclofenac delivery, Colloids Surf. B: Biointerfaces 114 (2014) 36–44.

[7] A.C. Balazs, T. Emrick, T.P. Russell, Nanoparticle polymer composites: where two small worlds meet, Science 314 (2006) 1107–1110.

[8] D.R. Paul, L.M. Robeson, Polymer nanotechnology: nanocomposites, Polymer 49 (2008) 3187–3204.

[9] X. Tang, S. Alavi, Recent advances in starch, polyvinyl alcohol based polymer blends, nanocomposites and their biodegradability, Carbohydr. Polym. 85 (2011) 85–87.

[10] A.K. Nayak, S. Kalia, M.S. Hasnain, Optimization of aceclofenac-loaded pectinate-poly (vinyl pyrrolidone) beads by response surface methodology, Int. J. Biol. Macromol. 62 (2013) 194–202.

[11] A.K. Nayak, S. Khatua, M.S. Hasnain, K.K. Sen, Development of diclofenac sodium-loaded alginate-PVP K 30 microbeads using central composite design, DARU J. Pharm. Sci. 19 (2011) 356–366.

[12] M.S. Hasnain, P. Rishishwar, S. Ali, Floating-bioadhesive matrix tablets of hydralazine HCL made of cashew gum and HPMC K4M, Int. J. Pharm. Pharm. Sci. 9 (2017) 124–129.

[13] M.S. Hasnain, P. Rishishwar, S. Ali, Use of cashew bark exudate gum in the preparation of 4% lidocaine HCL topical gels, Int. J. Pharm. Pharm. Sci. 9 (2017) 146–150.

[14] P.R. Guru, H. Bera, M.P. Das, M.S. Hasnain, A.K. Nayak, Aceclofenac-loaded *Plantago ovata* F. Husk mucilage-Zn^{+2}-pectinate controlled-release matrices, Starch-Stärke 70 (2018) 1700136.

[15] A.K. Nayak, D. Pal, J. Pradhan, M.S. Hasnain, Fenugreek seed mucilage-alginate mucoadhesive of metformin HCl: Design, optimization and evaluation, Int. J. Biol. Macromol. 54 (2013) 144–154.
[16] A.K. Nayak, D. Pal, M.S. Hasnain, Development, optimization and *in vitro-in vivo* evaluation of pioglitazone-loaded jackfruit seed starch-alginate beads, Curr. Drug Deliv. 10 (2013) 608–619.
[17] E.T. Thostenson, C. Li, T.W. Chou, Nanocomposites in context, Compos. Sci. Technol. 65 (2005) 491–516.
[18] P.M. Ajayan, Nanocomposite Science and Technology, Wiley-VCH Verlag, Weinheim, 2000.
[19] Y. Zheng, J. Monty, R.J. Linhardt, Polysaccharide-based nanocomposites and their applications, Carbohydr. Res. 405 (2015) 23–32.
[20] P.H.C. Camargo, K.G. Satyanarayana, F. Wypych, Nanocomposites: synthesis, structure, properties and new application opportunities, Mater. Res. 12 (2009) 1–39.
[21] S.K. Kumar, R. Krishnamoorti, Nanocomposites: structure, phase behavior, and properties, Annu. Rev. Chem. Biomol. Eng. 1 (2010) 37–58.
[22] E. Manias, Nanocomposites: stiffer by design, Nat. Mater. 6 (2007) 9–11.
[23] Y. Kasirga, A. Oral, C. Caner, Preparation and characterization of chitosan/montmorillonite-K10 nanocomposites films for food packaging applications, Polym. Compos. 33 (2012) 1874–1882.
[24] W. Hu, S. Chen, J. Yang, Z. Li, H. Wang, Functionalized bacterial cellulose derivatives and nanocomposites, Carbohydr. Polym. 101 (2014) 1043–1060.
[25] A. Ashori, S. Sheykhnazari, T. Tabarsa, A. Shakeri, M. Golalipour, Bacterial cellulose/silica nanocomposites: preparation and characterization, Carbohydr. Polym. 90 (2012) 413–418.
[26] M.D. Sanchez-Garcia, L. Hilliou, J.M. Lagaron, Nanobiocomposites of carrageenan, zein, and mica of interest in food packaging and coating applications, J. Agric. Food Chem. 58 (2010) 6884–6894.
[27] R. Zafar, K.M. Zia, S. Tabasum, F. Jabeen, A. Noreen, M. Zuber, Polysaccharide based bionanocomposites, properties and applications: a review, Int. J. Biol. Macromol. 92 (2016) 1012–1024.
[28] C.J. Grande, F.G. Torres, C.M. Gomez, M.C. Bañó, Nanocomposites ofbacterial cellulose/hydroxyapatite for biomedical applications, Acta Biomater. 5 (2009) 1605–1615.
[29] H. Ullah, F. Wahid, H.A. Santos, T. Khan, Advances in biomedical and pharmaceutical applications of functional bacterial cellulose-based nanocomposites, Carbohydr. Polym. 150 (2016) 330–352.
[30] V. Raj, G. Prabha, Synthesis, characterization and in vitro drug release of cisplatin loaded cassava starch acetate–PEG/gelatin nanocomposites, J. Assoc. Arab Univ. Basic Appl. Sci. 21 (2016) 10–16.
[31] S. Bala Subramanian, A.P. Francis, T. Devasena, Chitosan-starch nanocomposite particles as a drug carrier for the delivery of bis-desmethoxy curcumin analog, Carbohydr. Polym. 114 (2014) 170–178.
[32] S.G. Fullana, H. Ternet, M. Freche, J.L. Lacout, F. Rodriguez, Controlled release properties and final macroporosity of a pectin microspheres-calcium phosphate composite bone cement, Acta Biomater. 6 (2010) 2294–2300.
[33] R.I. Illiescu, E. Andronescu, C.D. Ghitulica, G. Voicu, A. Ficai, M. Hotetu, Montmorillonite-alginate nanocomposite as a drug delivery system-incorporation and in vitro release of irinotecan, Int. J. Pharm. 463 (2014) 184–192.

[34] M.S. Hasnain, A.K. Nayak, M. Singh, M. Tabish, M.T. Ansari, T.J. Ara, Alginate-based bipolymeric-nanobioceramic composite matrices for sustained drug release, Int. J. Biol. Macromol. 83 (2016) 71–77.

[35] P.L. Nayak, D. Sahoo, Chitosan-sodium alginate nanocomposites blended with cloisite 30b as a novel drug delivery system for anticancer drug curcumin, Int. J. Appl. Biol. Pharm. Technol. 2 (2011) 402–411.

[36] H. Schmitt, N. Creton, K. Prashantha, J. Soulestin, M.F. Lacrampe, P. Krawczak, Melt-blended halloysite nanotubes/wheat starch nanocomposites as drug delivery system, Polym. Eng. Sci. 55 (2015) 573–580.

[37] R. Khlibsuwan, F. Siepmann, J. Siepmann, T. Pongjanyakul, Chitosan-clay nanocomposite microparticles for controlled drug delivery: Effects of the MAS content and TPP crosslinking, J. Drug Deliv. Sci. Technol. 40 (2017) 1–10.

[38] J. Zhang, Q. Wang, A. Wang, In situ generation of sodium alginate/hydroxyapatite nanocomposite beads as drug-controlled release matrices, Acta Biomater. 6 (2010) 445–454.

[39] H. Hezaveh, I.I. Muhamad, The effect of nanoparticles on gastrointestinal release from modified K-carrageenan nanocomposite hydrogels, Carbohydr. Polym. 89 (2012) 138–145.

[40] A.K. Nayak, K.K. Sen, Hydroxyapatite-ciprofloxacin implantable minipellets for bone delivery: Preparation, characterization, in vitro drug adsorption and dissolution studies, Int. J. Drug Dev. Res. 1 (2009) 47–59.

[41] A.K. Nayak, A. Bhattacharya, K.K. Sen, Hydroxyapatite-antibiotic implantable minipellets for bacterial bone infections using precipitation technique: Preparation, characterization and in-vitro antibiotic release studies, J. Pharm. Res. 3 (2010) 53–59.

[42] A.K. Nayak, A. Bhattacharyya, K.K. Sen, *In vivo* ciprofloxacin release from hydroxyapatite-based bone implants in rabbit tibia: a preliminary study, ISRN Orthop. 2011 (2011) 1–4.

[43] A.K. Nayak, B. Laha, K.K. Sen, Development of hydroxyapatite-ciprofloxacin bone-implants using quality by design, Acta Pharma. 6 (2011) 25–36.

[44] A.K. Nayak, M.S. Hasnain, J. Malakar, Development and optimization of hydroxyapatite-ofloxacin implants for possible bone-implantable delivery in osteomyelitis treatment, Curr. Drug Deliv. 10 (2013) 241–250.

[45] G.A.X. Acasigua, G.M. Olyveira, L.M.M. Costa, D.I. Braghirolli, A. Fossati, A.C.M. Carlos Guastaldi, Novel chemically modified bacterial cellulose nanocomposite as potential biomaterial for stem cell therapy applications, Curr. Stem Cell Res. Ther. 9 (2014) 117–123.

[46] H. Ghanbari, D. Radenkovic, S.M. Marashi, S. Parsno, N. Roohpour, G. Burriesci, A.M. Seifalian, Novel heart valve prosthesis with self-endothelialization potential made of modified polyhedral oligomeric silsesquioxane-nanocomposite material, Biointerphases 11 (2016) 029801.

[47] H. Mohammadi, D. Boughner, L.E. Millon, W.K. Wan, Design and simulation of a poly(vinyl alcohol)-bacterial cellulose nanocomposite mechanical aortic heart valve prosthesis, Proc. Inst. Mech. Eng. H. 223 (2009) 697–711.

[48] E.E. Brown, D. Hu, N. Abu Lail, X. Zhang, Potential of nanocrystalline cellulose-fibrin nanocomposites for artificial vascular graft applications, Biomacromolecules 14 (2013) 1063–1071.

[49] B. Fang, Y.Z. Wan, T.T. Tang, C. Gao, K.R. Dai, Proliferation andosteoblastic differentiation of human bone marrow stromal cells on hydroxyapatite/bacterial cellulose nanocomposite scaffolds, Tissue Eng. Part A 15 (2009) 1091–1098.

[50] M. Peter, P.T.S. Kumar, N.S. Binulal, S.V. Nair, H. Tamura, R. Jayakumar, Development of novel a-chitin/nanobioactive glass ceramic composite scaffolds for tissue engineering applications, Carbohydr. Polym. 78 (2009) 926–931.
[51] Y. Wan, C. Gao, M. Han, H. Liang, K. Ren, Y. Wang, et al., Preparation and characterization of bacterial cellulose/heparin hybrid nanofiber for potential vascular tissue engineering scaffolds, Polym. Adv. Technol. 22 (2011) 2643–2648.
[52] G.M. Olyveira, D.P. Valido, L.M.M. Costa, P.B.P. Gois, L. Xavier Filho, P. Basmaji, First otoliths/collagen/bacterial cellulose nanocomposites as apotential scaffold for bone tissue regeneration, J. Biomater. Nanobiotechnol. 2 (2011) 239–243.
[53] A.L. Daniel-Da-Silva, A.B. Lopes, A.M. Gil, R.N. Correia, Synthesis and characterization of porous K-carrageenan/calcium phosphate nanocomposite scaffolds, J. Mater. Sci. 42 (2007) 8581–8591.
[54] D. Archana, B.K. Singh, J. Dutta, P.K. Dutta, In vivo evaluation of chitosan–PVP–titanium dioxide nanocomposite as wound dressing material, Carbohydr. Polym. 95 (2013) 530–539.
[55] D. Archana, J. Dutta, P.K. Dutta, Chitosan-pectin-titanium dioxide nanocomposite film: an investigation for wound healing applications, Asian. Chitin J. 6 (2010) 45–46.
[56] N.A. Kamel, S.L.A. El-messieh, N.M. Saleh, Chitosan/banana peel powder nanocomposites for wound dressing application: preparation and characterization, Mater. Sci. Eng. C. 72 (2017) 543–550.
[57] P.T.S. Kumar, S. Abhilash, K. Manzoor, S.V. Nair, H. Tamura, R. Jayakumar, Preparation and characterization of novel b-chitin/nanosilver composite scaffolds for wound dressing applications, Carbohydr. Polym. 80 (2010) 761–767.
[58] G. Sandri, M.C. Bonferoni, F. Ferrari, S. Rossi, C. Aguzzi, M. Mori, P. Grisoli, P. Cerezo, M. Tenci, C. Viseras, C. Caramella, Montmorillonite-chitosan-silver sulfadiazine nanocomposites for topical treatment of chronic skin lesions: In vitro biocompatibility, antibacterial efficacy and gap closure cell motility properties, Carbohydr. Polym. 102 (2014) 970–977.
[59] A.S. Montaser, A.M. Abdel-Mohsen, M.A. Ramadan, A.A. Sleem, N.M. Sahffie, J. Jancar, A. Hebeish, Preparation and characterization of alginate/silver/nicotinamide nanocomposites for treating diabetic wounds, Int. J. Biol. Macromol. 92 (2016) 739–747.
[60] A. Khalid, R. Khan, M.R. Ul-Islam, T. Khan, F. Wahid, Bacterial cellulose-zinc oxide nanocomposites as a novel dressing system for burn wounds, Carbohydr. Polym. 164 (2017) 214–221.
[61] K.T. Shalumon, K.H. Anulekha, S.V. Nair, K.P. Chennazhi, R. Jayakumar, Sodium alginate/poly(vinyl alcohol)/nano ZnO composite nanofibers for antibacterial wound dressings, Int. J. Biol. Macromol. 49 (2011) 247–254.
[62] B.L. Wang, X.H. Liu, Y. Ji, B.F. Ren, J. Ji, Fast and long-acting antibacterial properties of chitosan–Ag/polyvinylpyrrolidone nanocomposite films, Carbohydr. Polym. 90 (2012) 8–15.
[63] A.J. Nathanael, J.H. Lee, D. Mangalaraj, S.I. Hong, Y.H. Rhee, Multifunctional properties of hydroxyapatite/titania bio-nano-composites: bioactivity and antimicrobial studies, Powder Technol. 228 (2012) 410–415.
[64] I. Malagurski, S. Levic, M. Pantic, D. Matijasevic, M. Mitric, V. Pavlovic, S. Dimitrijevic-Brankovic, Synthesis and antimicrobial properties of Zn-mineralized alginate nanocomposites, Carbohydr. Polym. 165 (2017) 313–321.

[65] N.A. Abdelwahab, N. Shukry, Synthesis, characterization and antimicrobial properties of grafted sugarcane bagasse/silver nanocomposites, Carbohydr. Polym. 115 (2015) 276–284.
[66] M. Sneha, N.M. Sundaram, Preparation and characterization of an iron oxide-hydroxyapatite nanocomposite for potential bone cancer therapy, Int. J. Nanomedicine 2015 (2015) 1099–1106.
[67] S. Radhakrishnan, C. Sumathi, A. Umar, S.J. Kim, J. Wilson, V. Dharuman, Polypyrrole-poly (3,4-ethylenedioxythiophene)-Ag (PPy-PEDOT-Ag) nanocomposite films for label-free electrochemical DNA sensing, Biosens. Bioelectron. 47 (2013) 133–140.
[68] J. Wilson, S. Radhakrishnan, C. Sumathi, V. Dharuman, Polypyrrole polyaniline-Au (PPy-PANi-Au) nanocomposite films for label-free electrochemical DNA sensing, Sens. Actuators B 171–172 (2012) 216–222.
[69] M. Baghayeri, E.N. Zare, M.M. Lakouraj, A simple hydrogen peroxide biosensor based on a novel electromagnetic poly (p-phenylene diamine)@Fe_3O_4 nanocomposite, Biosens. Bioelectron. 55 (2014) 259–265.
[70] N. Ruecha, R. Rangkupan, N. Rodthongkum, O. Chailapakul, Novel paper-based cholesterol biosensor using graphene/polyvinylpyrrolidone/polyaniline nanocomposite, Biosens. Bioelectron. 52 (2014) 13–19.
[71] K. Radhapyari, P. Kotoky, M.R. Das, R. Khan, Graphene–polyaniline nanocomposite based biosensor for detection of antimalarial drug artesunate in pharmaceutical formulation and biological fluids, Talanta 111 (2013) 47–53.
[72] R. Jain, D.C. Tiwari, S. Shrivastava, Polyaniline–bismuth oxide nanocomposite sensor for quantification of anti-parkinson drug pramipexole in solubilized system, Mater. Sci. Eng. B 185 (2014) 53–59.
[73] A.K. Nayak, Hydroxyapatite synthesis methodologies: an overview, Int. J. Chemtech Res. 2 (2010) 903–907.
[74] P.R. Odgreen, T.J. Martin, Therapeutic approaches to bone diseases, Science 289 (2002) 1508–1514.
[75] A.L. Golob, M.B. Laya, Osteoporosis screening, prevention, and management, Med. Clin. North Am. 99 (2015) 587–606.
[76] H.H. Pham, P. Luo, F. Genin, A.K. Das, Synthesis and characterization of hydroxyapatite-ciprofloxacin delivery systems by precipitation and spray drying technique, AAPS PharmSciTech. 3 (2002). article 1.
[77] A.K. Nayak, K.K. Sen, Bone-targeted drug delivery systems, in: S. Maiti, K.K. Sen (Eds.), Bio-Targets & Drug Delivery Approaches, CRC Press, Boca Raton, London, 2016, pp. 207–231.
[78] C. Soundrapandian, B. Sa, S. Dutta, Organic-inorganic composites for bone drug delivery, AAPS PharmSciTech 10 (2009) 1158–1171.
[79] M.S. Hasnain, A.K. Nayak, R. Singh, F. Ahmad, Emerging trends of natural-based polymeric systems for drug delivery in tissue engineering applications, Sci. J. UBU 1 (2010) 1–13.
[80] A.K. Nayak, D. Pal, Natural polysaccharides for drug delivery in tissue engineering, Everyman's Sci. XLVI (2012) 347–352.
[81] M.A. Rauschmann, T.A. Wichelhaus, V. Stirnal, E. Dingeldein, L. Zichner, R. Schnettler, V. Alt, Nanocrystalline hydroxyapatite and calcium sulphate as biodegradable composite carrier material for local delivery of antibiotics in bone infections, Biomaterials 26 (2005) 2677–2684.

[82] A.C. Queiroz, J.D. Santos, F.J. Monteiro, I.R. Gibson, J.C. Knowles, Adsorption and release studies of sodium ampicillin from hydroxyapatite and glass-reinforced hydroxyapatite composites, Biomaterials 22 (2001) 1393–1400.
[83] S. Hesaraki, F. Moztarzadeh, N. Nezafati, Evaluation of a bioceramic-based nanocomposite material for controlled delivery of a non-steroidal anti-inflammatory drug, Med. Eng. Phys. 31 (2009) 1205–1213.
[84] H. Aghaei, A.A. Nourbakhsh, S. Karbasi, R. Kalbasid, M. Rafienia, N. Nourbakhsh, S. Bonakdar, K.J.D. Mackenzie, Investigation on bioactivity and cytotoxicity of mesoporous nano-composite MCM-48/hydroxyapatite for ibuprofen drug delivery, Ceram. Int. 40 (2014) 7355–7362.
[85] I. Izquierdo-Barba, L. Ruiz-Gonzalez, J.C. Doadrio, J.M. Gonzalez-Calbet, M. Vallet-Regi, Tissue regeneration: a new property of mesoporous materials, Solid State Sci. 7 (2005) 983–989.
[86] S. Itoh, M. Kikuchi, Y. Koyama, K. Takakuda, K. Shinomiya, J. Tanaka, Development of a hydroxyapatite-collagen nanocomposite as medical device, Cell Transplant. 13 (2004) 451–461.
[87] S. Itoh, M. Kikuchi, K. Takakuda, The biocompatibility and osteoconductive activity of a novel hydroxyapatite/collagen composite biomaterial, and its function as a carrier of rhBMP-2, J. Biomed. Mater. Res. 54 (2001) 445–453.
[88] M. Otsuka, R. Hirano, Bone cell activity responsive drug release from biodegradable apatite/collagen nano-composite cements—in vitro dissolution medium responsive vitamin K2 release, Colloids Surf. B: Biointerfaces 85 (2011) 338–342.
[89] G.D. Venkatasubbu, S. Ramasamy, V. Ramakrishnan, J. Kumar, Hydroxyapatite-alginate nanocomposite as drug delivery matrix for sustained release of ciprofloxacin, J. Biomed. Nanotechnol. 7 (2011) 759–767.
[90] R. Nithya, N. Meenakshi Sundaram, Biodegradation and cytotoxicity of ciprofloxacin-loaded hydroxyapatite-polycaprolactone nanocomposite film for sustainable bone implants, Int. J. Nanomedicine 10 (2015) 119–127.
[91] M. Jia, H. Li, H. Guo, A. Xie, S. Wang, F. Huang, S. Lia, Y. Shen, J. He, A novel porous aspirin-loaded $(GO/CTS\text{-}HA)_n$ nanocomposite films: synthesis and multifunction for bone tissue engineering, Carbohydr. Polym. 153 (2016) 124–132.
[92] K. Ariga, Y. Yamauchi, G. Rydzek, Q. Ji, Y. Yonamine, K.C.W. Wu, et al., Layer-by-layer nanoarchitectonics: invention, innovation, and evolution, Chem. Lett. 43 (2014) 36–68.
[93] E. Kolanthai, P.A. Sindu, K.T. Arul, V.S. Chandra, E. Manikandan, S.N. Kalkura, Agarose encapsulated mesoporous carbonated hydroxyapatite nanocomposites powder for drug delivery, J. Photochem. Photobiol. B Biol. 166 (2017) 220–231.
[94] V.C. Martin, G. Goissis, A.C. Rebeiro, E. Marcantinio Jr., M.R. Bet, The controlled release of antibiotic by hydroxyapatite: anionic collagen composites, Artif. Org. 22 (1998) 215–221.
[95] V.C. Martin, G. Goissis, Nanostoichiometric hydroxyapatite-anionic collagen composite, a support for the double sustained release of gentamicin and norfloxacin/ciprofloxacin, Artif. Org. 24 (2002) 224–230.
[96] M.T. Ethell, R.A. Bennett, R.A. Brown, M.P. Brown, K. Merritt, J.S. Davidson, T. Tran, In vitro elution of gentamicin, amikacin, ceftiofur from polymethyl-methacrylate and hydroxyapatite cement, Vet. Surg. 29 (2000) 375–382.
[97] B. Ben-Nissan, I. Macha, S. Cazalbou, A.H. Choi, Calcium phosphate nanocoatings and nanocomposites, part 2: thin films for slow drug delivery and osteomyelitis, Nanomedicine 11 (2016) 531–544.

[98] M. Kharaziha, M.H. Fathi, H. Edris, N. Nourbakhsh, A. Talebi, S. Salmanizadeh, PCL-forsterite nanocomposite fibrous membranes for controlled release of dexamethasone, J. Mater. Sci. Mater. Med. 26 (2015) 53–64.
[99] S. Hesaraki, F. Moztarzadeh, R. Nemati, N. Nezafati, Preparation and characterization of calcium sulfate-biomimetic apatite nanocomposites for controlled release of antibiotics, J. Biomed. Mater. Res. B Appl. Biomater. 91 (2009) 651–661.
[100] Y. Zhang, M. Zhang, Calcium phosphate/chitosan composite scaffolds for controlled in vitro antibiotic drug release, J. Biomed. Mater. Res. 62 (2002) 378–386.
[101] S. Saber-Samandari, S. Saber-Samandari, Biocompatible nanocomposite scaffolds based on copolymer-grafted chitosan for bone tissue engineering with drug delivery capability, Mater. Sci. Eng C. Mater. Biol. Appl. 75 (2017) 721–732.
[102] J.P. Cattalini, J. Roether, A. Hoppe, F. Pishbin, L. Haro Durand, A. Gorustovich, A.R. Boccaccini, S. Lucangioli, V. Mouriño, Nanocomposite scaffolds with tunable mechanical and degradation capabilities: co-delivery of bioactive agents for bone tissue engineering, Biomed. Mater. 11 (2016) 065003.
[103] A. Pistone, D. Iannazzo, S. Panseri, M. Montesi, A. Tampieri, S. Galvagno, Hydroxyapatite-magnetite-MWCNT nanocomposite as a biocompatible multifunctional drug delivery system for bone tissue engineering, Nanotechnology 25 (2014) 425701.
[104] H. Pacheco, K. Vedantham, Y.A. Aniket, I. Marriott, A. El-Ghannam, Tissue engineering scaffold for sequential release of vancomycin and rhBMP2 to treat bone infections, J. Biomed. Mater. Res. A 102 (2014) 4213–4223.
[105] T.S. Sun, K. Guan, S.S. Shi, et al., Effect of nano hydroxyapatite/collagen composite and bone morphogenetic protein-2 on lumbar intertransverse fusion in rabbits, chin, J. Traumatol. 7 (2004) 18–24.
[106] S. Sotome, T. Uemura, M. Kikuchi, J. Chen, S. Itoh, J. Tanaka, T. Tateishi, K. Shinomiya, Synthesis and in vivo evaluation of a novel hydroxyapatite/collagen–alginate as a bone filler and a drug delivery carrier of bone morphogenetic protein, Mater. Sci. Eng C 24 (2004) 341–347.
[107] Y. Zhang, Y. Hu, Z. Luo, X. Shen, C. Mu, K. Cai, Simultaneous delivery of BMP-2 factor and anti-osteoporotic drugs using hyaluronan-assembled nanocomposite for synergistic regulation on the behaviors of osteoblasts and osteoclasts in vitro, J. Biomater. Sci. Polym. Ed. 26 (2015) 290–310.
[108] X. Liu, K. Zhao, T. Gong, J. Song, C. Bao, E. Luo, J. Weng, S. Zhou, Delivery of growth factors using a smart porous nanocomposite scaffold to repair a mandibular bone defect, Biomacromolecules 15 (2014) 1019–1030.
[109] M. Farokhi, F. Mottaghitalab, M.A. Shokrgozar, J. Ai, J. Hadjati, M. Azami, Bio-hybrid silk fibroin/calcium phosphate/PLGA nanocomposite scaffold to control the delivery of vascular endothelial growth factor, Mater. Sci. Eng. C 35 (2014) 401–410.
[110] M. Farokhi, F. Mottaghitalab, J. Ai, M.A. Shokrgozar, Sustained release of platelet-derived growth factor and vascular endothelial growth factor from silk/calcium phosphate/PLGA based nanocomposite scaffold, Int. J. Pharm. 454 (2013) 216–225.
[111] P.X. Ma, Biomimetic materials for tissue engineering, Adv. Drug Deliv. Rev. 60 (2008) 184–198.
[112] R. Skalak, C.F. Fox, Tissue Engineering: Proceedings of a Workshop Held at Granlibakken, Lake Tahoe, New York, (1988), 26–29.
[113] B.S. Kim, C.E. Baez, A. Atala, Biomaterials for tissue engineering, World J. Urol. 18 (2000) 2–9.
[114] Y.-M. Kong, C.-J. Bae, S.-H. Lee, H.-W. Kim, H.-E. Kim, Improvement in biocompatibilty of zirconia-alumina nanocomposite by addition of HA, Biomaterials 26 (2005) 509–517.

[115] S.K. Swain, I. Gotman, R. Unger, E.Y. Gutmanas, Bioresorbable β-TCP-FeAg nanocomposites for load bearing bone implants: high pressure processing, properties and cell compatibility, Mater. Sci. Eng. C 78 (2017) 88–95.

[116] S. Nath, B. Basu, M. Mohanty, P.V. Mohanan, In vivo response of novel hydroxyapatite-mullite composites: results up to 12 weeks of implantation, J. Biomed. Mater. Res. B 90 (2009) 547–557.

[117] L.M. Famá, V. Pettarin, S.N. Goyanes, C.R. Bernal, Starch/multi-walled carbon nanotubes composites with improved mechanical properties, Carbohydr. Polym. 83 (2011) 1226–1231.

[118] C. Du, F.Z. Cui, Q.L. Feng, X.D. Zhu, K. de Groot, Tissue response to nano-hydroxyapatite/collagen composite implants in marrow cavity, J. Biomed. Mater. Res. 42 (1998) 540–548.

[119] C. Du, F.Z. Cui, Q.L. Feng, X.D. Zhu, K. de Groot, Three-dimensional nano-HAp/collagen matrix loading with osteogenic cells in organ culture, J. Biomed. Mater. Res. 44 (1999) 407–415.

[120] M. Kikuchi, S. Itoh, S. Ichinose, K. Shinomiya, J. Tanaka, Self-organization mechanism in a bone like hydroxyapatite/collagen nanocomposite synthesized in vitro and its biological reaction *in vivo*, Biomaterials 22 (2001) 1705–1711.

[121] H.W. Kim, H.E. Kim, V. Salih, Stimulation of osteoblast responses to biomimetic nanocomposites of gelatin–hydroxyapatite for tissue engineering scaffolds, Biomaterials 26 (2005) 5221–5230.

[122] H.-W. Kim, J.C. Knowles, H.-E. Kim, Gelatin/hydroxyapatite nanocomposite scaffolds for bone repair, J. Biomed. Mater. Res. 72A (2005) 136–145.

[123] D.A. Wahl, J.T. Czernuszka, Collagen-hydroxyapatite composites for hard tissue repair, Eur. Cell. Mater. 11 (2006) 43–56.

[124] A.K. Nayak, Controlled release drug delivery systems, Sci. J. UBU 2 (2011) 1–8.

[125] H. Ishikawa, T. Koshino, R. Takeuchi, T. Saito, Hydroxyapatite and gelatin composite foams processed via novel freeze-drying and crosslinking for use as temporary hard tissue scaffolds, Biomaterials 22 (2001) 1689–1694.

[126] S. Mohanapriya, M. Mumjithaa, K. Purna Sai, V. Raja, Fabrication and characterization of poly (vinyl alcohol)-TiO2 nanocomposite films for orthopedic applications, J. Mech. Behav. Med. Mater. 63 (2016) 141–156.

[127] L. Zhang, C. Ning, T. Zhou, X. Liu, K.W.K. Yeung, T. Zhang, Z. Xu, X. Wang, S. Wu, P.K. Chu, Polymeric nanoarchitectures on Ti-based implants for antibacterial applications, ACS Appl. Mater. Interfaces 6 (2014) 17323–17345.

[128] A. Bhowmick, N. Pramanik, P. Jana, T. Mitra, A. Gnanamani, M. Das, P.P. Kundu, Development of bone-like zirconium oxide nanoceramic modified chitosan based porous nanocomposites for biomedical application, Int. J. Biol. Macromol. 95 (2017) 348–356.

[129] S. Shi, S. Chen, X. Zhang, W. Shen, X. Li, W. Hu, Biomimetic mineralization synthesis of calcium-deficient carbonate-containing hydroxyapatite in a three-dimensional network of bacterial cellulose, J. Chem. Technol. Biotechnol. 84 (2009) 285–290.

[130] F. Xie, E. Pollet, P.J. Halley, L. Averous, Starch-based nano-biocomposites, Prog. Polym. Sci. 38 (2013) 1590–1628.

[131] B.S. Harrison, A. Atala, Carbon nanotube applications for tissue engineering, Biomaterials 28 (2007) 344–353.

[132] C.Q. Ning, Y. Zhou, *In vitro* bioactivity of a biocomposite fabricated from HA and Ti powder metallurgy method, Biomaterials 23 (2002) 2909–2915.

[133] M. Salarian, W.Z. Xu, Z. Wang, T.K. Sham, P.A. Charpentier, Hydroxyapatite-TiO(2)-based nanocomposites synthesized in supercritical CO(2) for bone tissue engineering: Physical and mechanical properties, ACS Appl. Mater. Interfaces 6 (2014) 16918–16931.

[134] M. Bernstein, I. Gotman, C. Makarov, A. Phadke, S. Radin, P. Ducheyne, E.Y. Gutmanas, Low temperature fabrication of β-TCP-PCL nanocomposites for bone implants, Adv. Eng. Mater. 12 (2010) B341–B347.

[135] X. Xie, K. Hu, D. Fang, L. Shang, S.D. Tran, M. Cerruti, Graphene and hydroxyapatite self-assemble into homogeneous, free standing nanocomposite hydrogels for bone tissue engineering, Nanoscale 7 (2015) 7992–8002.

[136] S. Saska, L.N. Teixeira, P.T. de Oliveira, A.M.M. Gaspar, S.J.L. Ribeiro, Y. Messaddeq, Bacterial cellulose-collagen nanocomposite for bone tissue engineering, J. Mater. Chem. 22 (2012) 22102–22112.

[137] S. Saska, H. Barud, A. Gaspar, R. Marchetto, S. Ribeiro, Y. Messaddeq, Bacterial cellulose-hydroxyapatite nanocomposites for bone regeneration, Int. J. Biomater. 2011 (2011). article ID 175362.

[138] I. Yamaguchi, K. Tokuchi, H. Fukuzaki, Preparation and mechanical properties of chitosan/hydroxyapatite nanocomposites, Key Eng. Mater. 192-195 (2011) 673–679.

[139] E. Boanini, P. Torricelli, M. Gazzano, R. Giardino, A. Bigi, Nanocomposites of hydroxyapatite with aspartic acid and glutamic acid and their interaction with osteoblast-like cells, Biomaterials 27 (2006) 4428–4433.

[140] S.H. Rhee, J. Tanaka, Synthesis of a hydroxyapatite/collagen/chondroitin sulphate nanocomposite by a novel precipitation method, J. Am. Ceram. Soc. 84 (2001) 459–461.

[141] M. Peter, N.S. Binulal, S. Soumya, S.V. Nair, T. Furuike, H. Tamura, R. Jayakumar, Nanocomposite scaffolds of bioactive glass ceramic nanoparticles disseminated chitosan matrix for tissue engineering applications, Carbohydr. Polym. 79 (2010) 284–289.

[142] A. Tampieri, G. Celotti, E. Landi, M. Sandri, G. Falini, N. Roveri, Biologically inspired synthesis of bone-like composite: self assembled collagen fibers/hydroxyapatite nanocrystals, J. Biomed. Mater. Res. 67A (2003) 618–625.

[143] H.W. Kim, J.H. Song, H.E. Kim, Bioactive glass nanofiber–collagen nanocomposite as a novel bone regeneration matrix, J. Biomed. Mater. Res. A 79 (2006) 698–705.

[144] H.W. Kim, H.H. Lee, G.S. Chun, Bioactivity and osteoblast responses of novel biomedical nanocomposites of bioactive glass nanofiber filled poly(lactic acid), J. Biomed. Mater. Res. A 85 (2008) 651–663.

[145] A.M. Gamal-Eldeen, S.A. Abdel-Hameed, S.M. El-Daly, M.A. Abo-Zeid, M.M. Swellam, Cytotoxic effect of ferrimagnetic glass-ceramic nanocomposites on bone osteosarcoma cells, Biomed. Pharmacother. 88 (2017) 689–697.

Tailoring surface properties from nanotubes and anodic layers of titanium for biomedical applications

8

Gabriel Goetten de Lima, Aline Rossetto da Luz†, Bruno Leandro Pereira†, Eduardo Mioduski Szesz†, Gelson Biscaia de Souza‡, Carlos Maurício Lepienski†, Neide Kazue Kuromoto§, Michael J.D. Nugent**

*Materials Research Institute, Athlone Institute of Technology (AIT), Athlone, Ireland, †Graduate Program in Engineering and Science of Materials—PIPE, Federal University of Paraná (UFPR), Curitiba, Brazil, ‡Departament of Physics, State University of Ponta Grossa (UEPG), Ponta Grossa, Brazil, §Department of Physics, Federal University of Paraná (UFPR), Curitiba, Brazil

8.1 Introduction

In the past decades, there has been considerable improvement in both quality of life of the population and life expectancy [1,2]. The increase in longevity contrasts with increasing numbers of traumatic accidents with resultant life-changing injuries. Biomaterials have been widely used for bone repair, fracture fixation, dental implants, and orthopedic prostheses to alleviate the effects of these traumas. The improvement of life quality is allied to the technological advances and new surgical techniques. The development of materials able to mimic human tissues is still in progress [3]; therefore, metallic prosthesis are the replacements most commonly used for routine bone surgeries [4,5].

There is ongoing progress in the techniques of characterization and development of novel materials to repair hard tissues, as the surfaces of these materials are generally bioinert. These characteristics are important because there are many situations where the surface of the biocompatible material must be bioinert, such as orthopedic bone screws. In this case, there is a formation of a fibrous tissue that isolates the material from the biological environment [1,4].

In more difficult settings, these metals can be used as partial or total bone replacement where it is necessary that the surface must be bioactive, as it must induce bone growth and connect with living tissues through chemical and/or biological bonds to promote osseointegration between the implant and living tissue [6]. Surface treatments are used to improve these surfaces, making them bioactive [4,7] and it is necessary that this interface is free of inflammatory processes with the formation of fibrous encapsulating tissues. Eventually, if the osseointegration of the implant is interrupted, it will result in loss of the primary stability of the material causing the implant to loosen [1].

Applications of Nanocomposite Materials in Orthopedics. https://doi.org/10.1016/B978-0-12-813740-6.00010-7

Implants undergoing cyclic stressing are subjected to stress shielding, a phenomenon responsible for bone resorption, loosening, and eventually replacement surgeries. The mechanical compatibility of the synthetic material and the bone is essential to avoid stress shielding, which can be attained by matching the Young's modulus. Hence, alloys containing β-stabilizer elements are being studied and the addition of elements Nb, Ta, V, Mo, Mg, and Cr to Ti decreases these alloys' elastic modulus to values as lower as 40 GPa, which is much closer to the range of values for bones (10–40 GPa) than pure metals (typically >100 GPa) [8,9], although the microstructure of these alloys depends on the %wt [4–6].

Among the various techniques of surface modification, anodic oxidation is an effective electrochemical method that has been successfully used in the surface treatment of implants in order to promote the bone integration of biomaterials. This technique allows the production of oxide films on the surface of a transition metal and its alloys, such as Ti, Ta, Zr, Nb, TiNb, TiNbTa, and TiNbTaZr [4,7,10,11]. Surface modification is essential to promote the bone integration of biomaterials [4,5,10] and with this technique, one can obtain compact, porous, and roughened oxides or nanotube structures. However, the morphology of these surfaces will depend on the voltage applied, with low voltages resulting in compact oxides and as the voltage (V) increases, it is possible to obtain sparking during the oxidation process (the V value depends on the electrolyte) leading to anodic films being more rough and porous. When this occurs, the process is known as plasma electrolytic oxidation (PEO) or micro arc oxidation (MAO). If fluoride ions are introduced into the electrolyte, nanostructured and self-organized surfaces are obtained because these fluorine ions attack the oxide surface giving rise to the nanotubular structure [12].

Ti and its alloys are the metals most used as dental (grade 2 Ti) and orthopedic (Ti-6Al-4V) implant materials due to their mechanical properties, suitable for the substitution of hard tissue, excellent corrosion resistance, and biocompatibility conferred by the natural oxide film. However, bone growth is a slow process onto metallic implants without superficial treatments when compared to healing times required for implants. One method to decrease the time for healing is a fast osseointegration between the metallic implant and the bone tissue. Thus, it is necessary to use surface-modification techniques to improve the biological activity of these materials and to promote bone integration [4,7,10].

The biocompatibility of titanium is closely related to the properties of the oxide surface layer in terms of its structure, morphology, and composition [7,13]. The surface of Ti when exposed to air is rapidly coated with a thin layer, in the order of tens of nanometers, of amorphous titanium oxide that has shown to have important values of corrosion resistance [4]. This layer also has been shown to be essential for an inert body response as it forms an interface layer between biological systems and implants [7]. In addition, it also acts as a barrier preventing metal ions from the interior from being released into the biological medium, and prevents corrosion of the metal [4,7]. This low tendency to corrode is necessary in order to avoid ions being released into the biological medium and it guarantees stability for the implant.

The success of the osseointegration process between the implant and bone tissue depends on surface chemical composition, porosity, surface topography, surface energy, wettability, and mechanical compatibility [13].

Surface chemical composition is a determinant for bioactivity. Anodic films rich in Ca and P elements produced by an anodizing process may be more effective due to the formation of calcium phosphates after being implanted in the human body [14]. On titanium-treated surfaces, it was observed that the presence of Ti-OH functional groups is directly related to the formation of apatite [15,16].

Porous interfaces increase fixation through bone growth toward the coating, which forms a mechanical bond. Results in vivo have shown that porous interfaces influence vascular formation on mesenchymal cell proliferation and the osteogenesis process [7]. Interconnected pores are an important characteristic of blood circulation, as it increases osteoconduction and the exchange of extracellular liquids [7,17,18].

Surface topography significantly influences cellular orientation, adhesion, proliferation, and differentiation of osteoblasts [1,19]; such a topography is defined by surface orientation and roughness, which is characterized by successive peaks and valleys [20]. Additionally, the most commonly used parameter for implant surfaces is R_a (arithmetic or average roughness) [21,22].

Surface energy can be measured from the contact angle of this surface with drops of liquids with different viscosities such as water [23], which is related to the chemical composition and surface geometry [18,22]. Hydrophilic surfaces exhibit better responses from the biological environment [18,22] as it can: enhance the wettability with blood; disseminate and bind fibrin proteins and matrix; induce a greater differentiation of osteoblasts; increase osteoconductivity [24,25]; aid in the adsorption of proteins [24]; improve initial bone apposition and bone/implant contact [26]; and facilitate the fixation of the cells and accelerate the healing of tissues, especially soon after the surgical placement of the implant [27].

Mechanical compatibility influences the recovery of damaged bone due to applied mechanical stress; also, the absence of mechanical stress can cause atrophy and bone resorption. The modulus of elasticity E of the Ti-6Al-4V alloy (110 GPa) is lower than other metals used for this application but it is still higher than bone tissue (10–40 GPa). If these values differ significantly in the relationship between the implant and the bone, the transfer of tension between them will not be homogeneous. A low value in the modulus of elasticity can result in a better distribution of stresses between the bone and the implant, thus avoiding its loss [6].

8.1.1 Film formation by electrochemical process

8.1.1.1 Anodic oxidation and plasma electrolytic oxidation (PEO)

Anodic oxidation can be performed in an electrochemical cell (Fig. 8.1) with two electrodes (titanium anode, platinum or titanium cathode), under galvanostatic (constant current density) or potentiostatic mode (potential constant) [5,28,29]. Oxidation and reduction reactions occur in the anode when a current or constant voltage is applied, establishing an electric field that guides the diffusion of ions present in the electrolyte, and leading to the formation of an oxide film on the surface of the anode. The main chemical reactions that occur during titanium anodization are [28,29]:

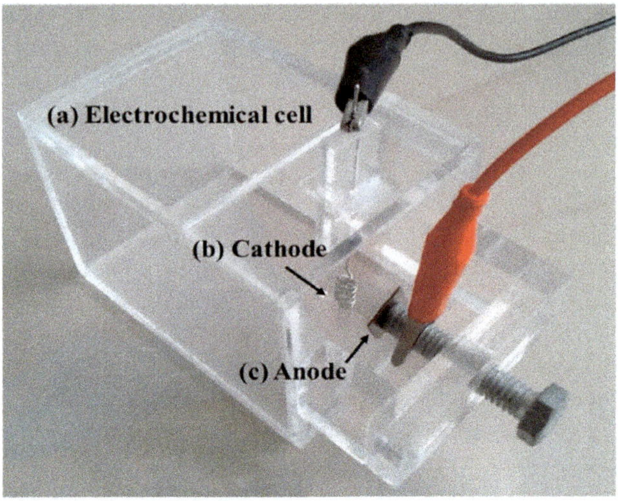

Fig. 8.1 Photo of the anodizing apparatus: (a) Electrochemical cell, (b) cathode, and (c) anode.

$$Ti \leftrightarrow Ti^{2+} + 2e^- \quad (8.1)$$

$$2H_2O \leftrightarrow 2O^{2-} + 4H^+ \quad (8.2)$$

$$2H_2O \leftrightarrow O_2(gas) + 4H^+ + 4e^- \quad (8.3)$$

$$Ti^2 + 2O^{2-} \leftrightarrow TiO_2 + 2e^- \quad (8.4)$$

Anodic titanium oxides have high resistivity with respect to electrolyte and metallic substrate; thus, the current/voltage drop occurs mainly through the oxide film. As the electric field is sufficiently high to conduct ions through the oxide, with a current in free flow, the oxide will continue to increase, creating a thicker layer [29]. In this case, the thickness, d, of the oxide layer is almost linearly dependent on the voltage U applied following the relation $d=\alpha U$, where α is a constant, which is usually within the range of 1.5–3.0 nm V^{-1} [29]. It has been reported that at 80 V and 100 V, using H_3PO_4 and H_2SO_4 electrolyte, anode oxide films are relatively thin and generally nonporous [30].

In conventional anodizing, with increasing voltage, if it exceeds the threshold of the dielectric rupture potential of the oxide barrier, a discharge phenomenon occurs. This phenomenon is an intermediate process between conventional anodizing and high energy plasma coating in a controlled gas pressure chamber without an aqueous solution [31].

The discharge phenomenon in an electrolytic process was described first in the 1880s and further developed in the 1930s [32–34]. Despite the technology being relatively old, the treatment has a wide range of technological applications mainly due to the ability to modulate superficial characteristics like hardness, elastic modulus, wettability, porosity, roughness, chemical composition, crystallinity, corrosion resistance, and tribological behavior [35–37]. Other features are process speed, ease of application, low safety hazard, relative geometrical independence of the sample surface, and incorporation of positive and negative ions into the oxide formed over the anode, even when using DC current [38,39].

The discharge phenomenon in an electrolyte is known as a PEO or MAO; however, this technique is also referred to as anode spark electrolysis, anodic oxidation under spark discharge, microplasma anodizing, anode spark electrolysis, anodic spark deposition, and anodic discharge anodizing [31].

There are different ways to perform PEO by modifying only the electrical source parameters. PEO may be conducted using DC or AC mode, electric current, or voltage ramps, in one step or two steps and the simplest PEO configuration uses two electrodes immersed in an electrolyte and a DC source with the possibility to apply the oxidation in potentiostatic or galvanostatic mode.

PEO consists of an electrolysis process, similar to anodizing with the formation of sparks or electric discharges [11]. Under relatively low voltages, the anodizing phenomenon can be equated by Faraday's law. When the applied voltage exceeds the dielectric rupture limit of the oxide, it will behave like a conductor that lead the process to the release gas and sparks could be generated [7,28,29], which contributes to the formation of a porous morphology film.

Fig. 8.2 exhibits niobium (Nb), titanium commercially pure, grade 2, (2cpTi), and Ti20% (wt) Nb alloy that were oxidized by PEO during 60s using electrolytes containing phosphoric acid (1 mol L^{-1} H_3PO_4), under 350V for Nb and 250V for Ti and Ti20Nb alloy. For the Nb condition, hydrogen peroxide H_2O_2 (80% vol 1 mol L^{-1} H_3PO_4 + 20% vol H_2O_2) was added.

Dielectric rupture, during the formation of the anodic film, produces high temperatures and pressures in the electric discharge channels over the anode, resulting in electrolyte thermolysis and incorporation of ions presented in solution during the formation of film structure [11]. One of the main characteristics is the formation of luminescent gas, which is attributed to electrolyte vaporization near the electrode (in this case anode) due to the Joule effect, combined with the electric voltage in the electrolytic cell.

The dielectric rupture can contribute to the crystalline transformation of the anodic films [11,28]. In general, when low voltage is applied, anodic oxide film is amorphous [11,28] and as the voltage increases, the oxide structure changes from an amorphous

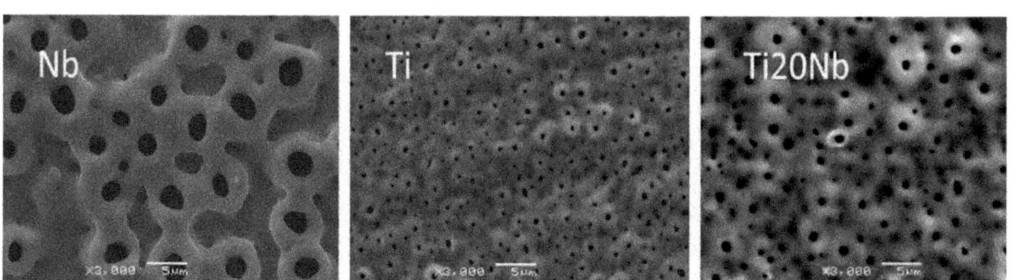

Fig. 8.2 SEM images of oxide coatings grown on Nb with 80% vol 1 mol L^{-1} H_3PO_4 + 20% vol H_2O_2 electrolyte under 350V/60s, cp2-Ti and Ti-20Nb with 1 mol L^{-1} H_3PO_4 under 250V/60s.

to a crystalline structure [40]. However, films grown on pure Ti at galvanostatic mode exhibit a degree of crystallinity at low applied potentials (10–20V) [30]. Additionally, the crystallization sites inside the films can be associated with the presence of anions of acid residues in the film [30].

The electrochemical parameters of anodic oxidation significantly affect growth behavior and chemical and structural properties of the films. Such parameters include the electrolyte solution, concentration, temperature, electrical parameters, anodizing time, and stirring speed of the solution [7,29]. In general, it is accepted that growth behavior of anodic oxide film is characterized by electrochemical dynamics between the oxide film formation capacity and oxide dissolution rate determined by the nature of the electrolyte [29,40].

Optimizing the experimental conditions of oxidation by electrolytic plasma allows the production of porous crystalline films with good adhesion to the substrate, high hardness, and with thicknesses up to 100 μm [5,7,28,36,38]. For anodic films obtained on Ti6Al4V, the KOH and Na_2WO_4 additions in aluminate-based electrolytes smooth the surface and reduce porosity of these films, improving the tribological properties [41].

8.1.2 Nanotube arrays

Many types of processes have been used to prepare nanotube arrays, such as anodizing, sol-gel chemistry, polymerization, and hydrothermal [10,42]. Anodization is the most effective method to obtain nanotube arrays as it can grow self-organized nanotubes, and when a fluoride-based electrolyte is used, their dimensions can be controlled [43,44].

The anodization technique allows growth of nanotubes on transition metals (Ti, Al, Zr, Hf, Nb, Ta, W) and its alloys [7]. Fig. 8.3 shows a typical nanotube morphology via top-view and cross-section grown on Nb, Ti, and Ti-40Nb alloy [45].

The formation mechanism of nanotubes is similar to anodic films as the electrochemical cell is the same [42]. Normally, the anodization process to grow nanotube arrays is usually carried out by applying a potential step (or ramp) at a constant voltage (1–30V) in aqueous electrolyte or in nonaqueous electrolytes (5–150V) containing approximately 0.05–0.5 m (0.1–1 wt%) of fluoride ions (and usually some background ion species) [44].

The process of nanotube formation occurs in accordance with reactions and stages illustrated in Fig. 8.4 [46]. The three steps (Fig. 8.4c–e) can be monitored by a typical curve of current vs time obtained during the anodizing process [42,46], corresponding to stages I–III of Fig. 8.5, respectively. At the stage I of anodization (Fig. 8.5), a rapid decrease in the current due to formation of a compact oxide layer is observed. At stage II, the current rises to a maximum value owing to the process of pore nucleation. Finally, at stage III, the current attains a constant value and at the oxide/electrolyte interface, the rate of oxide formation and dissolution are equal, allowing the formation of nanotube arrays [44,47].

Various parameters can influence the physical properties and morphology of nanotubes. The most important are composition, pH and concentration of electrolyte, applied

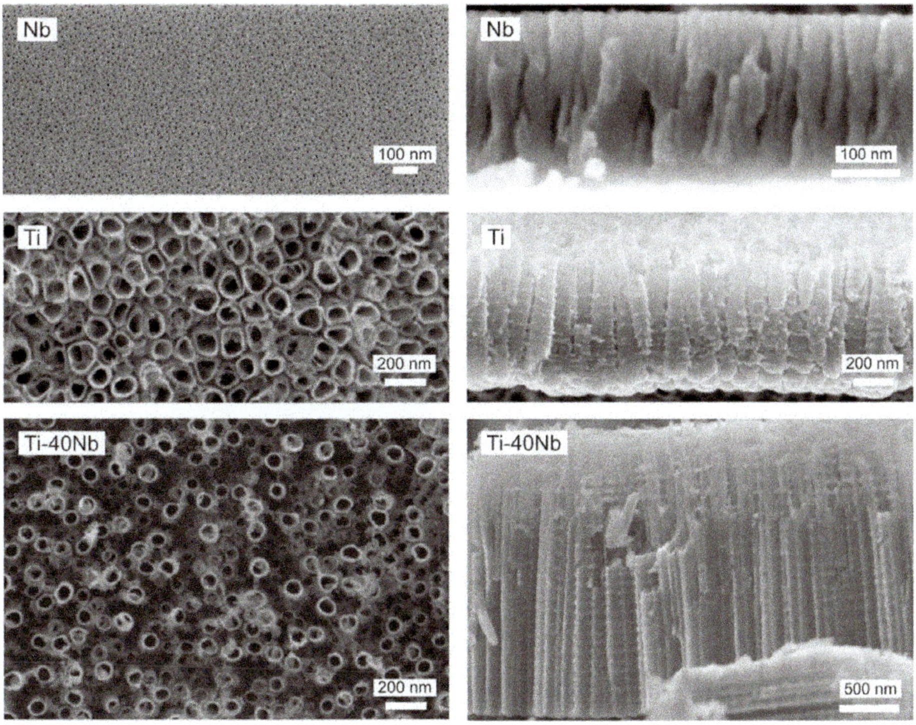

Fig. 8.3 SEM images of oxide coatings grown on Nb, cp2-Ti and Ti-40Nb at 20V in 1M H3PO4 with 0.16M HF solution: *left column*—top view after 2h anodization and *right column*—cross-sectional view after 4h anodization.
Reproduced with permission from A. Gebert, D. Eigel, P.F. Gostin, V. Hoffmann, M. Uhlemann, A. Helth, S. Pilz, R. Schmidt, M. Calin, M. Göttlicher, M. Rohnke, J. Janek, Oxidation treatments of beta-type Ti-40Nb for biomedical use, Surf. Coat. Technol. 302 (2016) 88–99. ©2016, Elsevier.

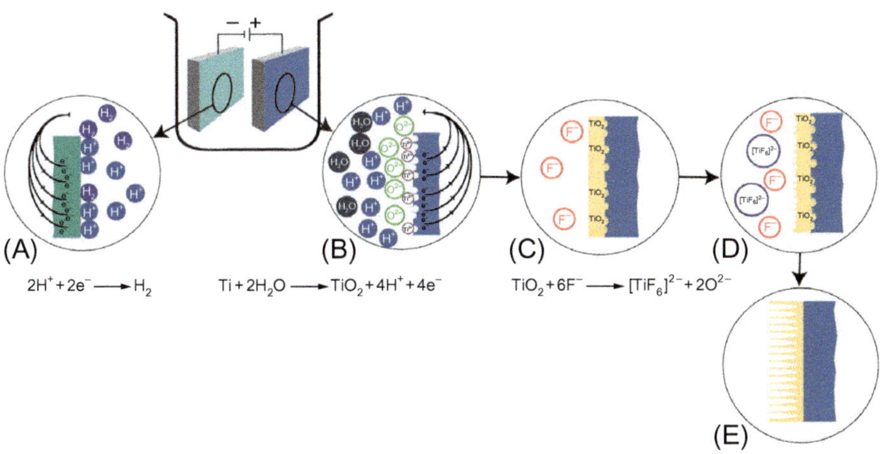

Fig. 8.4 Growth of regular TiO_2 nanotubes: (a) cathodic reaction, (b) anodic reaction, (c) transition state of TiO_2 layer, (d) starting of nanotube formation, and (e) titania nanotubes.
Reproduced with permission from S. Minagar, C.C. Berndt, J. Wang, E. Ivanova, C. Wen, A review of the application of anodization for the fabrication of nanotubes on metal implant surfaces, Acta Biomater. 8 (2012) 2875–2888. ©2012, Elsevier.

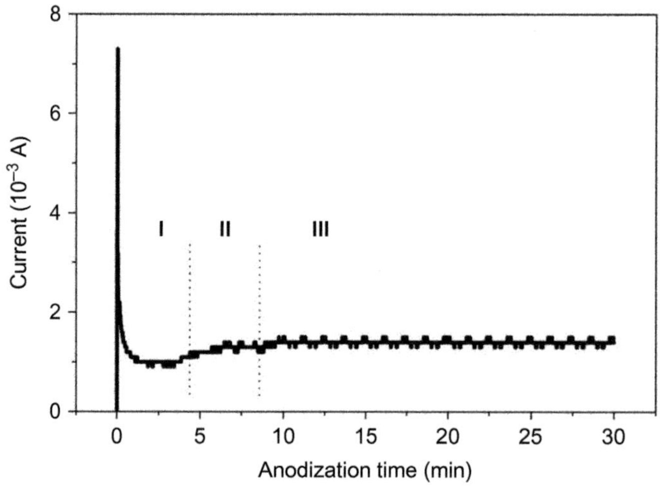

Fig. 8.5 Typical current-time anodizing curve to titanium alloy sample; I, II, and III are the different nanotube growth process steps.

voltage, current density, temperature, and anodization time, which can be controlled by anodizing techniques [46–51]. Researchers have not only analyzed the characteristics of the nanotube arrays, but also demonstrated biocompatibility, bioactivity, and bone cell/material interaction of nanotube surfaces using in vitro or in vivo tests [48–53].

Several researchers have shown that the average tube diameter increases with increasing anodizing voltage [42,46,48–50]. This behavior occurred because the electric field strength across the oxide is determined by the applied anodization voltage, thus affecting the migration of ions and consequently the dissolution rate of oxide layers [49,52]. As a result, the average internal tube diameter of nanotube arrays is directly proportional to the applied voltage [49,52].

The effects of temperature and applied voltage on nanotubes grown on pure Ti are showed in Fig. 8.6 [48]. The electrolyte used was a nonaqueous solution containing ethylene glycol + 4 wt% H_3PO_4 + 0.25 wt% HF at the range of voltages and temperatures of 60–120 V and 0–30°C, respectively. When increasing the applied voltage, the diameter increases as well. However, at 30°C the diameter was greater compared to 0°C; this occurs due to a low temperature that decreases the ionic exchange in the electrolyte; and at 0°C the tube diameter had a hexagonal shape, while at 30°C a circular shape. These geometries were due to the temperature effect, as at low temperature the role of fluorine ion etching becomes unnoticeable [48]. The nanotubes anodized at 120 V and 0°C showed better results for in vitro tests, because the nanotube arrays showed the higher surface area exposed to SBF (simulated body fluid).

The electrochemical process to obtain nanotubes or porous films has been shown to be efficient at improving the metal surfaces for biomedical applications. This knowledge regarding the experimental parameters of the electrochemical process is of significant interest, as the reactions of body fluids on the implant surface depend on the surface properties.

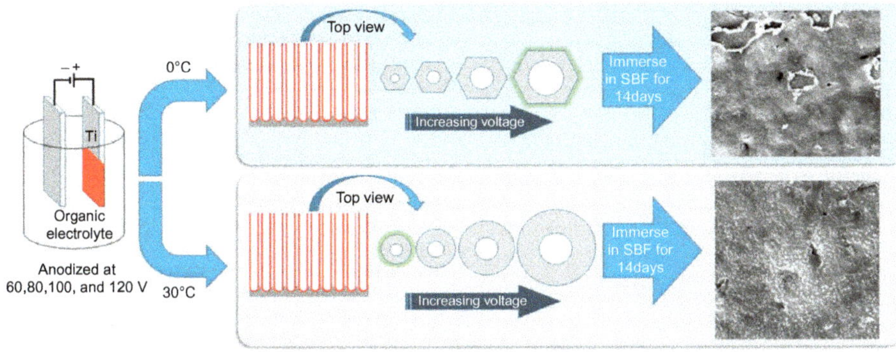

Fig. 8.6 Effects of temperature (0°C and 30°C) and voltage (60–120V) on diameter and in vitro tests of nanotube arrays.
Reproduced with permission from F. Nasirpouri, I. Yousefi, E. Moslehifard, J. Khalil-Allafi, Tuning surface morphology and crystallinity of anodic TiO_2 nanotubes and their response to biomimetic bone growth for implant applications, Surf. Coat. Technol. 315 (2017) 163–171. ©2017, Elsevier.

8.2 Commercial applications

Surface treatments have been extensively proposed and developed for the dental and orthopedic applications of titanium devices [1,54]. Nanofilms are one of the most prominent classes of surface modifications achieved through the use of several techniques aiming at this area of application [55]; among them, titanium oxide nanofilms are one of the most used in applications of commercially available medical devices.

Dental implants are generally intended to be used according to the osseointegration principle, which advocates a fusion between the implant and host bone. Surface modification techniques are often utilized to improve the osseointegration ability of titanium dental implants [54]; in these cases, osseointegration is highly desirable once these devices are permanent implants. The success of prosthetic devices strongly depends upon the implant stability and anchorage in bone and nanofilms obtained by anodic oxidation have been successively used to promote the osseointegration of dental implants.

Busquim et al. [56] investigate the surface chemistry and oxide crystal structure of machined dental implants and two types of commercially anodized titanium dental implants, Vulcano Actives® and TiUnite® (manufactured by Conexão Sistemas de Prótese (Brazil) and Nobel Biocare (Sweden), respectively). According to Busquim et al. [56], the implant anodized surfaces have the ability to promote osseointegration.

Liu et al. [57] compared the surface characteristics and cellular adhesion for several dental implants that are commercially available: 3i Nanotite (4×10mm, Biomet 3i, Palm Beach Gardens, Florida), Astra OsseoSpeed (3.1×11 mm, Astra Tech, Molndal, Sweden), Nobel Biocare TiUnite (3.5×10mm, Nobel Biocare, Goteborg, Sweden), and Straumann SLActive (3.3×10mm, Straumann, Waldenburg, Switzerland). Surface treatment technologies such as sand blasting, acid etching, and anodic oxidation were employed by the manufacturers on the implant threads, which should integrate with the host bone.

Table 8.1 Commercially available implants

Product	Surface treatment	Manufacturer
Allfit	Al_2O_3-blasted	Dr. Idhe Dental, Munich, Germany
Ice	Smooth machined	3i, Palm Beach Gardens, FL, United States
IMZ TPS	Titanium plasma-sprayed	Friedrichsfeld, Mannheim, Germany
Laser Lok	Tricalcium phosphate/ hydroxyapatite-blasted	Biohorizons, Birmingham, AL, United States
PrimaConnex	Calcium phosphate-blasted	Lifecore Biomedical, Chaska, MN, United States
Ospol	Calcium-anodized	Ospol, Malmö, Sweden
Osseospeed TX	Titanium oxide blasted/ fluoride-treated	Astra Tech, Mölndal, Sweden
Osseotite Full	Acid-etched (double)	3i, Palm Beach Gardens, FL, United States
Replace Select	Calcium phosphate-anodized	Nobel Biocare, Göteborg, Sweden
SLA	Al_2O_3-blasted/acid-etched	Institute Straumann, Basel, Switzerland
Trilobe	Al_2O_3-blasted	Southern Implants, Irene, South Africa

Reproduced with permission from M. Gaintantzopoulou, S. Zinelis, N. Silikas, G. Eliades, Micro-Raman spectroscopic analysis of TiO_2 phases on the root surfaces of commercial dental implants, Dent. Mater. 30 (2014) 861–867. ©2014, Elsevier.

Gaintantzopoulou et al. [58], studied phases of the root surface of commercially available titanium dental implants, subjected to various surface treatments. Table 8.1 shows the commercially available implants tested by Gaintantzopoulou et al. [58]. According to Gaintantzopoulou et al. [58] titanium oxide anatase, rutile, and amorphous phases may be present in commercially available dental implants. The anatase phase is more pronounced in implants submitted to anodic oxidation and can be accompanied with a rutile phase. Surface treatments such as double-etching, sand blasting, and blast-acid etching are related with the occurrence of rutile and/or amorphous phases. Thus, the identification of the TiO_2 phases in the surface layers of dental implants should be judged as mandatory by the manufacturers once variations in the biological activity of these polymorphs are possible.

In contrast to dental implants, orthopedic medical devices in many cases are not intended to osseointegrate, especially when plates and screws are used in the surgical internal fixation of fractured bones. Implants used in orthopedic trauma treatments generally are not permanent; they are used to provide stabilization of fracture focus during the time period necessary for the bone to heal; after that, the plates and screws should be removed [6].

Orthopedic implant manufacturers have used anodized surfaces (TiO_2 nanofilms) to improve corrosion resistance and for identification purposes [59]. Titanium anodization produces surfaces with different colors due to the interference phenomenon. Different colors have been used to identify differences in dimensions or other features in implants for a given application, for example, orthopedic plates for bone fractures that possess right and left models. The different implant surface colors are used to avoid confusion by the surgeon during the implantation procedure allowing easy identification of implant models and dimensions.

Orthopedic permanent implants such as those used in hip and knee prosthesis, however, depend upon osseointegration to achieve success in restoring the articulation function. These medical devices can be manufactured to adapt better to bone using several methods such as meshing, texturing, and porous or bioactive coatings. Depending on the design strategy, these coatings either should be dissolved gradually or should provide a stable, nondissolving interface for bone growth [60]. Commercially available implants for total hip arthroplasty typically have coatings applied using hydroxyapatite, calcium phosphate, and porous metal surface treatments. These approaches also have yielded good clinical results when premature delamination of the coatings is avoided by optimal adhesion to the implant [60,61].

Despite the clinical success of hydroxyapatite demonstrated after several years of patient follow-up [62], alternatives such as the use of nanofilms or the combination of different treatments have been proposed to obtain bioactive surfaces in total hip arthroplasty implants. As an example, Hall et al. [63] used hip arthroplasty femoral stems coated with Ti6Al4V beads that were treated by anodic oxidation for enhanced bioactivity. An in vivo animal study was conducted and according to Hall et al. [63] treatment of titanium alloy using anodic oxidation had a marked effect on bone ingrowth and soft tissue in the porous coating of a functioning total hip replacement femoral stem. The surface treatment significantly increased the amount of bone ingrowth and enhanced implant integration with bone and marrow, decreasing fibrous tissue within the porous coating of the component.

According to Zhang et al. [64], the design of coatings for permanent orthopedic implants must satisfy several important criteria: firstly, the coating must be biocompatible and not trigger significant immune or foreign-body response; secondly, it must be osteoconductive; thirdly, the implant must also be osteoinductive. Furthermore, the coating must have sufficient mechanical stability when under physiological stresses and should have antimicrobial properties minimizing the risk of prosthetic infection. However, none of the commercially available prostheses are able to satisfy all of the above criteria. In summary, there is enormous potential for new research and development of coatings and surface treatments for new dental and orthopedic medical applications.

8.3 Mechanical stability of anodic layers

Perhaps because of the complexity in obtaining effective functional and novel coatings on materials used in medicine, authors can concentrate mainly on specific aspects of interest: bioinertness, bioactivity, drug delivery, and corrosion resistance. However, the compatibility must be satisfied as a whole, preserving or even improving other critical properties expected for coatings on biomaterials, as thermal diffusivity [65], the absence of toxic elements to living tissues [66], and many others. Our research group has developed a tradition to contrast two or more functionalities, searching for appropriate ratios among them, notably addressing the *biocompatibility* and the *tribo-mechanical behavior* of layers and films produced on biomedical materials. The mechanical stability of a coating grown or deposited on a device is complementary to any other property it can exhibit. More than everything else, the functional coating must survive the surgical procedure—it cannot be lost by screwing or shearing [67]; after that, it is expected to not release debris or detach prematurely from the surface [68], assuring the long-term use of the prosthesis in the body.

Analyses of residual stresses σ allow predicting a threshold layer thickness h to prevent spontaneous detachment [69]:

$$h = \frac{EG}{Z\sigma^2} \tag{8.5}$$

In Eq. (8.5), Z is a dimensionless constant, and $G = {\left(1-\nu^2\right)K_{IC}}/{E}$ is the released energy rate for a layer with elastic modulus E, Poisson's ratio ν and stress intensity factor K_{IC}. However, we observed that such direct prediction is hardly effective for ceramic-like layers as those produced by PEO, because of its inherent porosity, allotropic transformations occurring in film growth (which results in material densification), and pore settlement after cooling [70]. These statements are important since the estimated localized temperature during PEO is >600°C, attained during the electric breakdown regime (see Section 8.1.1). As a result, PEO TiO_2 layers produced with H_2SO_4 electrolytes over Ti substrates detached spontaneously during growth when they reached ~3 μm thickness at 180 s anodization time, a much lower thickness than that predicted by Eq. (8.5)—45 μm. In nondetached coatings, the residual stresses in the interface substrate coatings also dictated the surface performance under plastic deformation. In this particular case [70], optimum layer thickness versus adequate tribo-mechanical features (low elastic modulus, good scratch resistance) was attained with 60 s anodizing time. The coating presented *tensile* residual stresses in opposition to compressive residual stresses in the titanium substrate beneath. Since the substrate residual stresses varied from −30 to −110 MPa in PEO processes carried out for 10–60 s, the critical load for detachment under severe scratching with an acute tip could be increased 10-fold in the same anodizing time range.

Another matter of concern in coatings is the stress shielding in the implant-bone interface, previously discussed in Section 8.1. The occurrence of such a phenomenon at the implant site can permanently limit the stability and performance of a dental or bone prosthesis, eventually resulting in unexpected second surgeries and extra medical costs [9]. In spite of the low thickness of a layer or film as compared to the prosthesis dimensions, the surface is inwardly connected with living tissues, through which mechanical stresses are transferred from the body to the implant during the biomechanical activity of the patient. In the case of bone or dental implants, assuming a well-established osseointegration and negligible influence of micromotions, the lack of a matching elastic modulus can lead to coating detachment either from the bone or from the implant bulk.

Our group has worked with PEO TiO_2 layers rich in Ca and P. These elements are incorporated into the oxide during the spark regime of the anodization, as discussed previously in Section 8.1.1.1, providing bioactive surfaces for dental and bone prosthesis. The process involves ion release in the material's surrounding with subsequent calcium phosphate growth on its surface and improved cellular response [71–73]. The good bioactivity was easily confirmed in our experiments through in vitro tests, carried out using a simulated body fluid with 14 days soaking time [74]. We first reported that these layers feature low elastic modulus values as compared to the titanium substrate, that is, about 40 GPa in TiO_2 layers in contrast with ~130 GPa in the bulk titanium [75]. The results indicate that PEO-modified surfaces presented conditions to be more mechanically compatible with the host bone, therefore facilitating load transfer to the biomaterial. Of course, the aforementioned elastic modulus of the Ca-P-containing TiO_2 layers are much lower from a dense TiO_2 bulk, ~200 GPa [76]. Such drastic reduction owes to the inherent porosity of layers anodically grown in the dielectric rupture regime, as depicted in Fig. 8.2.

One matter of concern could be the real effectiveness of a mechanically biocompatible coating, few micrometers thick, to minimize stress-shielding effects with a material with a higher elastic modulus than the host bone. In this case, one should consider that titanium and some Ti-alloys as Ti-6Al-4V present relatively low elastic modulus (~130 GPa) as compared to other typical metallic biomaterials, as stainless steel AISI 316L (200 GPa) and CoCr (220 GPa) [76]. Moreover, Ti(Nb,Ta,Zr)-based alloys present even lower elastic modulus (<100 GPa) [1]. By assuming mechanical compatibility as the only criterion, PEO TiO_2 layers could be adequate choices for Ti and Ti-based alloys, better than stiffer coatings as DLC or dense ceramic layers, with the additional advantage to present adequate conditions for the osseointegration. Moreover, PEO layers grown on low elastic modulus alloys presented similar features as those produced on commercially pure Ti, as discussed in Section 8.1. However, besides adhesion and elastic modulus, the tribo-mechanical compatibility also involves the layer strength during loading, as seen further.

As discussed above, as well as in other topics presented in this section, elastic moduli were assessed through the instrumented indentation technique, a method where force and displacements of a tip loaded against the surface are controlled with nN and nm resolutions, respectively, as schematized in Fig. 8.7. Typical indenters used present

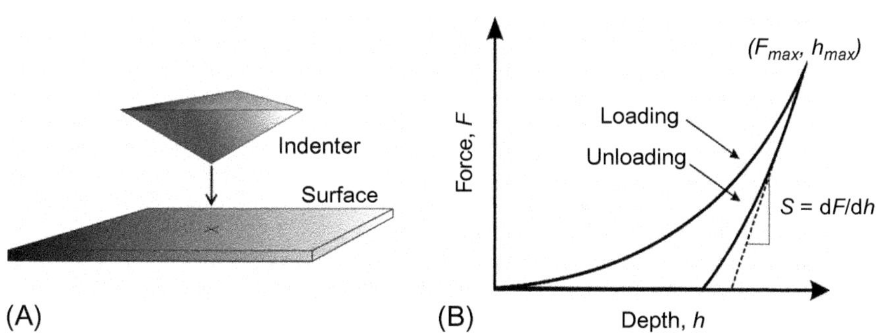

Fig. 8.7 (A) Schematic representation of a Berkovich tip reaching a surface, in an indentation test. (B) A typical force versus depth curve, from which hardness and elastic modulus are calculated through the Oliver and Pharr method. The analysis demands the knowledge of maximum force and depth, the contact stiffness *S*, and tip geometric factors.

the Berkovich geometry, a three-sided pyramid made of diamond. This method, also known as nanoindentation, is the technique suitable to investigate thin films and a few micrometer-thick layers [77], which are not assessed by traditional methods designed for bulk materials, say, the tensile strength testing, as a thin coating represents only a tiny fraction of the whole centimeter-size specimen. In instrumented indentation, hardness and elastic modulus are obtained from the force-displacement curve (Fig. 8.7B), calculated by the well-established Oliver and Pharr analytical method [78]. Some limitations occur in applying the Oliver and Pharr method if the surface is rough [79], or if the material presents some limited strain hardening [78].

In addition, it is possible to determine the effective elastic modulus, by subtracting the substrate effects from the surface elastic response. The knowledge of a layer's effective elastic modulus is important for design purposes. The obtained result correlates "only" to the PEO layer, instead of the layer + substrate (the composite). The Xu and Pharr proposition proved to be an efficient method for several types of PEO layers. It presents a solution for the effective elastic compliance $[(1-\nu)/\mu]$ of a layer over a half-space indented with a flat cylindrical punch, where μ and ν are the composite shear modulus and Poisson's ratio, respectively. The adaptation of the method to provide the elastic modulus E of a layer measured with a Berkovich tip demands some basic assumptions from the contact mechanics theory [80]. Fig. 8.8 shows the elastic modulus profile measured on a PEO TiO_2 layer over a titanium substrate, and the corresponding effective elastic modulus profile of the layer, calculated with the

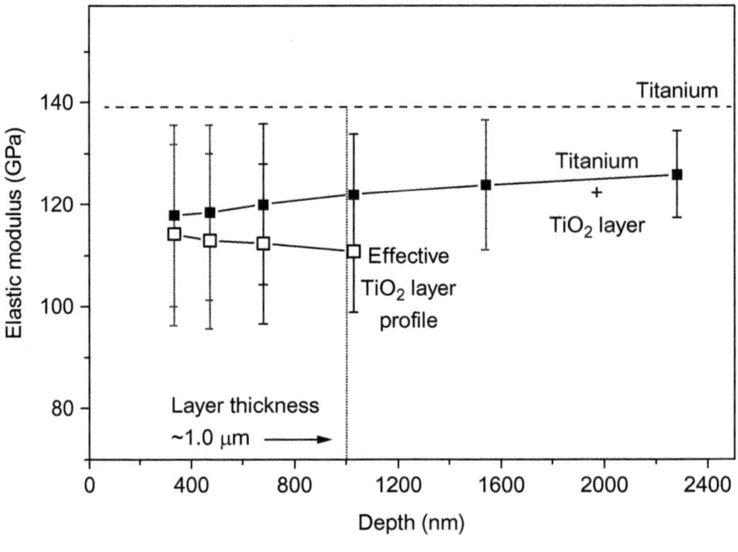

Fig. 8.8 Elastic modulus from instrumented indentation, of a 1.0-μm thick TiO_2 layer produced on titanium by PEO, using H_2SO_4 electrolyte. *Closed symbols* correspond to the layer + substrate profile, calculated directly with the Oliver and Pharr method. The *open symbols* ascribe to the effective layer elastic modulus, independent of the substrate influence, inferred with the Xu and Pharr method. The *dashed line* indicates the substrate elastic modulus for comparison purposes.

Xu and Pharr method [81]. As values of the composite profile increased toward the titanium elastic modulus with the penetration depth, due to the increased depth to the layer thickness ratio, they were approximately constant in the substrate-independent profile. We have employed other analytical methods in biomaterial coatings, such as the Doerner and Nix method [35] and the proposition by Hay and Crawford [82], also obtaining satisfactory results.

In spite of the interesting elastic response disclosed by the CaP-containing TiO_2 layers, whose elastic modulus is close to the bone values, layers were rather fragile [75] under scratching and indentation tests. Improvements to the structures of these layers were investigated using various methods, with the focus being preserving the good bioactivity response of these Ca-P rich layers.

One of the methods proposed by our group was heat treatment subsequent to PEO [83]. The wear rate of CaP-containing TiO_2 layers decreased one and two orders of magnitude after heat treatments in air at 400°C and 600°C, respectively, as compared to the only PEO sample, maintaining the integrity of the film at specific conditions. The enhanced wear strength was mainly due to the anatase-to-rutile allotropic transformation. However, such improvement was obtained at the expense of elastic modulus, which increased from 40 to ~140 GPa. On the other hand, these heat-treated layers presented very hydrophilic surfaces due to the production of vacancy sites favorable for chemical bonds of OH groups, leading to an increase in water adsorption.

Another method investigated to improve the tribo-mechanical response of PEO layers was the combination of two treatment techniques at the same surface, namely the use of shot blasting or plasma nitriding prior to PEO. In this configuration, instead of changing the layer properties by varying their crystalline structure, the modification of the substrate offers different conditions for the anodization process, at the same time that it provides a better load-bearing capacity to the layer under scratching or normal loading [84]. The shot-blasting technique consists of blasting the surface with hard particles, such as alumina (Al_2O_3), which can be chemically stable and not harmful to the body [29]. The CaP-rich TiO_2 layers grown over titanium surfaces, previously shot blasted with alumina particles of 280 μm diameter, presented low values of elastic modulus, similar to those of layers formed on the raw titanium (~40 GPa). Moreover, the shot blasted + PEO layer maintained its integrity under nanoscratch tests, disclosing critical loads for detachment three times higher than that presented by the only PEO layer [84]. Thus, combined shot blast and anodic oxidation processes can lead to some of the requirements for improved bioactivity on titanium surfaces. In a different approach [85], the plasma nitrided titanium surface was harder and stiffer than the substrate, consisting of a stratified case composed of nitride precipitates (TiN, Ti_2N) and nitrogen-expanded Ti matrix. PEO TiO_2 layers grown on these surfaces presented an elastic modulus of ~75 GPa, with reduced brittleness under normal loading and scratch tests.

Although sodium hydroxide is relatively known for modification of surface on Ti in alkali treatment, our group used sodium hydroxide as an electrolyte for PEO [80] exhibiting better mechanical properties compared to conventional alkali treatment [85]. The combination of the shot blasting of the substrate follow by the PEO treatment using sodium hydroxide as an electrolyte [86] improves the corrosion resistance when compared with the reference Ti (surface without treatments). Results of electrochemical measurements using the open-circuit potential (OCP) conditions showed that the protective effect to corrosion in the studied physiological environment increased in Ti and shot-blasted Ti after the oxidation process, due to the oxide layer.

8.4 Conclusions

Ideal materials that can be a substitute for the use of orthopedic applications are what drives the current research on titanium implants. The recent development on titanium surfaces assists in the biocompatibility, mechanical and chemical stability as well as the improvement in the healing process for targeted applications. Even though there has been considerable research on titanium surfaces, there is still need for improvement because of the specific properties of the bone and the wide use of commercial implants, which will always need evolution in the regenerative process and compatibility of the body.

References

[1] M. Geetha, A.K. Singh, R. Asokamani, A.K. Gogia, Ti based biomaterials, the ultimate choice for orthopaedic implants—a review, Prog. Mater. Sci. 54 (2009) 397–425, https://doi.org/10.1016/j.pmatsci.2008.06.004.

[2] United Nations, World Population Ageing 2015 Highlights, United Nations, New York, 2015.

[3] Y. Wang, Bioadaptability: an innovative concept for biomaterials, J. Mater. Sci. Technol. 32 (9) (2016) 801, https://doi.org/10.1016/j.jmst.2016.08.002.

[4] L. Kunčická, R. Kocich, T.C. Lowe, Advances in metals and alloys for joint replacement, Prog. Mater. Sci. 88 (2017) 232–280, https://doi.org/10.1016/j.pmatsci.2017.04.002.

[5] J.M. Cordeiro, V.A.R. Barão, Is there scientific evidence favoring the substitution of commercially pure titanium with titanium alloys for the manufacture of dental implants? Mater. Sci. Eng. C 71 (2017) 1201–1215, https://doi.org/10.1016/j.msec.2016.10.025.

[6] Q. Chen, G.A. Thouas, Metallic implant biomaterials, Mater. Sci. Eng. R Rep. 87 (2015) 1–57, https://doi.org/10.1016/j.mser.2014.10.001.

[7] X. Liu, P.K. Chu, C. Ding, Surface nano-functionalization of biomaterials, Mater. Sci. Eng. R Rep. 70 (2010) 275–302, https://doi.org/10.1016/j.mser.2010.06.013.

[8] H. Ikehata, N. Nagasako, T. Furuta, A. Fukumoto, K. Miwa, T. Saito, First-principles calculations for development of low elastic modulus Ti alloys, Phys. Rev. B 70 (2004) 174113https://doi.org/10.1103/PhysRevB.70.174113.

[9] M. Niinomi, Biologically and mechanically biocompatible titanium alloys, Mater. Trans. 49 (2008) 2170–2178, https://doi.org/10.2320/matertrans.L-MRA2008828.

[10] M. Abdullah, S.K. Kamarudin, Titanium dioxide nanotubes (TNT) in energy and environmental applications: an overview, Renew. Sustain. Energy Rev. 76 (2017) 212–225, https://doi.org/10.1016/j.rser.2017.01.057.

[11] I.V. Lukiyanchuk, V.S. Rudnev, L.M. Tyrina, I.V. Chernykh, Plasma electrolytic oxide coatings on valve metals and their activity in CO oxidation, Appl. Surf. Sci. 315 (2014) 481–489, https://doi.org/10.1016/j.apsusc.2014.03.040.

[12] A. Ghicov, H. Tsuchiya, J.M. MacAk, P. Schmuki, Titanium oxide nanotubes prepared in phosphate electrolytes, Electrochem. Commun. 7 (2005) 505–509, https://doi.org/10.1016/j.elecom.2005.03.007.
[13] S. Gupta, V. Dahiya, P. Shukla, Surface topography of dental implants: a review, J. Dent. Implant 4 (2014) 66, https://doi.org/10.4103/0974-6781.131009.
[14] X. Zhu, K.H. Kim, Y. Jeong, Anodic oxide films containing Ca and P of titanium biomaterial, Biomaterials 22 (2001) 2199–2206.
[15] T. Kokubo, H.-M. Kim, M. Kawashita, Novel bioactive materials with different mechanical properties, Biomaterials 24 (2003) 2161–2175, https://doi.org/10.1016/S0142-9612(03)00044-9.
[16] T. Miyazaki, Development of bioactive materials based on bone-bonding mechanism on metal oxides, J. Ceram. Soc. Jpn. 116 (2008) 260–264, https://doi.org/10.2109/jcersj2.116.260.
[17] N. Tamai, A. Myoui, T. Tomita, T. Nakase, J. Tanaka, T. Ochi, et al., Novel hydroxyapatite ceramics with an interconnective porous structure exhibit superior osteoconduction in vivo, J. Biomed. Mater. Res. 59 (2002) 110–117, https://doi.org/10.1002/jbm.1222.
[18] R. Gilbert Triplett, O. Budinskaya, New frontiers in biomaterials, Oral Maxillofac. Surg. Clin. North Am. 29 (2017) 105–115, https://doi.org/10.1016/j.coms.2016.08.011.
[19] D.M. Campos, E.S. Jr, N.K. Kuromoto, G.A. Soares, Preliminary results of osteoblast adhesion on titanium anodic films, Matéria (Rio J.) 12 (2007) 150–155, https://doi.org/10.1590/S1517-70762007000100019.
[20] A. Wennerberg, T. Albrektsson, Effects of titanium surface topography on bone integration: a systematic review, Clin. Oral Implants Res. 20 (2009) 172–184, https://doi.org/10.1111/j.1600-0501.2009.01775.x.
[21] D.A. Puleo, M.V. Thomas, Implant surfaces, Dent. Clin. North Am. 50 (2006) 323–338, https://doi.org/10.1016/j.cden.2006.03.001.
[22] K.M. Hotchkiss, G.B. Reddy, S.L. Hyzy, Z. Schwartz, B.D. Boyan, R. Olivares-Navarrete, Titanium surface characteristics, including topography and wettability, alter macrophage activation, Acta Biomater. 31 (2016) 425–434, https://doi.org/10.1016/j.actbio.2015.12.003.
[23] L. Bacakova, E. Filova, M. Parizek, T. Ruml, V. Svorcik, Modulation of cell adhesion, proliferation and differentiation on materials designed for body implants, Biotechnol. Adv. 29 (2011) 739–767, https://doi.org/10.1016/j.biotechadv.2011.06.004.
[24] J.H. Park, C.E. Wasilewski, N. Almodovar, R. Olivares-Navarrete, B.D. Boyan, R. Tannenbaum, et al., The responses to surface wettability gradients induced by chitosan nanofilms on microtextured titanium mediated by specific integrin receptors, Biomaterials 33 (2012) 7386–7393, https://doi.org/10.1016/j.biomaterials.2012.06.066.
[25] M. Zuldesmi, A. Waki, K. Kuroda, M. Okido, Enhancement of valve metal osteoconductivity by one-step hydrothermal treatment, Mater. Sci. Eng. C 42 (2014) 405–411, https://doi.org/10.1016/j.msec.2014.05.049.
[26] D. Buser, N. Broggini, M. Wieland, R.K.K. Schenk, a.J.J. Denzer, D.L.L. Cochran, et al., Enhanced bone apposition to a chemically modified SLA titanium surface, J. Dent. Res. 83 (2004) 529–533, https://doi.org/10.1177/154405910408300704.
[27] K. Lin, C. Wu, J. Chang, Advances in synthesis of calcium phosphate crystals with controlled size and shape, Acta Biomater. 10 (2014) 4071–4102, https://doi.org/10.1016/j.actbio.2014.06.017.
[28] Y. Wang, H. Yu, C. Chen, Z. Zhao, Review of the biocompatibility of micro-arc oxidation coated titanium alloys, Mater. Des. 85 (2015) 640–652, https://doi.org/10.1016/j.matdes.2015.07.086.

[29] X. Liu, P.K. Chu, C. Ding, Surface modification of titanium, titanium alloys, and related materials for biomedical applications, Mater. Sci. Eng. R Rep. 47 (2004) 49–121.
[30] C. Jaeggi, M. Parlinska-Wojtan, P. Kern, Correlation of electrolyte-derived inclusions to crystallization in the early stage of anodic oxide film growth on titanium, Thin Solid Films 520 (2012) 1804–1808, https://doi.org/10.1016/j.tsf.2011.08.092.
[31] F.C. Walsh, C.T.J. Low, R.J.K. Wood, K.T. Stevens, J. Archer, A.R. Poeton, et al., Plasma electrolytic oxidation (PEO) for production of anodised coatings on lightweight metal (Al, Mg, Ti) alloys, Trans. IMF 87 (2009) 122–135, https://doi.org/10.1179/174591908X372482.
[32] N.P. Sluginov, On light phenomena observed in liquids during their electrolysis, J. Russ. Phys. Chem. Soc. 12 (1880) 193–203.
[33] A. Gunterschultze, H. Betz, Electrolytkondensatoren, vol. 1 Krayn, Berlin, 1937.
[34] A.L. Yerokhin, X. Nie, A. Leyland, A. Matthews, S.J. Dowey, Plasma electrolysis for surface engineering, Surf. Coat. Technol. 122 (1999) 73–93, https://doi.org/10.1016/S0257-8972(99)00441-7.
[35] P. Soares, A. Mikowski, C.M. Lepienski, E. Santos, G.A. Soares, V. Swinka Filho, et al., Hardness and elastic modulus of TiO_2 anodic films measured by instrumented indentation, J. Biomed. Mater. Res. B Appl. Biomater. 84 (2008) 524–530, https://doi.org/10.1002/jbm.b.30900.
[36] B.L. Pereira, C.M. Lepienski, I. Mazzaro, N.K. Kuromoto, Apatite grown in niobium by two-step plasma electrolytic oxidation, Mater. Sci. Eng. C 77 (2016) 1235–1241, https://doi.org/10.1016/j.msec.2016.10.073.
[37] A. Lugovskoy, M. Zinigr, Plasma electrolytic oxidation of valve metals, in: Materials Science—Advanced Topics, InTech, 2013, https://doi.org/10.5772/54827.
[38] J.A. Curran, T.W. Clyne, Porosity in plasma electrolytic oxide coatings, Acta Mater. 54 (2006) 1985–1993, https://doi.org/10.1016/j.actamat.2005.12.029.
[39] S.V. Dorozhkin, Calcium orthophosphate deposits: preparation, properties and biomedical applications, Mater. Sci. Eng. C 55 (2015) 272–326, https://doi.org/10.1016/j.msec.2015.05.033.
[40] J.H. Liu, G.L. Wu, M. Yu, L. Wu, Y. Zhang, S.M. Li, Influence of incremental rate of anodising current on roughness and electrochemical corrosion of oxide film on titanium alloy Ti-10V-2Fe-3Al, Surf. Eng. 28 (2012) 406–411, https://doi.org/10.1179/1743294411Y.0000000091.
[41] D. Quintero, M. Gómez, J. Castaño, E. Tsuji, Y. Aoki, F. Echeverría, et al., Anodic films obtained on Ti6Al4V in aluminate solutions by spark anodizing: effect of OH− and WO4−2 additions on the tribological properties, Surf. Coat. Technol. 310 (2017) 180–189, https://doi.org/10.1016/j.surfcoat.2016.12.063.
[42] M. Kulkarni, A. Mazare, E. Gongadze, S. Perutkova, V. Kralj-Iglic, I. Milosev, et al., Titanium nanostructures for biomedical applications, Nanotechnology 26 (2015). 62002 (1–18) https://doi.org/10.1088/0957-4484/26/6/062002.
[43] S. Minagar, C.C. Berndt, J. Wang, E. Ivanova, C. Wen, A review of the application of anodization for the fabrication of nanotubes on metal implant surfaces, Acta Biomater. 8 (2012) 2875–2888, https://doi.org/10.1016/j.actbio.2012.04.005.
[44] P. Roy, S. Berger, P. Schmuki, TiO_2 nanotubes: synthesis and applications, Angew. Chem. Int. Ed. 50 (2011) 2904–2939, https://doi.org/10.1002/anie.201001374.
[45] A. Gebert, D. Eigel, P.F. Gostin, V. Hoffmann, M. Uhlemann, A. Helth, et al., Oxidation treatments of beta-type Ti-40Nb for biomedical use, Surf. Coat. Technol. 302 (2016) 88–99, https://doi.org/10.1016/j.surfcoat.2016.05.036.

[46] D. Khudhair, A. Bhatti, Y. Li, H.A. Hamedani, H. Garmestani, P. Hodgson, et al., Anodization parameters influencing the morphology and electrical properties of TiO_2 nanotubes for living cell interfacing and investigations, Mater. Sci. Eng. C 59 (2016) 1125–1142, https://doi.org/10.1016/j.msec.2015.10.042.

[47] K. Indira, U.K. Mudali, T. Nishimura, N. Rajendran, A review on TiO_2 nanotubes: influence of anodization parameters, formation mechanism, properties, corrosion behavior, and biomedical applications, J. Bio- Tribo-Corros. 1 (2015) 1–28, https://doi.org/10.1007/s40735-015-0024-x.

[48] F. Nasirpouri, I. Yousefi, E. Moslehifard, J. Khalil-Allafi, Tuning surface morphology and crystallinity of anodic TiO_2 nanotubes and their response to biomimetic bone growth for implant applications, Surf. Coat. Technol. 315 (2017) 163–171, https://doi.org/10.1016/j.surfcoat.2017.02.006.

[49] D. Khudhair, H. Amani Hamedani, J. Gaburro, S. Shafei, S. Nahavandi, H. Garmestani, et al., Enhancement of electro-chemical properties of TiO_2 nanotubes for biological interfacing, Mater. Sci. Eng. C 77 (2017) 111–120, https://doi.org/10.1016/j.msec.2017.03.112.

[50] G. Liu, K. Du, K. Wang, Surface wettability of TiO_2 nanotube arrays prepared by electrochemical anodization, Appl. Surf. Sci. 388 (2016) 313–320, https://doi.org/10.1016/j.apsusc.2016.01.010.

[51] V.B. Damodaran, D. Bhatnagar, V. Leszczak, K.C. Popat, Titania nanostructures: a biomedical perspective, RSC Adv. 5 (2015) 37149–37171, https://doi.org/10.1039/C5RA04271B.

[52] A. Roguska, M. Pisarek, A. Belcarz, L. Marcon, M. Holdynski, M. Andrzejczuk, et al., Improvement of the bio-functional properties of TiO_2 nanotubes, Appl. Surf. Sci. 388 (2016) 775–785, https://doi.org/10.1016/j.apsusc.2016.03.128.

[53] L. Salou, A. Hoornaert, G. Louarn, P. Layrolle, Enhanced osseointegration of titanium implants with nanostructured surfaces: an experimental study in rabbits, Acta Biomater. 11 (2015) 494–502, https://doi.org/10.1016/j.actbio.2014.10.017.

[54] X. Liu, P. Chu, C. Ding, Surface modification of titanium, titanium alloys, and related materials for biomedical applications, Mater. Sci. Eng. R Rep. 47 (2004) 49–121, https://doi.org/10.1016/j.mser.2004.11.001.

[55] F. Variola, F. Vetrone, L. Richert, P. Jedrzejowski, J.H. Yi, S. Zalzal, et al., Improving biocompatibility of implantable metals by nanoscale modification of surfaces: an overview of strategies, fabrication methods, and challenges, Small 5 (2009) 996–1006, https://doi.org/10.1002/smll.200801186.

[56] T.D.P. Busquim, C.N. Elias, J.E. May, S.E. Kuri, P.A. Nascente, in: Titanium oxide layer on the surface of anodized dental implants, Med Device Mater V; Proc from Mater Process Med Devices Conf, 2009, pp. 60–65.

[57] R. Liu, T. Lei, V. Dusevich, X. Yao, Y. Liu, M.P. Walker, et al., Surface characteristics and cell adhesion: a comparative study of four commercial dental implants, J. Prosthodont. 22 (2013) 641–651, https://doi.org/10.1111/jopr.12063.

[58] M. Gaintantzopoulou, S. Zinelis, N. Silikas, G. Eliades, Micro-Raman spectroscopic analysis of TiO_2 phases on the root surfaces of commercial dental implants, Dent. Mater. 30 (2014) 861–867, https://doi.org/10.1016/j.dental.2014.05.030.

[59] A.P. Ross, T.J. Webster, Anodizing color coded anodized Ti6Al4V medical devices for increasing bone cell functions, Int. J. Nanomed. 8 (2013) 109–117.

[60] R. Lappalainen, S.S. Santavirta, Potential of coatings in total hip replacement, Clin. Orthop. Relat. Res. 430 (2005) 72–79, https://doi.org/10.1097/01.blo.0000150000.75660.ff.

[61] W.N. Capello, J.A. D'Antonio, M.T. Manley, J.R. Feinberg, Hydroxyapatite in total hip arthroplasty. Clinical results and critical issues, Clin. Orthop. Relat. Res. 355 (1998) 200–211.
[62] J.-A. Epinette, M.T. Manley, Fifteen Years of Clinical Experience with Hydroxyapatite Coatings in Joint Arthroplasty. vol. 1 Springer Paris, Paris, 2004https://doi.org/10.1007/978-2-8178-0851-2.
[63] D.J. Hall, R.M. Urban, R. Pourzal, T.M. Turner, A.K. Skipor, J.J. Jacobs, Nanoscale surface modification by anodic oxidation increased bone ingrowth and reduced fibrous tissue in the porous coating of titanium-alloy femoral hip arthroplasty implants, J. Biomed. Mater. Res. Part B Appl. Biomater. 105 (2017) 283–290, https://doi.org/10.1002/jbm.b.33554.
[64] B. Zhang, D. Myers, G. Wallace, M. Brandt, P. Choong, Bioactive coatings for orthopaedic implants—recent trends in development of implant coatings, Int. J. Mol. Sci. 15 (2014) 11878–11921, https://doi.org/10.3390/ijms150711878.
[65] A. Somer, F. Camilotti, G.F. Costa, A.R. Jurelo, A. Assmann, G.B. De Souza, et al., Effects of thermal oxidation on the effective thermal diffusivity of titanium alloys, J. Phys. D. Appl. Phys. 47 (2014) 385306.
[66] C.L. Chu, C. Guo, X.B. Sheng, Y.S. Dong, P.H. Lin, K.W.K. Yeung, et al., Microstructure, nickel suppression and mechanical characteristics of electropolished and photoelectrocatalytically oxidized biomedical nickel titanium shape memory alloy, Acta Biomater. 5 (2009) 2238–2245.
[67] G.B. de Souza, C.M. Lepienski, C.E. Foerster, N.K. Kuromoto, P. Soares, A. Ponte H de, Nanomechanical and nanotribological properties of bioactive titanium surfaces prepared by alkali treatment, J. Mech. Behav. Biomed. Mater. 4 (2011) 756–765.
[68] P. Korkusuz, F. Korkusuz, Hard tissue—biomaterial interactions, in: M.J. Yaszemski, D.J. Trantolo, K.-U. Lewandrowski, V. Hasirci, D.E. Alto-Belli, D.L. Wise (Eds.), Biomaterials in Orthopedics, vol. 1, Marcel Dekker Inc, New York, 2004, pp. 1–40.
[69] J.W. Hutchinson, Z. Suo, Mixed mode cracking in layered materials, Adv. Appl. Mech. 29 (1991) 63–191.
[70] E. Santos, G.B. de Souza, F.C. Serbena, H.L. Santos, G.G. de Lima, E.M. Szesz, et al., Effect of anodizing time on the mechanical properties of porous titania coatings formed by micro-arc oxidation, Surf. Coat. Technol. 309 (2017) 203–211.
[71] J. Chen, Y. Shi, L. Wang, F. Yan, F. Zhang, Preparation and properties of hydroxyapatite-containing titania coating by micro-arc oxidation, Mater. Lett. 60 (2006) 2538–2543.
[72] J.-H. Lee, S.-E. Kim, Y.-J. Kim, C.-S. Chi, H.-J. Oh, Effects of microstructure of anodic titania on the formation of bioactive compounds, Mater. Chem. Phys. 98 (2006) 39–43.
[73] W. Simka, A. Iwaniak, G. Nawrat, A. Maciej, J. Michalska, K. Radwański, et al., Modification of titanium oxide layer by calcium and phosphorus, Electrochim. Acta 54 (2009) 6983–6988.
[74] G.B. de Souza, G.G. de Lima, N.K. Kuromoto, P. Soares, C.M. Lepienski, C.E. Foerster, et al., Tribo-mechanical characterization of rough, porous and bioactive Ti anodic layers, J. Mech. Behav. Biomed. Mater. 4 (2011) 796–806, https://doi.org/10.1016/j.jmbbm.2010.09.012.
[75] G.B. de Souza, G.G. de Lima, N.K. Kuromoto, P. Soares, C.M. Lepienski, C.E. Foerster, et al., Tribo-mechanical characterization of rough, porous and bioactive Ti anodic layers, J. Mech. Behav. Biomed. Mater. 4 (2011) 796–806.
[76] M.A. Meyers, K.K. Chawla, Mechanical Behavior of Materials, vol. 1 Prentice-Hall, Upper Saddle River, 1999.
[77] C.M. Lepienski, C.E. Foerster, Nanomechanical properties by nanoindentation, in: H.S. Nalwa (Ed.), Encyclopedia of Nanoscience and Nanotechnology, vol. 6, American Scientific Publishers, Stevenson Ranch, 2004, pp. 1–20.

[78] W.C. Oliver, G.M. Pharr, Measurement of hardness and elastic modulus by instrumented indentation: advances in understanding and refinements to methodology, J. Mater. Res. 19 (2004) 3–20.
[79] G.B. de Souza, A. Mikowski, C.M. Lepienski, C.E. Foerster, Indentation hardness of rough surfaces produced by plasma-based ion implantation processes, Surf. Coat. Technol. 204 (2010) 3013–3017, https://doi.org/10.1016/j.surfcoat.2010.03.001.
[80] E.M. Szesz, G.B. de Souza, E. Santos, N.K. Kuromoto, Nanomechanical properties of bioactive Ti surfaces obtained by NaOH-based anodic oxidation and alkali treatment, Key Eng. Mater. 493 (2012) 524–529.
[81] H. Xu, G.M. Pharr, An improved relation for the effective elastic compliance of a film/substrate system during indentation by a flat cylindrical punch, Scr. Mater. 55 (2006) 315–318.
[82] J. Hay, B. Crawford, Measuring substrate-independent modulus of thin films, J. Mater. Res. 26 (2011) 727–738.
[83] G.G. De Lima, G.B. De Souza, C.M. Lepienski, N.K. Kuromoto, Mechanical properties of anodic titanium films containing ions of Ca and P submitted to heat and hydrothermal treatment, J. Mech. Behav. Biomed. Mater. 64 (2016) 18–30, https://doi.org/10.1016/j.jmbbm.2016.07.019.
[84] E.M. Szesz, G.B. de Souza, G.G. de Lima, B.A. da Silva, N.K. Kuromoto, C.M. Lepienski, Improved tribo-mechanical behavior of CaP-containing TiO_2 layers produced on titanium by shot blasting and micro-arc oxidation, J. Mater. Sci. Mater. Med. 25 (2014) 2265–2275, https://doi.org/10.1007/s10856-014-5238-9.
[85] G.B. de Souza, B.A. da Silva, G. Steudel, S.H. Gonsalves, C.E. Foerster, C.M. Lepienski, Structural and tribo-mechanical characterization of nitrogen plasma treated titanium for bone implants, Surf. Coat. Technol. 256 (2014) 30–36.
[86] E.M. Szesz, B.L. Pereira, N.K. Kuromoto, C.E.B. Marino, G.B. de Souza, P. Soares, Electrochemical and morphological analyses on the titanium surface modified by shot blasting and anodic oxidation processes, Thin Solid Films 528 (2013) 163–166.

Zirconia-alumina composite for orthopedic implant application

9

Vasanth Gopal[*,†], *Geetha Manivasagam*[†]
[*]Department of Physics, School of Advanced Sciences, Vellore Institute of Technology, Vellore, India, [†]Centre for Biomaterials, Cellular and Molecular Theranostics (CBCMT), Vellore Institute of Technology, Vellore, India

9.1 Introduction

The evolution of ceramics in the medical field dates back to the 18th century where porcelain ceramics were used as a dental crown. Not surprisingly, ceramics also find a place in orthopedic applications in the late 19th century due to their outstanding properties such as high hardness, low wear, and superior chemical stability. Boutin in the 1970s was the first person to introduce alumina hip joints in orthopedic surgery as an alternate to metal-on-polyethylene hip bearings [1]. The invention of metal-on-ultrahigh molecular weight polyethylene (UHMWPE) (M-o-P) hip bearing by Sir John Charnley led to the widespread use in hip arthroplasty and indeed it is seen as one of the most successful orthopedic operations today [2]. However, the M-o-P is susceptible to high wear, mainly the wear of UHMWPE in the form of wear debris triggers osteolysis and aseptic loosening and consequently results in revision surgeries [3–6]. The metal-on-metal (M-o-M) hip prosthesis, another combination of hip bearing, was introduced in the market by orthopedic companies. The wear rate of M-o-M was significantly lower compared to the M-o-P. Nevertheless, the complications associated with M-o-M such as metal ion release and metal wear debris causes metal hypersensitivity called "Metallosis," which restricted their further uses [7–9]. All the above factors concomitantly reduce the lifetime of the hip prosthesis. Further, the demand for hip replacement is increasing drastically every year. Estimation shows that the market for hip replacement is projected to grow by 174% in the United States by the year 2030 [10]. Also, an alarming increase in a younger patient (<65 years) for hip arthroplasty demands the hip prosthesis to function for >30 years (the lifetime of current implants is 15–20 years), thus leading to a higher demand for an implant to be more reliable and durable [11].

The ceramic composite hip prosthesis can be a promising alternative to the M-o-P, M-o-M, and monolithic ceramic implants due to their extremely low wear rate, enhanced fracture toughness, and excellent chemical stability, which are prerequisites for the longevity of the hip prosthesis. In the following section, a detailed discussion on the evolution of ceramic matrix composites and its performance is presented.

Applications of Nanocomposite Materials in Orthopedics. https://doi.org/10.1016/B978-0-12-813740-6.00011-9

9.1.1 Evolution of ceramic composite hip prostheses

First-generation alumina with low friction, high wear resistance, and excellent wettability has led to the implantation of >3.5 million alumina components worldwide since 1990 [12]. Despite the low wear rates, the first-generation alumina hip prosthesis experienced higher fracture rates. The intrinsic brittleness of alumina led to the unexpected catastrophic failure of the femoral head in vivo [13]. This devastating failure led to the development of second- (Biolox) and third-generation alumina bearing (Biolox Forte). Biolox Forte was commercially available in 1990, and it contains high-purity alumina (>98%) with a small amount of magnesium oxide (MgO) to inhibit the grain growth of alumina during sintering. Further, Biolox Forte, which was manufactured by the HIPing process (discussed in detail in Section 9.3.1.2) has fetched a more homogeneous and dense microstructure with improved properties over first-generation alumina bearings. Even with this enhanced features, still, the alumina bearings have been restricted to a limited number of designs of hip components. For instance, 22-mm alumina femoral heads are not currently manufactured for reliability reasons [14].

Yttria stabilized zirconia (Y-TZP) was introduced in the 1990s as an alternative to alumina hip prosthesis. In fact, the zirconia bearings have gained a lot of attention due to their superior fracture toughness and strength compared to brittle alumina. The excellent fracture toughness of zirconia is associated with phase transformation toughening (tetragonal-monoclinic phase). More than 600,000 zirconia femoral heads have been implanted worldwide, mainly in the United States and in Europe. Unfortunately, the zirconia bearings also faced a catastrophic failure due to the phenomenon called premature low-temperature degradation (LTD) or aging of the material in vivo accompanied by a phase transformation [13,15].

As an alternative to the above ceramic hip prosthesis, the ceramic composite hip prosthesis was introduced in 2000, and since then it has been used widely. In general, the ceramic composites were fabricated by mixtures of alumina and zirconia. The most common ceramic matrix composites are yttria stabilized zirconia (YSZ)-toughened alumina (ZTA) where YSZ inclusion is upto 25 wt%, and alumina-toughened YSZ (ATZ) ceramic composites which are suitable for both ceramic-on-polyethylene (C-o-P) and ceramic-on-ceramic (C-o-C) applications, respectively. ATZ comprises 80% tetragonal zirconia polycrystals (ZrO_2-TZP) and 20% alumina (Al_2O_3) and is reported to have superior mechanical and tribological properties compared to alumina [16]. Bio-Hips (Metoxit AG, Thayngen, Switzerland) and Ceramys (Mathys Ltd., Bettlach, Switzerland) have developed the ATZ components. However, Bio-Hip was not commercialized whereas Ceramys was commercialized in 2007. The ZTA prosthesis such as Biolox® Delta and AZ209 are currently available in the market. Biolox® Delta was developed by Ceramtech Medical Products (Plochingen, Germany) and AZ209 by KYOCERA Medical (Osaka, Japan), respectively [17]. Fig. 9.1 shows the microstructure of Biolox® Delta [13]. This ceramic composite prosthesis is successful due to its excellent mechanical properties, and the underlying mechanism is discussed in detail in the below section.

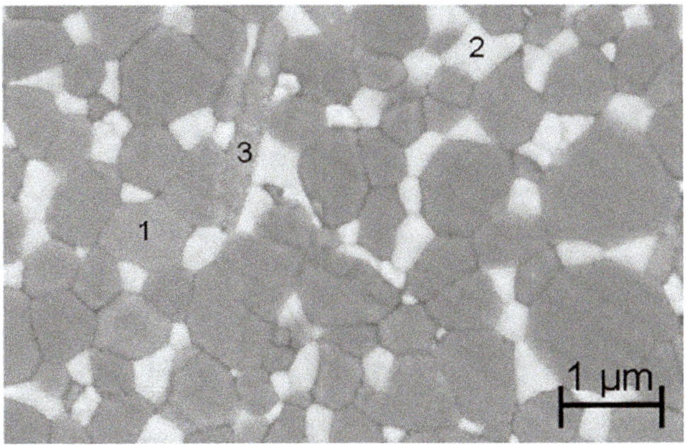

Fig. 9.1 Microstructure of Biolox Delta showing (1) alumina grains, (2) yttria stabilized zirconia grains, (3) platelet-shaped crystal of strontium aluminate.
Source: http://www.mdpi.com/1996-1944/10/6/569.

9.2 The toughening mechanism in ceramic composite

The toughening mechanisms in ceramic composite, particularly zirconia-toughened alumina, are related to volume expansion and shear strain associated with the transformation from tetragonal to monoclinic phase. Two well-accepted toughening mechanisms are (1) stress-induced transformation toughening and (2) micro-cracking toughening.

It is well known that on the application of external tensile stress, the metastable tetragonal zirconia in the alumina matrix transforms to its thermodynamically stable monoclinic phase. This stress-induced phase transformation results in a volume expansion of 3%–4% and shear strain of approximately 6% and provides compressive stress that inhibits crack propagation [18]. A schematic of phase transformation toughening is shown in Fig. 9.2.

Another toughening mechanism is the micro-crack. In general, micro-cracks in ZTA can be categorized as residual micro-cracks and stress-induced micro-cracks. The residual micro-cracks are due to the volume expansion and shear strain associated

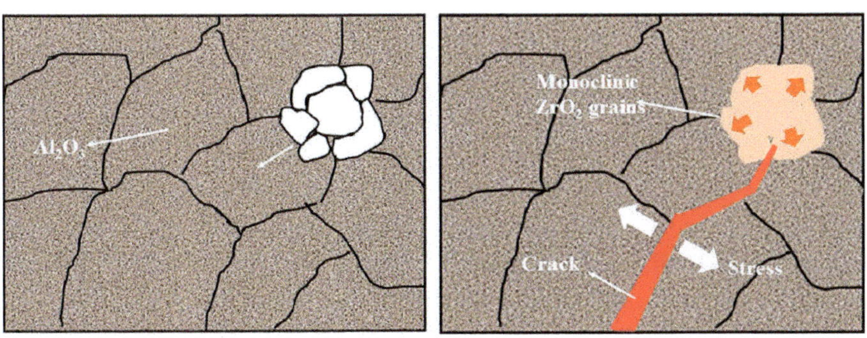

Fig. 9.2 Phase transformation toughening mechanism in ZTA.

with the tetragonal—monoclinic transformation that takes place on cooling from the sintering temperature. The stress-induced micro-cracks are those caused by the volume expansion and shear strain associated with the subsequent stress-induced transformation during the fracture process. In fact, the residual stress micro-cracks are the most commonly occurring toughening mechanism in ZTA. However, a higher density of micro-cracks results in reduced fracture toughness rather than toughening [17,18].

Further, it is perceived that the toughening mechanism in monolithic zirconia and zirconia-toughened alumina has a significant difference. In monolithic zirconia, the stress induced by the single transformed grain makes the neighboring tetragonal zirconia to transform as well and consequently, spread the transformation throughout the material. This transformation results in LTD and deterioration under long-term usage. In the case of the composite, the zirconia is contained in the stable alumina matrix, thereby preventing the transformation of the adjacent grains. Hence, the ZTA has the better tetragonal retention rate compared to the monolithic zirconia when exposed to an in vitro hydrothermal condition [13,17–19].

Also, it is noteworthy to mention that the phase transformation has dual properties; it can enhance the toughness, but at the same time, it can be detrimental also. For instance, if the monoclinic phase transformation occurs at the bearing surface it results in increased surface roughness due to the volume expansion, which in turn leads to higher wear rate. Similarly, if it occurs at the head-trunnion interface, it can initiate fracture.

The manufacturing process of zirconia-toughened alumina is of crucial importance, and it should retain the metastable tetragonal ZrO_2 phase that usually transforms to a stable monoclinic ZrO_2 phase at or near room temperature when exposed to stress. Further, it is suggested to add yttria or magnesia to monoclinic ZrO_2 to retard phase transformation when it is used as orthopedic components [17].

9.2.1 Influence of platelets to inhibit crack propagation

Another toughening mechanism is the introduction of platelets to encounter crack propagation. In fact, Biolox® Delta and AZ209 utilize platelet structures to block or deflect crack growth. A strontium oxide crystal was included in their material. During the sintering process, strontium oxide leads to the formation of strontium aluminate platelets ($SrAl_{12}$-Cr_xO_{19}). The presence of platelets reduces cracking and crack propagation by neutralizing the crack energy, that is, whenever the crack reaches the strontium aluminate platelets, it is unable to propagate further, and it needs extra energy to go around them. If sufficient energy is not available, then the crack does not propagate [17,20,21]. The underlying platelet-toughening mechanism is schematically shown in Fig. 9.3.

9.2.2 Strengthening additives

Stabilizers and additives are commonly used to maintain the zirconia grains in the tetragonal phase at ambient temperature and also to enhance the toughness. Biolox® Delta contains 1.3 M% of yttria as a stabilizer, which is usually lower

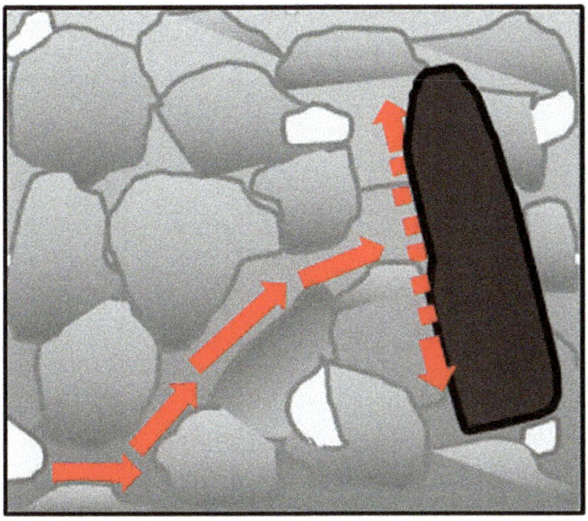

Fig. 9.3 Influence of platelets on inhibiting crack growth.

than the content used for monolithic zirconia due to the stabilization effect of the alumina matrix [17]. In addition to stabilizers, Biolox® Delta contains chromium oxide as additives to increase the hardness and wear characteristics. It has been reported that the addition of chromium oxide enhances the toughness with no change in hardness. Further, chromium oxide inclusion helps in slow down the hydrothermal degradation that occurs in zirconia. It is proposed that in ZTA ceramics, the alumina is a self-sacrificing phase that traps moisture on its surface and protects the zirconia from undergoing phase transformation [22]. The addition of chromia further enhances this protective effect. Particularly in cases of zirconia with yttria and chromia co-doping, the hydrothermal attack is reduced because of the strong interaction between the chromia and zirconia phase, which prevents the diffusion of oxygen into the zirconia phase. The Y-TZP phase is protected from hydrothermal attack at the expense of fast formation of oxygen vacancies in the chromium oxide-stabilized alumina matrix [17,23].

9.3 Fabrication of ceramic composites

The fabrication process of a ceramic composite comprises of different stages. The first stage is the preparation of the ceramic composite powder through various methods such as mechanical mixing, sol-gel technique, hydrothermal oxidation, etc. The second stage is the green forming of the powders; this process can be done by various methods such as die pressing, wet pressing, uniaxial pressing, and cold isostatic pressing to form a green compacted body. Finally, the green compacted powders are sintered for densification. The powder and green forming process are explained in detail in the literature [18]. In this chapter, the various densification processes have been described in detail.

9.3.1 Densification process

The densification of ceramic composites is attained through the sintering process. Sintering is considered as the final process in the fabrication of the ceramic materials. Sintering is a diffusional process that proceeds at relatively high temperatures, usually between 50% and 75% of the melting temperature of the ceramic to convert the porous powder compact that formed early by casting or pressing into a highly dense structure. Sintering is commonly categorized into four types with different densification mechanisms [24],

- Solid-state sintering: During this sintering process, the shaped green body is heated to a high temperature. No liquid remains throughout the sintering process. The densification is achieved by the atomic diffusion in the solid state leading to binding of the particles and reduction in the porosity.
- Liquid-state sintering: A small amount of liquid present throughout the sintering process. The volume of liquid is relatively lower than the original solid mixture. This type of sintering enhances the densification at lower temperatures.
- Vitrification: A relatively more significant amount of liquid that is sufficient to fill the volume of the remaining pores (typically >25% of the solid mixture) is present in the sintering process. Densification occurs by the flow of liquid into the pores and upon cooling, it is either crystallized or vitrified (glass formation).
- Viscous sintering: A consolidated mass of glass particles is heated near to or above its softening temperature. Densification occurs by the viscous flow of the glass particles under the influence of surface tension.

Sintering process can be divided into two types: (1) pressureless sintering and (2) pressure-assisted sintering.

9.3.1.1 Pressureless sintering

The sintering process in the ceramic composite is considered to be a solid state sintering process [18]. Several authors extensively studied the sintering of ZTA composites. Indeed, the ZTA composites offer advantages over the monolithic alumina during sintering. In general, coarsening or grain growth occurs during sintering. Notably, the monolithic alumina undergoes coarsening at higher temperatures [25]. However, in the case of the ZTA composite, the inclusion of zirconia inhibits the grain growth of the alumina matrix. Further, Lange et al. [26], confirmed that the incorporation of zirconia exhibits sufficient self-diffusion into the alumina four-grain junction during grain growth. This diffusion exerts a dragging force at these junctions and limits the grain growth of the alumina matrix. In fact, grain growth hindrance in ZTA is mutual during sintering, that is, the grain size of both alumina and zirconia is reduced compared to monolithic alumina or zirconia materials.

9.3.1.2 Pressure-assisted sintering

From the application point of view, the densification achieved through pressureless sintering might be inadequate. For example, load-bearing applications such as orthopedic bearings need a highly dense ceramic, which should be close to the theoretical density value. In such cases, high densification is achieved by the application of external pressure during sintering. This process is referred to as pressure-assisted sintering. In pressure-assisted sintering, the temperature can be decreased to almost half of the melting point of the ceramic. Further, differential sintering can be avoided using pressure-assisted sintering, which is considered to be critical for maintaining the microstructure and hence the mechanical properties [18]. The most common methods are the uniaxial pressing (hot pressing) and isostatic pressing (hot isostatic pressing, HIPing).

Hot pressing involves pressing (uniaxial) and sintering the ceramic powder in a single-step process. In general, the heating is achieved using indirect resistance heating, induction heating, and field-assisted sintering technique. Pressure and temperature up to 50 MPa and 2400°C can be applicable in a hot press, respectively. As stated above, hot pressing produces a dense, fine-grained material with significant improvement in physical properties compared to the conventional pressureless sintering. Further, the complications associated with powder preparation, dry pressing, and grinding process are eliminated. Despite marked improvement over conventional sintering, hot pressing possesses limitations such as the build-up of residual stresses during pressing, which influences the tetragonal-monoclinic transformation and results in less isotropic mechanical properties [18].

Another class of pressure-assisted sintering is hot isostatic pressing (HIPing). In the HIPing process, no rigid tool is used to transmit the pressure to the powder; instead, it uses insert gases (argon) with a very high pressure level (typically 100–320 MPa or 15–50 ksi) to compact the powder. The gas pressure is applied perpendicularly to and with the same magnitude on all accessible surfaces of the body. The temperature in the HIP can be up to ~2200°C. Based on the flow of gas through the voids in the powder body, it can be divided into two types: direct HIP and post-HIP.

In direct HIP, the powder is poured into a preshaped container that has a similar shape to the product to be made. In the case of post-HIP, the powder is presintered to a density typically 91%–97% of theoretical density. Therefore, the high-pressure gas is allowed to act directly on the surface of the part, and good effect on the closure of internal pores was found. The frequency of appearance of a certain size of pores was reduced by two orders of magnitude after HIP processing [18,27].

Overall, the strength achieved through HIP is higher than those of hot press and pressureless sintering processes. HIPed ZTA exhibited improved strength over sintered ZTA. Further, HIPing produces improved microstructure and isotropic mechanical properties compared to hot press where it gives less isotropic mechanical properties. Bernal et al. studied the effect of HIPing on as-sintered α-alumina for total hip arthroplasty. They achieved a near-theoretical density value without any significant grain growth. Also, they observed an increase in fracture toughness and hardness [23].

The advantage of HIPing led to the development of third-generation alumina hip bearing (Biolox Forte) by CeramTec Medical Products. The Biolox Forte femoral heads showed excellent mechanical properties over the first- and second-generation Biolox femoral heads. The rate of component fracture was significantly reduced in the case of Biolox Forte femoral heads. The schematic depicting the manufacturing process of Biolox Forte is shown in Fig. 9.4.

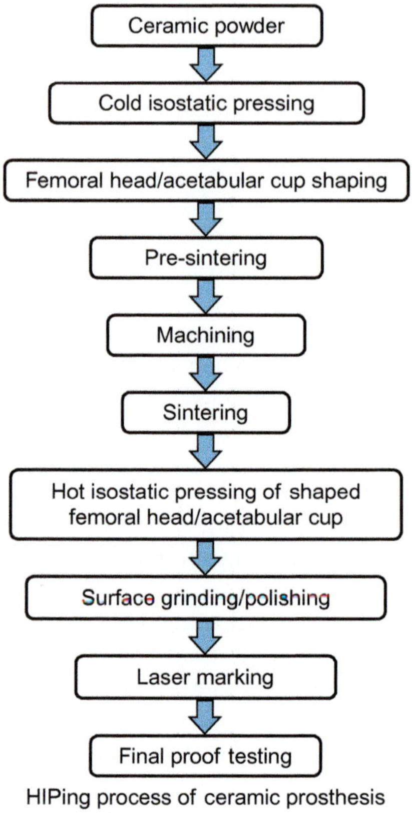

Fig. 9.4 HIPing process of Biolox Forte.

9.4 Wear of ceramic composite hip prosthesis

The wear debris from UHMWPE causes adverse tissue reaction that leads to osteolysis and often requires revision surgeries. This has led to renewed interest in alternative bearings with a lesser wear rate. Ceramic-on-ceramic hip implants have been introduced into the market to encounter the THR wear. Though M-o-M bearings reduce the THR wear to a greater extent when compared to M-o-P bearings, the metal ion release and metal wear debris due to the long-term contact with the corrosive physiological solution cause an allergic reaction called metal hypersensitivity, or metallosis, and are also associated with the formation of pseudotumors. This has restricted their uses, and in fact, it has been withdrawn from the market. As an alternative to M-o-P and M-o-M, ceramic on ceramic (C-o-C) hip joints were developed, and the wear rates were shown to be extremely low.

9.4.1 In vitro wear under standard conditions

The well-positioned C-o-C implants perform remarkably well with low friction, lubrication, and low wear when tested under loads and motions expected during the normal gait cycle. The contact mechanics under these ideal conditions shows that the predicted contact stress is relatively low, due to the close conforming geometries between the femoral head and the acetabular cup [28]. Various lubricants such as carboxymethyl cellulose (CMC) and bovine serum solutions of different viscosities have been used to study the friction behavior of the C-o-C joints. Scholes et al. studied the wear behavior of alumina-on-alumina joints in a hip simulator test under standard conditions (i.e., normal gait cycle). A CMC-Na solution was used as a lubricant as it replicates the shear thinning behavior of synovial fluid, which is a natural lubricating fluid in the human joints. Under this testing condition, the coefficient of friction was found to be 0.002, which suggests that it operates near to full fluid film lubrication. In the full fluid film lubrication regime, the surfaces are completely separated by the lubricant and the friction generated is due solely to the shearing of the lubricant film [2].

The natural lubricating synovial fluid contains proteins and it plays a vital role in the wear process. To mimic the natural synovial fluid, a bovine serum solution is often used for measuring the friction and also to validate the wear rates of the hip implants. In general, a 25% bovine serum that contains approximately 16 g/L protein concentration is used. Often, the bovine serum solution is mixed with 0.1 wt% sodium azide to retard bacterial growth during the wear test [5,29–31]. The proteins in the serum solution tend to form a tribolayer that dramatically influences the friction and wear behavior of the ceramic joints. In contrast to the CMC solution, the serum solution exhibits higher COF values 0.03 and a mixed lubrication regime. In mixed lubrication, the load is carried in part by the contact between the asperities of the bearing surfaces and also by the pressure generated within the lubricant. Despite the serum solution exhibiting higher COF and a mixed lubrication regime, there was no surface damage on the joints after the test. It was hypothesized that the protein that adheres to the surface of ceramic joints formed a thick layer and resulted in protein-to-protein contact and shearing [2].

Similarly, Ma et al. [32], study the effect of lubrication on the friction and wear of Biolox® Delta under a reciprocating sliding wear test with different lubrications such as water, 1 wt% CMC-Na and 25% new-born calf serum. They observed that the serum solution produces a worn surface that exhibited the greatest extent of pitting but with the lowest wear rate. In contrast, the water lubrication gave the highest wear rate but the least pitting. This evidences that the proteins in the serum solution play a vital role in the wear process. Further, a Stribeck curve was constructed to study the friction behavior of Biolox® Delta under three lubricant conditions. The Stribeck curve is a plot used to measure the friction coefficient against the Sommerfeld number (a dimensionless parameter dependent on the lubricant viscosity, the entraining velocity of the bearing surfaces, the joint radius, and the load applied). Mixed lubrication and full fluid film lubrication were obtained in both water and CMC-Na solution, while full fluid film lubrication occurred only in serum solution for the load/speed conditions used. Under the normal circumstances, the wear volume loss of the ceramic joints was shown to be very less and found to be a promising alternative to M-o-P and M-o-M hip joints.

9.4.2 In vitro wear under adverse conditions

It is well proven that the wear rate of ceramic joints is extremely low under normal gait condition. However, these ideal conditions do not always occur during steady-state walking. The retrieval studies of some of the C-o-C joints shown higher wear rates and the phenomenon of stripe wear (a line contact between the femoral head and the edge of the liner) was observed on both the femoral head and acetabular cup. In general, stripe wear can be seen visually as a less polished, lunar shape on the femoral head and a circumferential stripe adjacent to the cup level [20,33]. The stripe wear phenomenon in C-o-C joints has raised concerns and also it has been reported in first- and second-generation alumina bearings [34]. Initially, the stripe wear was thought to be associated with steep cup angles where the loose acetabular components migrated into steep orientations, and the subsequent high contact stress between the femoral head and the edge of the liner causes the stripe wear. Further, it is believed that the edge loading wear could be minimized by improving the bearing material properties, better positioning, and stability of the acetabular component. However, stripe wear was also observed on third-generation alumina C-o-C bearing (Biolox Forte), which has improved properties with well-fixed and well-positioned acetabular components. This suggests that the steep cup inclination does not produce stripe wear [34]. Recently, researchers proposed that micro-separation is responsible for the stripe wear.

Fluoroscopy studies have shown different motions of normal and implanted hip joints. In normal joints, the femoral head is retained within the acetabulum by supporting soft-tissue structures including the fibrous capsule, acetabular labrum, ligament of the head of the femur (LHF), and the iliofemoral, ischiofemoral, pubofemoral, and transverse acetabular ligaments. During surgery, the LHF is found disrupted or degenerative and is surgically removed. Soft-tissue disruption supports the fluoroscopic evidence that the femoral head slides away from the medial aspect of the acetabular component during gait, with active hip abduction-adduction and activity. This phenomenon is known as micro-separation [28,32,35–37]. After hip arthroplasty, during the swing phase of normal walking in the direction toward the cup axis, the femoral head can separate to a small distance from the acetabulum insert. At stance phase (heel strike), the femoral head moves vertically to relocate in the cup. This leads to edge loading where the femoral head contacts the superolateral rim of the acetabular cup resulting in stripe wear [2]. Fig. 9.5 shows the phenomenon of micro-separation during the gait cycle.

Micro-separation in hip joints plays a major role in the wear process and it is thought to be detrimental. Lombardi et al. [38] demonstrated micro-separation in metal-on-polyethylene (2 mm) during walking and the abduction/adduction leg-lift maneuver resulted in premature polyethylene wear. Also, they suggested that the

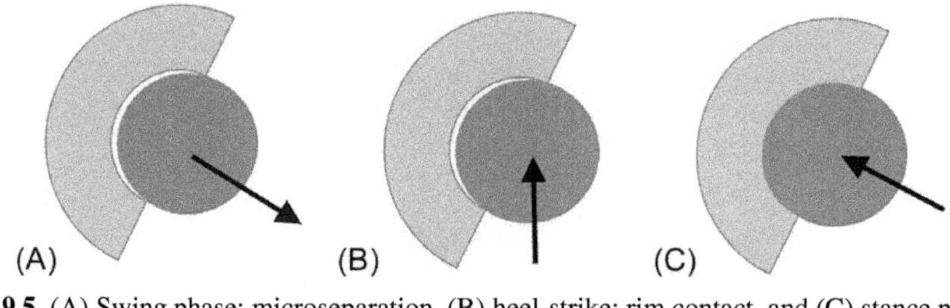

Fig. 9.5 (A) Swing phase: microseparation, (B) heel-strike: rim contact, and (C) stance phase: relocation.

introduction of micro-separation in a hip simulator test provides better replicated wear patterns observed in retrieval analysis. Nevelos et al. [37] hypothesized that the initiation of stripe wear in ceramic–ceramic hip prostheses could be due to micro-separation and they implemented the micro-separation in the hip simulator during the swing phase. Fig. 9.6 shows the micro-separation-assisted hip simulator test [41]. For the first time, they reproduce clinically relevant wear patterns, wear rates, and mechanism for alumina–alumina hip prostheses. Further, the 500-μm separation shows that the steady-state wear of alumina-on-alumina bearing is increased from 0.05 mm^3/million cycles in the standard simulation mode to 1.3 mm^3/million cycles in the severe separation mode.

As stripe wear was observed on third-generation alumina bearings, researchers were keen to study the wear mechanism of composite ceramic bearings under micro-separation and edge-loading conditions. Stewart et al. [39] studied the long-term wear of Biolox Forte (AL) and Biolox® Delta (AMC) under severe swing phase micro-separation. Different combinations of hip joints, the Biolox® Delta head/Biolox® Delta inserts, and Biolox® Delta head/Biolox Forte inserts were used in the study. The stripe wear was formed on all combinations and consequently increased the surface roughness. AMC on alumina produced wear stripes 4–5 mm wide and 17–53 μm deep. Severe micro-separation of AMC on AMC produced wear stripes similar in width to AMC on AL (4–5 mm), but, much shallower (1.5–16 μm). This compares to a typical stripe width of 5 mm and depth of 90 μm for HIPed alumina tested under the same severe conditions. The wear rate of AMC/AMC was 0.16 ± 0.08 mm^3/million cycles, and AMC/AL was 0.61 ± 0.21 mm^3/million cycles, which are 3 times and 12 times lower than the AL/AL ceramic bearing under severe micro-separation conditions. The results suggest that AMC heads articulating against AMC inserts and AMC heads articulating against alumina inserts may have better resistance to micro-separation wear compared to alumina on alumina bearings.

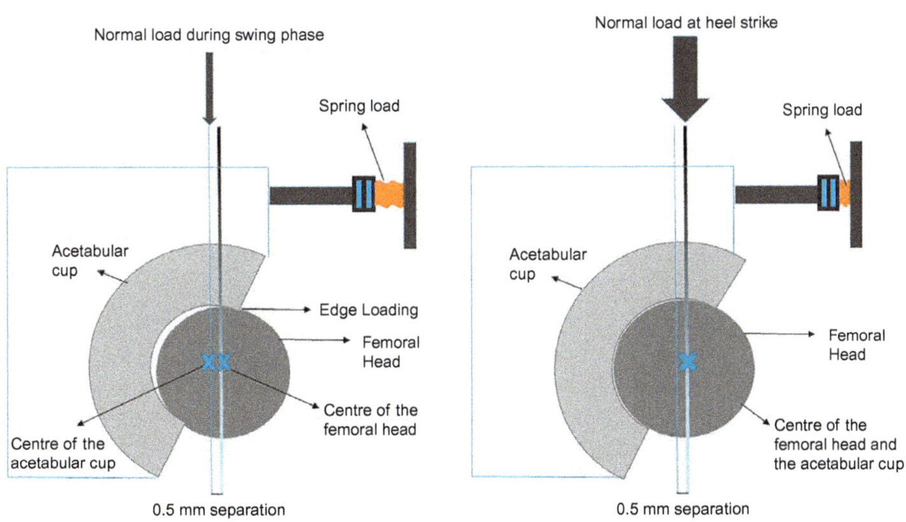

Fig. 9.6 Schematic showing the microseparation of femoral head and acetabular in hip simulator.

In another study, two novel ceramic composite materials, zirconia-toughened alumina (ZTA) and alumina-toughened zirconia (ATZ), developed by Mathys Orthopdie GmbH were tested under adverse micro-separation and edge-loading conditions using the Leeds II physiological anatomical hip joint simulator. Two bearing combinations, ATZ-on-ATZ and ZTA-on-ZTA, were tested and compared with alumina-on-alumina (Al_2O_3-on-Al_2O_3) bearings under severe micro-separation conditions. The overall wear rate of Al_2O_3-on-Al_2O_3 was $0.74 \pm 1.73\,mm^3$/million cycles whereas for the composite bearings the wear rates of ZTA-on-ZTA and ATZ-on-ATZ were $0.14 \pm 0.10\,mm^3$/million cycles and $0.06 \pm 0.004\,mm^3$/million cycles, respectively. The bearing ZTA-on-ZTA and ATZ-on-ATZ showed 5-fold and a 12-fold reduction in wear rate compared to the Al_2O_3-on-Al_2O_3 bearing. Further, all the bearing combination showed stripe wear on the lateral/superior position of the head with a corresponding wear area on the lateral edge of the cup. However, the stripe wear on Al_2O_3-on-Al_2O_3 showed the highest wear penetration depth compared to ZTA-on-ZTA and ATZ-on-ATZ bearings. Also, the monoclinic fraction of ZrO_2 on the ceramic bearing was measured using X-ray analysis, and it was found to be consistent with the International Standard ISO 13356 [40].

Similarly, the wear of composite ceramics in mixed material combination was tested under adverse edge loading conditions. The different head-on-cup configurations were *ATZ-on-ZTA*, *ZTA-on-ATZ*, alumina BIONIT-on-ATZ (*Al_2O_3-on-ATZ*), ATZ-on-alumina BIONIT (*ATZ-on-Al_2O_3*), alumina BIONIT-on-ZTA (*Al_2O_3-on-ZTA*), and ZTA-on-alumina BIONIT (*ZTA-on-Al_2O_3*). The wear rates of the mixed material combination have shown to be slightly higher compared to the like-on-like configuration, that is, ATZ-on-ATZ. Among the combinations, ZTA-on-ATZ had low wear rate similar to ATZ-on-ATZ. It is important to mention that the wear rates of all the combinations were significantly lower than that of Al_2O_3-on-Al_2O_3 bearings [41]. The bar diagram shows the different wear rates of ceramic composites under adverse testing conditions (Fig. 9.7) [40,41]. Overall, the combination ATZ-on-ATZ showed the lowest wear rate compared to all the combinations.

Green et al. [42] compared Biolox® Delta vs Biolox Forte under micro-separation mode to investigate the stripe wear and wear rates. They observed a lower magnitude of wear on the Biolox® Delta ball over the Biolox Forte ball, which demonstrates that the Biolox® Delta offers greater resistance to stripe wear compared to Biolox Forte.

9.5 Fracture—an ultimate challenge

The in vivo fracture of first-generation alumina prostheses pose a severe threat among the patients and surgeons due to the poor microstructure such as micropores, low density, and larger grain sizes that lead to slow crack growth. The advent of modern manufacturing techniques and the advantage of ceramic composites led to the development of improved ceramic prostheses and marketed as Biolox Forte and Biolox® Delta. As per the manufacturer's report (CeramTec GmbH, Plochingen, Germany), the fracture rate of the Biolox® Delta ceramic head was 0.002% compared to 0.02% of alumina ceramic, that is, 30 heads were fractured out of 1500,000 femoral heads sold in 2012 [43]. However, the fracture rate of the Biolox® Delta ceramic liner was not reduced

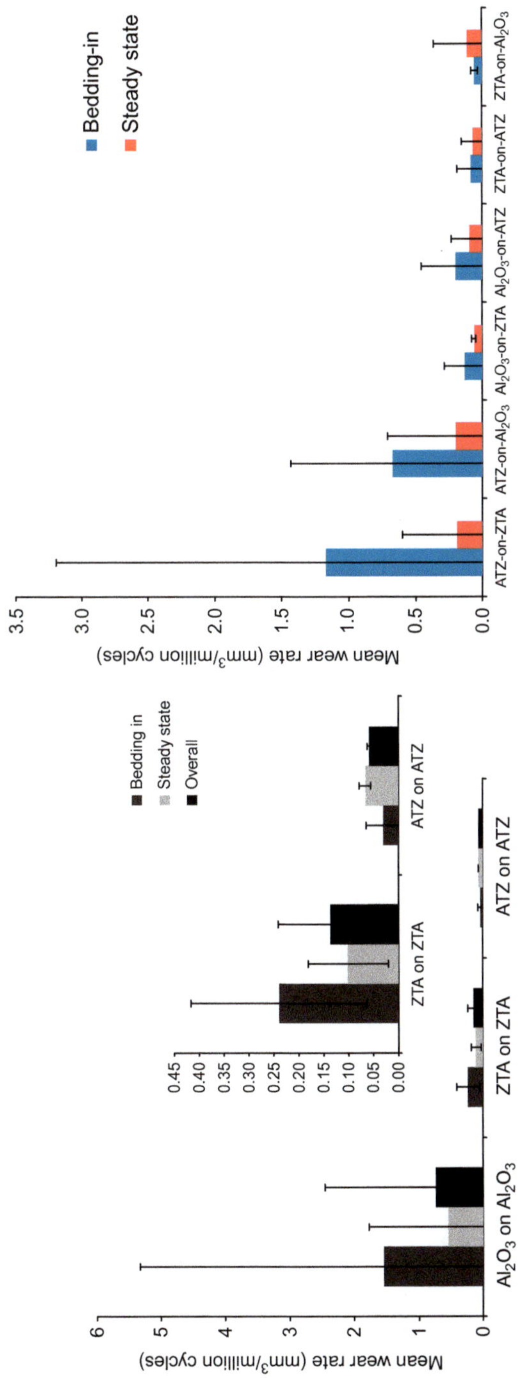

Fig. 9.7 Wear rates of different combination of ceramic composite hip joints.
Source: https://onlinelibrary.wiley.com/doi/abs/10.1002/jbm.b.33671 and https://onlinelibrary.wiley.com/doi/abs/10.1002/jbm.b.32965.

significantly. Howard et al. [44] reported that the fracture of Biolox® Delta liners was 101 of 80,170 (0.126%) compared to 35 of 31,258 (0.112%) of Biolox Forte liners. Hamilton et al. [45] reported that the fracture rate of the perioperative liner was 1.1% and that of the postoperative liner was 1%. The fracture of the ceramic head could be due to the quality of impaction, cleanliness of the Morse taper during the impaction, and smaller diameter femoral head whereas the liner fracture could be due to the diameter and thickness of the cup, cup positioning, and cup design. Though the fracture toughness was significantly enhanced by the introduction of the fourth-generation ceramic prosthesis, the insertion maneuvers and component design still decide the fracture rate. Also, the surgeons should scrutinize the specific technical maneuvers particularly for the impaction of a head component or liner (Morse taper) [43].

9.6 Squeaking—a noise or concern

Another major concern of the C-o-C hip joints is the incidence of postoperative squeaking noise. The postoperative audible squeaking of a hip replacement is a long-term concern, and it was observed as early as the 1950s in Judet acrylic hemiarthroplasty. The squeaking of Boutin C-o-C hip joints was first noted by Charnley when he tested them in his pendulum friction comparator [46]. Squeaking in THR is defined as an audible sound that emanates from hip movement. Sounds such as grinding, clicking or grating that are reproducible or which can be heard by other people are the characteristic sound of squeaking. In a study recording noise from some different THR bearing types, a variety of sounds including pops, snaps and thuds, knocking, crunching, grating, cracking, and squeaking were also reported [47].

The occurrence of squeaking is multifactorial in origin. Patient factors, surgical factors, and implant factors are responsible for squeaking. For instance, the patients who experience squeaking are on average younger (56 compared to 65 years), taller (179 cm compared to 169 cm), and have higher body mass index (90 kg compared to 76 kg) than patients whose C-o-C hips do not squeak [33]. However, a retrospective metaanalysis by Stanat and Capozzi demonstrated that squeaking was solely based on body mass index (BMI); patients with a higher BMI were at higher risk for squeaking while no significant relevance was found between squeaking and patient's age, gender, height, weight, or procedural laterality [48].

Another factor that involves squeaking is surgical procedures involving implant position and orientation. Walter et al. [33] demonstrate that high or low anteversion and inclination of the acetabular component were associated with squeaking. In addition to patient and surgical factors, implant design and materials used are also a contributing factor for squeaking.

Several studies reported the squeaking phenomenon of C-o-C bearings. However, the studies on squeaking in the ceramic composite bearing are scarce. Hamilton et al. [45] and Lombardi et al. [49] observed no squeaking phenomenon in ceramic composite bearings (Biolox® Delta).

One study conducted by Lee et al. [50] reported a total of 32 patients who had experienced postoperative noise in the hip. Out of 32 patients (34 hips), 20 patients experienced grinding noise, seven patient had squeaking, snapping was seen in three patients, clicking in two, cracking in one hip and popping in one hip. Of the 32 patients, 29 reported that they experienced noise when rising from a squatting position or from sitting on the floor. Of the 25 patients who had grinding or squeaking, 24 had it when rising from squatting or sitting on the floor. Six patients (seven hips) reported occasional squeaking and were asked to make the noise at the out-patient clinic. However, the squeaking was not audible. Thus, no patient had reproducible squeaking.

9.7 Clinical performance

The advent of new materials with improved properties is always desirable. However, one major shortcoming of new materials is their clinical follow-up. Clinical studies are of great importance to ensure that the new materials are safe for the patient. As in the case of ceramic composite bearing, the laboratory results are favorable in terms of wear and fracture resistance compared to first-generation alumina bearings. However, only a handful of clinical studies is available for ceramic composite bearings. Some of the clinical studies are discussed below to give a general overview of the performance of ceramic composite implants.

Hamilton et al. [45] reported the early outcomes and complication of a prospective, randomized, multicenter trial of 263 patients (264 hips) at eight centers, comparing a Delta C-o-C articulation with a Delta ceramic head-cross-linked polyethylene-bearing combination (C-o-P). There were 177 C-o-C hips and 87 C-o-P hips. The survivorship estimate at 3.2 years' follow-up was 97.6% in the C-o-C group and 97.7% in the C-o-P group. There were six revisions, four of 177 (2.3%) in the C-o-C group and two of 87 (2.3%) in the C-o-P group. There were no radiographic failures, cup migration, or change in inclination during the first 24 months. All cups were classified as stable, and no osteolysis was observed in either of the two groups. Follow-up of the C-o-C bearings at 3.2 years showed 1.1% insertional (operative) liner fracture and 1.1% postoperative liner fracture. Improper seating of the ceramic liner within the cup is the primary reason for the liner fracture. Further, the phenomenon of squeaking has not been reported for both C-o-C and C-o-P groups.

In another study, Lombardi et al. [49] compared the performance of 65 ZTA-on-alumina (Biolox® Delta-on-Biolox Forte) articulations inset in polyethylene (sandwich type) with 45 zirconia-on-polyethylene C-o-P bearings. The average follow-up was 6.1 years, and the femoral head diameters ranged from 28 to 32 mm. In this single center, single-surgeon study, the survivorship of the ZTA group was 95%. No cases of osteolysis were observed in the ZTA-on-alumina group, while there was one instance for the zirconia-on-polyethylene group. There was also no squeaking reported for the C-o-C Delta components.

Lee et al. [50] conducted the mid-term results of Biolox® Delta. A total of 274 patients were enrolled in the study. Only 252 patients (286 hips) were followed for a mean of 66.5 months (60–84). There were 144 men (166 hips) and 108 women (120 hips) with an average age of 49.7 years (16–83) at surgery. The mean preoperative Harris Hip Score of 47.1 points improved to 93.8 points at final follow-up. Six patients reported squeaking in seven hips; however, none were audible. Radiolucent lines involving Gruen zones one and seven were seen in 52 hips (18.2%). No hip had detectable wear, focal osteolysis, or signs of loosening. One hip was revised because of fracture of the ceramic liner, which occurred due to an undetected malseating of the ceramic liner at the time of surgery. One hip was revised for a periprosthetic fracture of the femur, and one hip was treated for periprosthetic joint infection. The 6-year survivorship with re-operation for any reason as the endpoint was 99.0% (95% confidence interval 97.8%–100%).

9.8 Conclusions

The incorporation of zirconia in alumina or vice versa significantly reduces the fracture rate, which ultimately overcomes the major setback of the monolithic ceramic prosthesis. The advent of modern processing techniques such as HIPing helps to achieve closer theoretical density value and the addition of platelets and stabilizers significantly enhances the mechanical properties. The wear rate of the ceramic composite hip prosthesis is extremely low compared to M-o-P and M-o-M implants. Even under the adverse micro-separation simulator condition, the wear rate was found to be low for the ceramic composite implants compared to the monolithic ceramic implants. Overall, the ceramic composite hip implant is found to be a promising alternative to M-o-P, M-o-M, and monolithic C-o-C implants.

9.9 Future aspects

The major limitation for the ceramic composite implants is the short-term clinical follow-up. As ceramic implants are suggested for a younger patient who can live for >30 years, a systematic long-term clinical follow-up is required to ensure the future application of this potential ceramic composite implants. The long-term clinical follow-up decides whether the ceramic composite hip implants outlast the monophasic ceramic hip prosthesis in the future.

References

[1] C. Piconi, G. Maccauro, F. Muratori, E.B.D.L. Prever, Alumina and zirconia ceramics in joint replacements, J. Appl. Biomater. Biomech. 1 (2003) 19–32.
[2] S.C. Scholes, T.J. Joyce, Ceramic-On-Ceramic Joints: A Suitable Alternative Material Combination? in: Advances in Biomaterials Science and Biomedical Applications, Rosario Pignatello, IntechOpen, 2013, https://doi.org/10.5772/53237.
[3] A. Buford, T. Goswami, Review of wear mechanisms in hip implants: paper I – general, Mater. Des. 25 (2004) 385–393, https://doi.org/10.1016/j.matdes.2003.11.010.
[4] Y. Abu-Amer, I. Darwech, J.C. Clohisy, Aseptic loosening of total joint replacements: mechanisms underlying osteolysis and potential therapies, Arthritis Res. Ther. 9 (Suppl. 1) (2007) S6, https://doi.org/10.1186/ar2170.

[5] S. Suñer, C.L. Bladen, N. Gowland, J.L. Tipper, Investigation of wear and wear particles from a UHMWPE/multi-walled carbon nanotube nanocomposite for total joint replacements, Wear 317 (2014) 163–169.

[6] M. Geetha, A.K. Singh, R. Asokamani, A.K. Gogia, Ti based biomaterials, the ultimate choice for orthopaedic implants – a review, Prog. Mater. Sci. 54 (3) (2009) 397–425, https://doi.org/10.1016/j.pmatsci.2008.06.004.

[7] T. Hanawa, Metal ion release from metal implants, Mater. Sci. Eng. C 24 (6–8) (2004) 745–752 (special issue). https://doi.org/10.1016/j.msec.2004.08.018.

[8] P. Campbell, E. Ebramzadeh, S. Nelson, K. Takamura, K. De Smet, H.C. Amstutz, Histological features of pseudotumor-like tissues from metal-on-metal hips, Clin. Orthop. Relat. Res. 468 (9) (2010) 2321–2327, https://doi.org/10.1007/s11999-010-1372-y.

[9] S. Mischler, A.I. Muñoz, Wear of CoCrMo alloys used in metal-on-metal hip joints: a tribocorrosion appraisal, Wear 297 (2013) 1081–1094, https://doi.org/10.1016/j.wear.2012.11.061.

[10] S. Kurtz, Projections of primary and revision hip and knee arthroplasty in the United States from 2005 to 2030, J. Bone Joint Surg. (American) 89 (4) (2007) 780, https://doi.org/10.2106/JBJS.F.00222.

[11] S.M. Kurtz, E. Lau, K. Ong, K. Zhao, M. Kelly, K.J. Bozic, Future young patient demand for primary and revision joint replacement: national projections from 2010 to 2030, Clin. Orthop. Relat. Res. 467 (10) (2009) 2606–2612, https://doi.org/10.1007/s11999-009-0834-6.

[12] J. Chevalier, L. Gremillard, Ceramics for medical applications: a picture for the next 20 years, J. Eur. Ceram. Soc. 29 (2009) 1245–1255, https://doi.org/10.1016/j.jeurceramsoc.2008.08.025.

[13] A. Perrichon, B.H. Liu, J. Chevalier, L. Gremillard, B. Reynard, F. Farizon, J.-D. Liao, J. Geringer, Ageing, shocks and wear mechanisms in ZTA and the long-term performance of hip joint materials, Materials 10 (2017) 569, https://doi.org/10.1016/10.3390/ma10060569.

[14] S. Affatato, P. Taddei, Wear phenomena of ceramic joints, in: Wear of Orthopaedic Implants and Artificial Joints, Woodhead Publishing Series in Biomaterials, 2013, pp. 278–297, Chapter 10.

[15] J. Chevalier, What future for zirconia as a biomaterial? Biomaterials 27 (2006) 535–543, https://doi.org/10.1016/j.biomaterials.2005.07.034.

[16] S.M. Kurtz, S. Kocagöz, C. Arnholt, R. Huet, M. Ueno, Advances in zirconia toughened alumina biomaterials for total joint replacement, NIH Public Access. 31 (2015) 107–116, https://doi.org/10.1016/j.jmbbm.2013.03.022.Advances.

[17] S.M. Kurtz, S. Kocago, C. Arnholt, R. Huet, M. Ueno, W.L. Walter, Advances in zirconia toughened alumina biomaterials for total joint replacement, J. Mech. Behav. Biomed. Mater. 31 (2014) 107–116.

[18] J. Wang, R. Stevens, Zirconia-toughened alumina (ZTA) ceramics—Review, J. Mat. Sci. 24 (1989) 3421–3440.

[19] H. Schubert, F. Frey, Stability of Y-TZP during hydrothermal treatment: neutron experiments and stability considerations, J. Eur. Ceramic Soc. 25 (2005) 1597–1602, https://doi.org/10.1016/j.jeurceramsoc.2004.03.025.

[20] F. Benazzo, F. Falez, M. Dietrich, in: Ceramics in orthopaedics—bioceramics and alternative bearings in joint arthroplasty, 11th BIOLOX® Symposium, Rome, 2006.

[21] BIOLOX® delta, Scientific information and performance data, CeramTec Medical Products.

[22] G. Pezzotti, M.C. Munisso, A.A. Porporati, K. Lessnau, On the role of oxygen vacancies and lattice strain in the tetragonal to monoclinic transformation in alumina/zirconia composites and improved environmental stability, Biomaterials 31 (27) (2010) 6901–6908, https://doi.org/10.1016/j.biomaterials.2010.05.035.

[23] M.H. Bocanegra-bernal, C. Domínguez-rios, A. Garcia-reyes, A. Aguilar-elguezabal, Hot isostatic pressing (HIP) of a -Al_2O_3 submicron ceramics pressureless sintered at different temperatures: improvement in mechanical properties for use in total hip arthroplasty (THA), Int. J. Refract. Metals Hard Mater. 27 (2009) 900–906, https://doi.org/10.1016/j.ijrmhm.2009.05.004.
[24] C.D.J. Lutgard, M.N. Rahaman, Sintering of ceramics, in: Handbook of Advanced Ceramics, Elsevier Inc., Waltham, MA, USA, 2003.
[25] R.H.R. Castro, Overview of conventional sintering, in: R. Castro, K. van Benthem (Eds.), Sintering, Engineering Materials, Vol. 35, Springer, Berlin, Heidelberg, 2012, https://doi.org/10.1007/978-3-642-31009-6.
[26] F.F. Lange, M.M. Hirlinger, Hindrance of grain growth in $A1_2O_3$ by ZrO_2 inclusions, J. Am. Ceram. Soc. 67 (2006) 164–168.
[27] H.T. Larker, R. Larker, Hot isostatic pressing, Mater. Sci. Technol. (2006), https://doi.org/10.1002/9783527603978.mst0201.
[28] M.M. Mak, A.A. Besong, Z.M. Jin, J. Fisher, Effect of microseparation on contact mechanics in ceramic-on-ceramic hip joint replacements, Proceedings of the Institution of Mechanical Engineers, Part H: Journal of Engineering in Medicine 216 (2002) 403, https://doi.org/10.1243/095441102321032193.
[29] M. Roba, Interaction of synovial fluid components with artificial hip-joint materials (Dissertation), Politecnico di Milano, ETH Zurich, 2009, Diss. ETH No. 18273.
[30] V. Thomas, B.A. Halloran, N. Ambalavanan, S.A. Catledge, Y.K. Vohra, In vitro studies on the effect of particle size on macrophage responses to nanodiamond wear debris, Acta Biomater. 8 (5) (2012) 1939–1947, https://doi.org/10.1016/j.actbio.2012.01.033.
[31] S. Williams, J.L. Tipper, E. Ingham, M.H. Stone, J. Fisher, In vitro analysis of the wear, wear debris and biological activity of surface-engineered coatings for use in metal-on-metal total hip replacements, Proc. Inst. Mech. Eng. H 217 (2003) 155–163.
[32] L. Ma, W.M. Rainforth, The effect of lubrication on the friction and wear of Biolox Ò delta, Acta Biomater. 8 (2012) 2348–2359, https://doi.org/10.1016/j.actbio.2011.12.037.
[33] W.L. Walter, P.J. Lusty, A. Watson, G. O"Toole, M.A. Tuke, B. Zicat, W.K. Walter, Stripe wear and squeaking in ceramic total hip bearings, Semin. Arthroplasty (2006) 13–15, https://doi.org/10.1053/j.sart.2006.09.015.
[34] W.L. Walter, G.M. Insley, W.K. Walter, M.A. Tuke, Edge loading in third generation alumina ceramic-on-ceramic bearings: stripe wear, J. Arthroplasty 19 (4) (2004) 402–413, https://doi.org/10.1016/j.arth.2003.09.018.
[35] M. Al-hajjar, I.J. Leslie, J. Tipper, S. Williams, J. Fisher, L.M. Jennings, Effect of cup inclination angle during microseparation and rim loading on the wear of BIOLOX V delta ceramic-on-ceramic total hip replacement, J. Biomed. Mater. Res. B Appl. Biomater. (2010) 263–268, https://doi.org/10.1002/jbm.b.31708.
[36] S. Affatato, F. Traina, A. Toni, Microseparation and stripe wear in alumina-on-alumina hip implants, Int. J. Artif. Organs 34 (2011) 506–512, https://doi.org/10.5301/IJAO.2011.8457.
[37] J. Nevelos, E. Ingham, C. Doyle, et al., Microseparation of the centers of alumina-alumina artificial hip joints during simulator testing produces clinically relevant wear rates and patterns, J. Arthroplasty 15 (6) (2000) 793–795, https://doi.org/10.1054/arth.2000.8100.
[38] A.V. Lombardi, T.H. Mallory, D.A. Dennis, R.D. Komistek, R.A. Fada, E.J. Northcut, An in vivo determination of total hip arthroplasty pistoning during activity, J. Arthroplasty 15 (6) (2000) 702–709, https://doi.org/10.1054/arth.2000.6637.

[39] T.D. Stewart, J.L. Tipper, G. Insley, R.M. Streicher, E. Ingham, J. Fisher, Long-term wear of ceramic matrix composite materials for hip prostheses under severe swing phase microseparation, J. Biomed. Mater. Res. B Appl. Biomater. 66 (2002) 567–573.

[40] M. Al-hajjar, L.M. Jennings, S. Begand, T. Oberbach, D. Delfosse, J. Fisher, Wear of novel ceramic-on-ceramic bearings under adverse and clinically relevant hip simulator conditions, J. Biomed. Mater. Res. B Appl. Biomater. (2013) 1456–1462, https://doi.org/10.1002/jbm.b.32965.

[41] M. Al-hajjar, S. Carbone, L.M. Jennings, et al., Wear of composite ceramics in mixed-material combinations in total hip replacement under adverse edge loading conditions, J. Biomed. Mater. Res. B Appl. Biomater. (2016) 1–8, https://doi.org/10.1002/jbm.b.33671.

[42] D.D. Green, P.A. Williams, T.K. Donaldson, I.C. Clarke, Biolox-forte vs. biolox-delta under micro separation test mode in the USA, 2004.

[43] P. Massin, R. Lopes, B. Masson, D. Mainard, French Hip & Knee Society, Does Biolox® delta ceramic reduce the rate of component fractures in total hip replacement? Orthop. Traumatol. Surg. Res. 100 (2014) 317–321.

[44] D.P. Howard, P.D.H. Wall, M.A. Fernandez, H. Parsons, P.W. Howard, Ceramic-on-ceramic bearing fractures in total hip arthroplasty, Bone Joint J. 99B (8) (2017) 1012–1019, https://doi.org/10.1302/0301620X.99B8.BJJ20170019.R1.

[45] W.G. Hamilton, J.P. Mcauley, D.A. Dennis, J.A. Murphy, T.J. Blumenfeld, J. Politi, THA with delta ceramic on ceramic results of a multicenter investigational device exemption trial, Clin. Orthop. Relat. Res. (2010) 358–366, https://doi.org/10.1007/s11999-009-1091-4.

[46] W.L. Walter, G.C.O. Toole, F. Tr, W.K. Walter, A. Ellis, B.A. Zicat, Squeaking in ceramic-on-ceramic hips the importance of acetabular component orientation, J. Arthroplasty 22 (4) (2007) 496–503, https://doi.org/10.1016/j.arth.2006.06.018.

[47] D.H. Owen, N.C. Russell, P.N. Smith, W.L. Walter, An estimation of the incidence of squeaking and revision surgery for squeaking in ceramic-on-ceramic total hip replacement, Bone Joint J. 96 (2) (2014) 181–187, https://doi.org/10.1302/0301-620X.96B2.32784.

[48] S.J.C. Stanat, J.D. Capozzi, Squeaking in third- and fourth-generation ceramic-on-ceramic total hip arthroplasty meta-analysis and systematic review, J. Arthroplasty 27 (3) (2012) 445–453, https://doi.org/10.1016/j.arth.2011.04.031.

[49] A.V. Lombardi, K.R. Berend, B.E. Seng, I.C. Clarke, J.B. Adams, Delta ceramic-on-alumina ceramic articulation in primary THA prospective, randomized FDA-IDE study and retrieval analysis, Clin. Orthop. Relat. Res. (2010) 367–374, https://doi.org/10.1007/s11999-009-1143-9.

[50] Y.K. Lee, Y.C. Ha, J. Yoo, K. Kim, K.H. Koo, Mid-term results of the BIOLOX delta ceramic-on-ceramic total hip arthroplasty, Bone Joint J. 99 (6) (2017) 12–14, https://doi.org/10.1302/0301-620X.99B6.BJJ-2016-0486.R3.

Nanocomposites in total hip joint replacements

10

Mohsin T. Mohammed
Nanotechnology and Advanced Materials Research Unit, Faculty of Engineering, University of Kufa, Najaf, Iraq

10.1 Introduction

Today, bone and joint inflammatory are the main clinical troubles of millions of people worldwide. Surgical intervention usually occurs to heal these diseases and in some severe cases, total joint replacements are necessary. It is well known that hip and knee joints are the most common classes of joint replacements. Total hip replacement (THR), or total hip arthroplasty (THA), is judged to be one of the most successful and frequently performed surgical operations worldwide, since they can offer an efficient solution for chronic pain and diseases such as osteoarthritis, rheumatoid arthritis, hip fractures, or tumors [1]. It was declared that THR is the operation of the century [2], since more than one million operations are carried out every year worldwide with prediction to be double within the coming years [3]. However, due to increasing elderly and active population in our world with concerning growth of health problems, it becomes urgent need to increase the longevity of the implants. It was reported that brilliant technical findings of THR survival can be gotten with considerable advantages for pain, mobility, and physical function [4–6]. On the other hand, THR presents an exceptional effectiveness/cost ratio by developing the functional state and quality of life of the patients in a reliable manner [7].

As is well known, biomaterials must have specific structures and properties depending on the type of the tissue to be replaced and its functions. These materials have to offer excellent wear resistance, mechanical strength and impact, as well as low friction in human body environment [8,9]. In particular, the main issue is to develop biomaterials for hip joint replacements with long lifetimes for younger and more active patients. Over the past decades, different kinds of biomaterials were introduced for replacing, permanently or temporarily, damaged hard tissues. Several various material combinations applied in design of femoral head and liner of hip joints involve metal on polymer (MoP), ceramic on polymer (CoP), metal on metal (MoM), and ceramic on ceramic (CoC). However, all these developed materials have advantages with some drawbacks for the patient using prosthesis. For example, high wear degradation occurs in CoC bearings due to their high sensitivity to cup position and edge loading. Moreover, the high contact stresses settled in very thin liners upon neck impingement may cause subsequent failure to component made of polymer [10]. Consequently, characteristics such as high toughness, wear and corrosion resistances, biocompatibility, and low Young's modulus are very necessary for THR.

Applications of Nanocomposite Materials in Orthopedics. https://doi.org/10.1016/B978-0-12-813740-6.00012-0

Among other applications, THR has been extensively investigated and analyzed in order to improve the properties and performance of the hip replacement materials and in turn reduce any drawbacks may be occurred. Therefore, serious attempts have been devoted through the past time to develop the technology of hip materials, and still these endeavors are continuous. In recent decades, total joint replacements have relatively long success periods, almost for 10–15 years. This comes from the progress in manufacturing techniques, materials, and implant designs [11]. The new design of materials provides several potential material combinations and large area of choices. It is well known that many kinds of biocompatible materials are employed for THR. Hence, numerous generations of metals, ceramics, highly cross-linked polyethylenes (HXLPEs), and nanocomposites (NCs) have presented for designing the artificial hip joints.

Indeed, there is an urgent need to develop innovative materials that can expand the life of the artificial joints more and more, at least more than 20 years. Nowadays, fabrication of composite materials is a possible way and promising technique for improving considerably various required properties such as wear, mechanical, and thermal properties using appropriate fillers [12]. Composite materials are defined as "naturally or engineered occurring materials which contain two or more distinct constituents with significantly different chemical, physical and mechanical properties" [13]. These materials are either micro-composites or NCs depending on the size of filler to be used. Hence, additional success with great properties is made using micro-composites [12] and NCs [14–16]. The production of nanostructural materials with exceptional properties in atomic scale has become a promising interdisciplinary field involving solid-state physics, chemistry, biology, and materials science [17]. The rapid innovation in nanotechnology permitted progress of superior NCs for various structural engineering applications especially for functional biomedical field. The research of functional NCs has become the research of present decade, since NCs with superior properties allow various novel applications.

Different unique physicochemical properties can be achieved once the size of low-dimensional materials approaches the nanometer scale. These properties differ significantly from those of an isolated atom and bulk materials. Nanoscale composites present high option of coupling different multifunctional properties such as high specific strength, biocompatibility, in addition to enhanced corrosion and wear resistances. However, the type of the reinforcing materials or fillers is vital and complicated factor in the manufacturing of the NCs as it affects directly on the considerations of the mechanical strength and the biocompatibility.

The aim of this chapter is to review the main types of NCs, with particular focus on the specific research area of THR. The chapter presents NCs based on a number of choices for their components of various matrixes and reinforcements. The main required properties of NCs used in THR such as mechanical and wear performances are also discussed. Some aspects regarding the fabrication techniques of NCs and their main parameters are also reported. Introducing this chapter will answer the question of whether these NCs could be an alternative to the current other micro materials used in total joint replacements.

10.2 Biomaterials and their essential characteristics

Biomaterials have been elaborately defined by several subject experts. The National Institutes of Health Consensus Development Conference, United States defines a biomaterial as any substance (other than drugs) or combination of substances, synthetic or natural in origin, which can be used for any period of time, as a whole or as a part of a system which treats, augments, or replaces any tissue, organ, or function of the body [18]. Recently, Guelcher and Hollinger [19] assumed that a biomaterial is a material which matches well with living tissue and demonstrates appropriate structural and mechanical stability when put to use. The field of biomedical materials is one of the fastest growing area of research for the contemporary materials scientist and engineers, as these materials can enhance the fineness and longevity of human life, and ameliorate patient health care. As the average age and life expectancy of human beings have continuously increased since 1950s, the demand for implantable materials has also increased since then. For instance, according to one report, over 500,000 THR surgeries are performed worldwide each year [20].

The development of materials for different biomedical applications is currently an area of active research in the globe and many serious attempts are achieved every year to improve different stringent requirements in this field. Over the past years, several implants with new biomaterials were developed to replace permanently damaged tissues. Depending on the replaced tissue and its function, biomaterials must keep particular characteristics. Hence, the selection of biomaterial to be employed as a load-bearing orthopedic device should be based on a reliable analysis of relevant materials properties. Nowadays, requirements of biomaterials are well understood and the credit for this goes to the efforts of doctors, biologists, and engineers. In short, the most important requirements for any material to be successful as implant in human body may be outlined as follows:

1. excellent biocompatibility;
2. better osseointegration (bone ingrowth);
3. improved mechanical properties such as specific strength, fatigue resistance, impact strength, ductility, and Young's modulus;
4. superior corrosion behavior in human body fluid;
5. outstanding tribological characteristics such as low friction and high wear resistance;
6. long-term dimensional stability;
7. ease of processing (casting, plastic deformation, powder metallurgy, machinability, and welding); and
8. availability (low price).

A number of comprehensive review articles and books extensively dealing with various aspects related to biomaterials are available [10,21–25]. Recently, great researches for design alternative new materials with high engineering characteristics have increased.

10.3 Tribological characteristics, the main issue for joint implant materials

Tribological performance such as wear resistance and coefficient of friction (COF) is an important factor for the biomaterials especially in load-bearing applications. Even though there are great developments in hip implants, wear of the articulating surfaces continues as one of the critical problems influencing the performance of THR [26]. It directly addresses the safety and efficacy for the patients. In general, the tribology of the bearing surfaces in hip implants plays a great role in verifying the performance of biomaterials since it is associated with the biocompatibility and the clinical success of the implant.

Wear has been defined as "the damage of a solid surface involving progressive loss of material when in contact with another surface in relative motion" [27]. The relative motion between the components of the joint bearing can cause wear. Hence, wear is considered as a significant issue since the longevity of joint replacements is related strongly to the wear of their components. Wear is still a serious problem which can lead to the definitive failure of the joint implant. In other words, wear is the main reason of implant's failure, causing severe pain to patients and emergency surgery may be required to replace or remove the implants [28]. Moreover, wear debris from hip implants may transfer to the liver, spleen, and lymph nodes where they cause clinical problems with inflammatory responses and are critical to identify [29]. The significance of wear issue becomes clear when the interested knows that only in the United States, there are more than 250,000 knee and hip implant surgeries each year and the failure of the artificial joints is always associated with the wear of the material [28]. Enormous biological concerns have constructed from the release of large numbers of wear debris and metal ions due to the effect of wear. The evaluation and prediction of the wear rate of the hip implants before their use in THR are of the main issues which should be considered. Consequently, reduction the wear rate of medical components is a major field in biotribology research of artificial joints.

Regardless of their articulating well against an ultrahigh molecular weight polyethylene (UHMWPE) acetabular cup function in most applications of THR, metallic or ceramic femoral heads suffer from high wear rate of the soft cup [30]. The wear debris from UHMWPE acetabular or tibial liner, as a normal result of wear effect, can cause cytotoxicity and hypersensitivity [31], and then decrease the functional life of a joint prosthetic [32]. As a result of excessive wear, particles distributing into the human body, aseptic loosening and then osteolysis with eventual failure of the prosthesis, may be induced by the biological response to wear debris [30,33,34]. The reduction in wear is critical to hip implants where small volume of wear debris can be harmful to patients and to implant performance. Therefore, a wear resistant implantable material with very less wear particles that resorb in vivo can essentially expand the lifetime of an implant. Furthermore, several factors can be selected as effected parameters on the tribological performance of bearings materials in hip replacements such as chemical composition, microstructural features, and the conditions of fabrication technique used. Therefore, understanding of the tribological characteristics of biomaterials by verifying the conditions to attain the best tribological behavior are becoming an imperative issue for such applications [35].

10.4 Morphology and importance of hip joint replacements

The hip implant as a kind of diarthrodial joint is one of the main joints in the human body. It is well known as a ball-in-socket joint since it consists of a ball and socket [36]. The top of the thighbone is a largest bone of human body, called femur joins with the horizontal pelvic coxal bone and lower end of that is fixed at the knee. The ball (femoral head) at the top of the thighbone fits into a portion of the pelvic bone forms the cup (acetabulum) or socket [37]. Accordingly, the articulation between the coxal bone and the femur identifies the hip joint [36]. A smooth glassy substance called cartilage lies between joining surfaces of the acetabulum and femoral. It provides frictionless cushion for constrained motion to femoral head within acetabular socket [37].

Up to date, hip implant involves primarily of three parts: stem, head, and socket, replacing the femoral stem and head, and pelvic acetabulum, respectively [38]. Many leading tasks may be achieved by hip implant such as sustaining the weight, stabilizing the body in static and dynamic conditions, and protecting the reproductive and digestive systems. Permitting an excessive range of motion is a major property of hip which made it the second most flexible joint after shoulder joint [25]. There are three directions of movement: flexion/extension, abduction/adduction, and intra-extra rotation can be attained as a result of the anatomical structure of a hip joint [36]. This optimum result comes from the configuration of ball-and-socket synovial joint where the roughly spherical femoral head is mostly contained within the acetabulum [25].

Over the last years, numerous numbers of patients worldwide, may be able to reach millions, suffer from end-stage joint disease. Unfortunately, the number of THRs and their own revisions have been found to increase with time. Hence, orthopedic application research has been focused on the development of total joint replacement (hard on soft and hard on hard hip bearings) and on the potential materials used for this biomedical field. The hip implants can be considered as the main accomplishment in orthopedic surgery since they have supported many of people with joints damaged by disease or trauma. In other words, THR has the ability to enhance the quality of patient's life and restore the function of his damaged natural joint and in turn significantly reduce the pain.

The number of patients suffering primary THR and revision surgery is anticipated to increase more and more because of several reasons like an aging population, decreasing average age at the first operation, and the limited lifetime of implants. It is reported that hip joints are most frequently replacements and the total number of them is anticipated to increase to about 272,000 by the year 2030 in the United States only [39], indicating that there is still a great need for advanced synthetic hip implants [40]. At the present time, stoical life analysis refers to increasing of a younger, more active, heavier, and longer-lived patient population which placing more and more necessities to develop hip joint bearings with high structures and characteristic. This will lead absolutely to enlarge the elongation of implant survivorship. In particular, THR surgery has permitted those patients with damaged joints to live independently without pain. Today, hip and knee, the main joints surgical implants, suffer degradation after 10–15 years of service [41]. This needs a successful, safe, and cost-effective medical intervention with biocompatible materials for reinstating the function of the hip joint and recovering pain-free mobility in patients suffering from severe joint disease or trauma. Therefore, the implantable materials must be designed for long-term applications with lifetime of more than 30 years [42]. However, the lifespan of artificial joints is limited and it is not unusual for young patients to outlive a total joint implant. The empirical outcomes of total hip implants surgery have discovered many clinical problems occurred such as wear debris and localized aseptic loosening of the prosthesis, which are associated with osteolysis [43–46].

Here, selecting an appropriate material with enhanced properties became a vital requisite to increase the service life of joint implants. The artificial hip joints are innovative, high-quality biomedical engineering products. Great attempts from the scientists and researchers are focused every year to develop new materials with high performance for hip prostheses and reduce the potential of complication and revision surgery. This has reflected a fast advance in the field of THR with a large number of hip implants on the market.

10.5 Implantable material systems for THR

The field of THR is a commonplace procedure and one of the most successful biomedical area in modern times, consequently, continuous developments with notable results have been attained [47]. It passed with long historical years since 1890s, and several serious researching attempts were made to improve the clinical performance of artificial total hip joints (THJs) with unique biomaterial's design changes in its structure and properties. The historical developments of THR are summarized in Refs. [36,37]. The manufacturing of artificial hip implants has suffered through numerous early stages of failure which afterward enhances extremely with applying fitting biomaterials.

Different biomaterials have been employed in modern THR [48]. Nowadays, the stem portions of most hip implants are made of orthinox-, cobalt (Co)/chromium (Cr)-, or titanium (Ti)-based alloys. Co-Cr-based alloys or ceramic materials are used in making the ball portions, which are polished smooth to allow easy rotation within the acetabular socket. This can be made of metal, UHMWPE, or a combination of polyethylene (PE) backed by metal. The metal-on-UHMWPE bearing surfaces are still accounted as the largest proportion of all THR parts. Ceramic-on-UHMWPE joints and MoM are employed with less quantities [38]. Moreover, ceramic-on ceramic solutions are also recently used for artificial hip joints [30]. Some details of main bearing systems for THR may be outlined as follows.

10.5.1 Metal-on-polymer

Over the past years, MoP bearing combination material, such as Co-Cr alloy on UHMWPE, has been widely used as a bearing couple in a variety of total joint replacements [49]. Co-Cr/UHMWPE is a standard bearing pair with approximately lifetime from 15 to 20 years [50]. However, UHMWPE wear debris may cause an inflammatory response, leading to bone resorption and finally aseptic loosening of the implant with completed failure [33,51]. Moreover, the metal ions released can also direct to the toxicity [44]. Therefore, alternative bearing combinations such as MoM, CoC, and ceramic-on-PE have been developed in order to decrease wear debris [52,53].

10.5.2 Metal on metal

Over the past 30 years, MoM hip replacements have been used as alternative bearing combination with great interest [54]. It is well known that MoM demonstrates higher wear resistance comparing with MoP bearings [26]. Owing to their suitable biocompatibility with high mechanical and wear performances, different kinds of metallic alloys have been effectively used for joint replacements especially stainless steel (SS), titanium (Ti)-, and cobalt (Co)-based alloys [55]. SS was applied as first biomaterial in THR by Philip Wiles in 1938. Afterward, Co-Cr alloy was chosen by Austin Moore in the 1950s. The poor corrosion, fatigue, and wear resistances, in addition to their toxicity, are the main drawbacks of SS. This expelled SS from the high applications as permanent implants. Consequently, SS is infrequently used in permanent implant devices. However, with MoM combination, long-term effects of the released metallic ion impurities are the main biological concerns [56]. Furthermore, metallic implants are associated with negative effects, including unpredicted pain, hypersensitive response, soft tissue necrosis, and pseudotumor [57–60]. Hence, reducing the amount of wear debris in human body and avoiding loosening of the implant are the significant functions of the most recent generation of MoM hip implants [61,62]. These critical aims can be achieved using Co-based alloys since they offer advanced wear resistance comparing with other metallic biomaterials [63]. In addition, these alloys have superior mechanical properties especially fatigue strength and hardness. The addition of the alloy element like Cr to the composition of Co alloys also enhances the corrosion behavior due to the formation of an adherent and protective oxide film on the surface [64].

For several decades, Co-Cr-based alloys are usually used as MoM bearing systems for hip implants [65,66]. The most common type of Co-based alloys used in orthopedic implants as femoral and acetabular components is ASTM F-75 (Co-Cr-Mo) alloy system [67]. However, these bearings materials are suffering from wear of the components used over 30 years [68]. In particular, there is a vital aim to increase the lifespan of MoM hip joints to more than 30 years for younger and more active patients. Therefore, understanding the tribological behavior of these materials with identifying its optimal conditions becomes a very essential issue for arthroplasty application [35]. Many significant concerns are associated with improving the wear behavior of the materials such as chemical composition, manufacturing process conditions, and microstructural constituents that achieve optimum tribological performance.

Some researchers investigated the tribological response of Co-Cr alloys modified with different boron additions for hip bearings. A considerably modification in the microstructure of alloys, followed by increasing the secondary hard phase content/higher precipitation of interdendritic compounds, has been seen as a result of increasing boron additions. Furthermore, wear resistance and hardness were also improved with increasing boron content without any inflammatory response at any boron content [62,69].

10.5.3 Ceramic on ceramic

As is well known, CoC hip implants have higher performance over polymer, metal, and polymer-metal hip implants [70,71]. The articulation of this materials system demonstrates very low wear rates of the order of 1 μm per year, compared to 100 μm per year for a standard ceramic-on-UHMWPE bearing [72], even lower than ceramic versus HXLPE [71]. This is a foremost benefit of CoC implant bearings. Moreover, ceramic hip implants have excellent strength, scratch resistance, wettability, and biocompatibility [70,71,73]. However, despite these exceptional characteristics, some dissatisfied clinical results still occur which may lead lastly to instability, loosening of the implants, and abrasive wear debris [71,73,74]. CoC bearings undergo crucial wear degradation as a result of their high sensitivity to cup position and edge loading [10]. This may be attributed to incorrect surgical techniques or prosthetic designs [75]. Furthermore, it was reported that these materials induce mechanical problems in the physiological fluid such as noise and squeaking [76], or may occur catastrophic failure via rapid crack propagation due to mechanical loading during service [77,78]. Therefore, developing of biocompatible ceramic matrix composites with enhanced fracture toughness and wear resistance becomes a dynamic concern.

10.6 Nanotechnology, the innovative approach

Nanotechnology has been defined as the progress of materials, devices, and systems revealing considerably modified and new properties using specific control at the atomic, molecular, and supramolecular levels [79]. Nanomaterial is the material that has grain size under ~100 nm, providing different characteristics such as mechanical and physical comparing with the micro-grained material of the same composition [80]. Over the last 15 years, materials with nanostructures have progressively attained interest in the world and motivated scientific research on sometimes fancy thoughts for future applications [81]. Different methods and techniques have been used to develop biomaterials with nanostructure such as sever plastic deformation [82,83], nanocrystallization surface modification [84,85], and by fabrication of NC materials [78,86–88]. Nanostructural materials, such as NCs, are considered for several essential applications in orthopedic implants industry. The incorporation of a high mechanical properties nanofiller with various matrix materials presents a better choice for high-performance composites.

10.7 Nanocomposites

In general, a NC can be defined as the material which consists of two or more materials, in which at least one of the materials is of a nanometer-scale [40]. In deep definition, Fajardo et al. [89] have defined a NC as "a multiphase solid material in which one of the phases has one, two, or three dimensions less than 100 nm." Depending on the matrix or continuous phase, NCs are typically characterized into three main categories; ceramic-, metallic-, and polymer-based materials. Moreover, higher technology of carbon nanotube (CNT) and graphene nanoparticles is a new approach for constructing NCs. The properties of the nanofillers as particles are depending directly on their size. Nanofiller materials require high biocompatibility and also strong interfacial bonding between filler and matrix in order to achieve maximum performance. However, hard and rigid fillers have two main roles, firstly they share in the compressive load distribution of the polymer matrix without strong interfacial bonding, and secondly they create nucleation sites to modify the crystallization process [12].

As is well known, the human bone, as a hard tissue, has special mechanical properties which make it strong and stiff with high modules of elasticity. The natural mineral of bone is a hierarchical NC of biological origin as it consists of flexible collagen fibers reinforced by nano-sized blade-like crystals of biological hydroxyapetite [90–92]. The dry bone weight contains almost 70% of high modulus hydroxyapetite in order to support the stiffness of the bone. However, bone can be affected by different types of applied stresses such as tension, compression, and torsion during activity routines. With high loading conditions, these stresses can lead to weaken and then break the bone. Therefore, NC material, where advantages of different material types can be combined, is a better choice for artificial implants due to its good mechanical performance and excellent biocompatibility.

10.8 Types of NCs used in hip implants

Some details of main types of NCs can be outlined as follows.

10.8.1 Polymer matrix NC

Polymers with outstanding characteristics such as lightweight and lower cost are widely used in different industries. Polymers of natural or synthetic origin with lower modulus are popular materials as alternative materials for orthopedic application. Due to the high strength and low modulus, composite materials with polymeric matrix are considered as the best choice. In general, they are variable in diverse compositions, properties, and performances. PE is the most common nondegradable polymer material used for orthopedic applications. It has various modifications like low-density polyethylene (LDPE), high-density polyethylene (HDPE), and UHMWPE [93,94]. Micron-sized fillers have been employed for many years for increasing mechanical performance of polymeric matrix. In contrast, using of nanofillers with particles size below 100 nm is complex approach due to size effects. The resultant properties of the composite material correlate with the volume of filler content in the matrix. The incorporation of nanofillers into a polymer matrix could present a NC with superior performances due to their high physical, mechanical, and others properties compared to the original matrix. Here, further aspects must be considered in the manufacturing of the polymer matrix composites [95]:

- particle shape, agglomeration, and size distribution,
- particle-specific surface area and related surface chemistry,
- particle-polymer matrix interface and interaction, and
- compounding method and related shear forces.

This type of NC is a classical category, where agglomerated nanoparticles are dispersed in a polymeric matrix. The addition of inorganic nanoparticles to polymeric matrix will certainly modify its physical and mechanical properties in addition to achieve new features in polymer matrix. Today, polymer matrix NCs have found a large area of research, mainly on nanophase ceramic reinforcements, owing to their ability to mimic the natural bone.

10.8.1.1 Ultrahigh molecular weight polyethylene

Over 60 years, UHMWPE is one of the most important polymer materials used as the acetabular cup liner in orthopedic hip implant, aiming to inhibit the metal-metal abrasion between femoral head and acetabular cup and reduce the amount of cytotoxic metallic debris. UHMWPE was considered as material of choice for total joint arthroplasty owing to its interesting clinical properties like biocompatibility, excellent toughness, good chemical and impact resistance, low friction, and high wear resistance [96–99]. However, in longer-term in vivo, conventional PE displayed low strength and drawback of generating large amounts of wear debris leading to the failure of the liner [100]. Moreover, UHMWPE suffers from oxidative degradation which may cause failure of joint implants in arthroplasty [101]. Currently, the life probability for an effectively implanted hip joint may be estimated to be approximately 15 years for a standard articulation of conventional PE liner [10]. Extensive attempts, by clinicians and scientists in the field of orthopedic, were devoted for reducing the osteolysis and aseptic loosening due to insufficient wear resistance [102–104]. Consequently, it is very important to develop the tribological and mechanical properties of UHMWPE, which can be attained by different materials and technologies. For instance, HXLPE has been developed through gamma irradiation and thermal or antioxidant stabilization [105], to enhance especially wear performance [106] with compromised mechanical characteristics [12]. The HXLPE stabilized with antioxidant vitamin-E (α-tocopherol) is nowadays recognized as one of the new standard materials in THA [102,107–110]. However, degradation of the polymer chains with resultant loss of wear and mechanical performances can be obtained through a cyclic process owing to the reaction between free radicals generated during cross-linking process and oxygen [94,111]. As a result, UHMWPE with outstanding properties has been developed recently by second nanophase reinforcements.

10.8.1.2 UHMWPE-based composites

Developed researches with new materials such as nano-reinforced UHMWPE composites have been detected for replacing the artificial joint implants [14]. UHMWPE is a polymeric material that has been most widely investigated as a matrix for different nanofillers composites in total joint replacements. Currently, UHMWPE components have limited thickness in their application for total joint arthroplasty because they may be fractured due to higher stresses generated. Thus, great interest in UHMWPE composites was made to increase the mechanical strength, with maintaining its other outstanding characteristics like biocompatibility, lubricity, and wear resistance [88]. The advanced scientific works have suggested UHMWPE-based NCs using different reinforcing materials as an alternative to HXLPE. There are many factors that may directly affect on the required properties of resultant NC like the fabrication method used and its parameters as well as the type, shape, and distribution of the reinforcements phase. However, this suggests that the structure of the UHMWPE has some defects in the microstructure, which certainly will lead to loss in the mechanical properties of the composite, regardless of the presence of the fillers and their exceptional mechanical properties. Enhancement of the viscosity of the UHMWPE is an important factor that influences on the dispersion of the reinforced filler in the matrix.

10.8.1.3 Advanced NCs using graphene and nanocarbon reinforcements

The development of innovative materials and techniques has shown great interest in order to extend the life of the implant with higher wear resistance, oxidation resistance, etc. In particular, state-of-the-art nanoparticles techniques such as graphene and CNTs are recently taken an extensive attention for developing the final performance of the biomaterials [112,113]. The carbon nanostructures have been widely investigated, owing to their high aspect dimensional ratio compared to other nanoparticles and their carbonaceous nature, which provides them exceptional properties. These materials suggest new alternatives for joint applications with more focus on the material's fabrication techniques to achieve homogeneous NCs [114]. Nanocarbon materials have been widely used as reinforcements of UHMWPE, the most common polymer used for bearing systems. A comprehensive review covering the evaluation of CNTs and graphene as reinforcements for UHMWPE-based composites in arthroplastic applications was published by Puértolas and Kurtz [88].

Many issues must be clearly determined and analyzed such as reinforcing mechanisms involved, as well as size, volume, morphology, and distribution of the particle debris and their influence on the required properties. However, more and more investigations are requisite to cover most biological issues related to these new materials in joint implant components [115].

An evaluation of the biocompatibility and human's health risk of graphene and CNTs reinforcements in biomedical NCs is very necessary to deal and analyze the overall concerns related to this field. However, an aspect associated with the cytotoxicity of carbon materials remains unexplained [116]. Indeed, the toxicity is a complicated action as many factors affect the interaction between the cells and the carbon materials such as shape, size, distribution, surface defects, and the chemical composition of the reinforcements. It should be considered that NCs reinforced with graphene or CNT have different biological responses compared to their constituents. This biological response is normal result of shielding the biological interactions due to the encapsulation of the reinforcements by the matrix [88].

Graphene/UHMWPE NCs

Graphene has been widely employed as promising material in the field of biotechnology and biomedicine [117]. The exceptional mechanical properties of graphene such as elastic modulus and tensile strength [118,119], in addition to its better biocompatibility, have motivated the specialist to use this material as a reinforcement for developing novel NCs appropriate to different structural applications. The use of graphene oxide (GO) has been proposed for reinforcing the polymer matrix since this material can introduce outstanding mechanical characteristics with high surface area [120,121]. The bidimensional GO particles, offer potentially superior benefits for reinforcing UHMWPE, owing to their improved actions in dispersion and lubrication, causing larger interface with more sufficient load-transfer sites between the matrix and reinforcement. Different techniques have been used to produce these UHMWPE composites with graphene fillers such as liquid-phase ultrasonication [122] or ball milling dispersion [123].

In general, using graphene as a filler in UHMWPE is an essential way to improve significantly the mechanical properties of UHMWPE matrix which depend strongly on the filler content. The filler content should be selected to award optimum values of the mechanical properties as these values increase at certain amounts of fillers and then decrease continuously with increasing filler content. Ren et al. [124] investigated the mechanical properties of UHMWPE reinforced with up to 2 wt% of graphene nanosheet (GNS). The authors observed this general result of the mechanical behavior as a substantial enhancement in mechanical properties is obtained at lower concentrations. The best concentrations for tensile modulus and strength appeared at 0.5 and 0.25 wt%, respectively. However, some studies showed unsuccessful mechanical performance, especially ductility and toughness, of UHMWPE after reinforcing it with various concentrations of graphene-based nanofiller [122,125].

The net mechanical performance of graphene/UHMWPE NCs is ultimately a compromise outcome between the excellent properties of the graphene fillers and the segregated structure formed around the particles of UHMWPE. The agglomeration of the graphene particles is the main structural defect which leads inevitably to decrease the interfacial bonding with the polymer matrix. As a result, microstructural flaws such as micro-cracks might appear, causing loss of toughness in these NCs. Many techniques have been adopted to improve the dispersion of the graphene fillers in the polymer matrix. For instance, Stuerzel et al. [126] used polymerization filling technique to create novel graphene/UHMWPE NCs with high mechanical performances. The optimum results of mechanical properties, that is strength, stiffness, and elongation, were observed at graphene content of 1 wt%. According to Ref. [120], great mechanical developments have been achieved, almost 54% and 71% in fracture toughness and tensile strength, respectively, using 0.1 wt% graphene nanoparticles reinforcement in UHMWPE.

As is well known, higher degree of tribological performance in frictional environment is most necessary for any bearing material such as UHMWPE. It was pointed out that graphene may offer a great promise and good lubrication in bearing applications [127,128]. This is due to the weak van der Waal force between the two-dimensional (2D) layers, resulting easy interlayer sliding in multilayer graphene and reduction in COF [129]. Furthermore, the high hardness of GO/UHMWPE would decrease the influence of the plastic deformation on the bearing surfaces against metal or ceramic counterparts. The great reductions in COF and wear loss are the major results of tribological performance when graphene reinforcement used for polymer matrix NCs [129–133].

Tai et al. [134] have found out an increase in the wear resistance using up to 1 wt% of nanofiller content in GO/UHMWPE NCs under reciprocating friction tests. The best reduction of wear rate (close to 50%) is obtained when 3 wt% GO nanosheet added to the matrix but with increased COF. Lahiri et al. [129] reinforced UHMWPE with graphene nanoplatelet (GNP) in different contents. The results showed that this type of NC presents lowering of the COF with increasing the content of GNP from 0.1 to 1 wt% owing to easy shear of GNP. Moreover, the authors pointed out that combined effect of lubrication and toughening can be presented by GNPs reinforcement due to the additional enhancement of the wear resistance by more than four times. Numerous latest reviews regarding graphene-based nanomaterials in biotechnology and biomedicine have been introduced [88,117,135–137]. Recently, Suner et al. [115] successfully prepared UHMWPE/GO NCs with various GO wt% amounts, aiming to investigate thermal stability, oxidative resistance, mechanical properties, and wettability of this biomaterial for joint replacement. The authors pointed out that all above requirements were improved as a result of GO additions. In addition, UHMWPE/GO NCs prepared with up to 0.5 wt% GO own better characteristics compared to virgin UHMWPE and NCs prepared with higher GO amounts. On the other hand, better biocompatibility is anticipated for GO/UHMWPE NCs, promoting the corporation of GO in orthopedic applications [122]. However, Lahiri et al. [120] pointed out that the presence of 1 wt% GNPs in UHMWPE matrix results in inferior biocompatibility with an obvious deterioration in the survival rate of the osteoblasts comparing with neat UHMWPE. Nowadays, ever more investigations on the biocompatibility of GO/UHMWPE NCs come to be an imperative concerns to understand and analyze the cytotoxicity and the bioactive response of these materials and their relations with essential tribological and mechanical properties. In short, the above results and discussion reveal clearly that graphene is an attractive reinforcement of UHMWPE for orthopedic hip implants that require essential developments in mechanical and wear properties.

CNTs/UHMWPE NCs

The dispersion of CNTs with polymer matrix is widely employed to increase the mechanical and thermal properties of the polymer systems, while retaining the structural capabilities of the polymer matrix. The intrinsic mechanical properties of CNTs made them attractive reinforcement materials for polymeric matrixes [87]. In recent years, multiwalled carbon nanotubes (MWCNTs) have found much interest for producing NCs because of their excellent mechanical properties and ultralightweight [112,138]. For example, MWCNTs have been succeeded as a reinforcement material of UHMWPE to develop the mechanical properties [139], wear and oxidative resistance [1,140], without recording any negative effect on its biocompatibility. Therefore, MWCNTs offer the potential to enhance properties of polymer systems owing to their high aspect ratio, with diameters of 10–40 nm and lengths of 10–100 μm. The dispersion of CNTs in polymer matrix must be well uniformed in order to increase the interfacial bonding between the CNTs and polymer matrix, consequently, getting optimal results of mechanical and thermal properties [141]. MWCNTs/UHMWPE NCs might be a better choice for bearing applications.

In general, the main route for manufacturing bulk UHMWPE/CNTs NCs consists of three steps: (a) solution or dry dispersion for CNTs, (b) blending CNTs with UHMWPE powder, and (c) hot compression press of the mixed materials [142–147]. The volume fraction of CNTs and some other main factors such as length-to-diameter aspect ratio, waviness, curliness, or orientation are considered as the most relevant factors affecting the mechanical performances of CNT-composites [88]. It seems that the addition of CNTs fillers to UHMWPE matrix will improve some main mechanical properties like modulus and yield strength [144,145,148].

As is well known, polymethyl methacrylate (PMMA) is the main material for acrylic bone cement. Potentially, cement mantle failure occurs when implants are fixed with PMMA bone cement in active or overweight patients. Recently, Ormsby et al. [149] prepared NCs by incorporating 0.1 wt% MWCNT to PMMA bone cement using different processing techniques. The results showed an enhancement in mechanical properties due to well-dispersed MWCNTs within the PMMA cement with arresting/retarding crack propagation through the cement. However, reductions in mechanical properties were observed in this kind of NCs due to the agglomerations of MWCNTs within the cement microstructure, and the degree of the agglomerations depends mainly on the method used to incorporate the MWCNTs into the cement. Moreover, the authors established that the level of heat generated within exothermic polymerization reaction of the PMMA bone cement and the thermal necrosis index value of the respective NC cements were significantly minimized as a result of MWCNTs addition, which potentially reduces the hyperthermia experienced in vivo.

A novel CNTs/UHMWPE composite with segregated and double-percolated structure using hot compaction between UHMWPE and HDPE particles was made by Pang et al. [150]. The new structure remarkably improved the toughness from 21.8 kJ/m^2 in CNT/UHMWPE to 58.1 kJ/m^2 in HDPE-CNT/UHMWPE composites due to the deformation of the HDPE with CNTs, getting higher ductility to the frontiers of UHMWPE grains. According to Ref. [148], different values of mechanical properties (yield stress, fracture strain, and toughness) have significantly increased by adding 2 wt% MWCNT to UHMWPE matrix. The results showed the antioxidant effect of CNTs by limiting the degradation of the mechanical properties after long postirradiation periods. However, it is also reported in literature that the obtained ductility and toughness of NCs were lower compared to the mechanical values of UHMWPE matrix [144,145]. Furthermore, a toughness loss in CNTs/UHMWPE NC was reported regardless of an increase of wear resistance [151]. The drawbacks of these vital mechanical characteristics may be related to the formation of segregated structure with induced defects during consolidation process, which in turn weakens the interface between the particles and the PE matrix. The superior adhesion in the interface of CNTs-matrix is an imperative parameter to develop the transmission of the load from the matrix to the filler, which in turn awards an excellent performance required such as mechanical and wear [152]. The lack in adhesion comes from the effect of CNT layers around the particles which may hinder molecular diffusion between UHMWPE grains. To obtain superior interfacial bonding between the CNT and the polymeric matrix and then compensating the loss of mechanical performance, various methods have been used such as physical functionalization method using gamma irradiation [145,153,154], chemical functionalization of CNTs [148,155], and polymerization-filling [156]. Moreover, the diverse structures and morphologies of CNTs associated with the different synthesis techniques might be considered as leading parameters involved in the filler-matrix interface. The incorporation of CNTs with the matrix is an effective way to improve wear behavior. It was reported that MWCNTs played a foremost role in enhancement of the wear resistance of conventional UHMWPE [157]. The literature [151,156,158–160] showed a noticeable decrease in wear effect with respect to neat UHMWPE, depending on the technique used and the concentration of CNTs. In general, wear resistance was improved when the filler content increased. The main mechanism of this enhancement of wear life of CNTs/UHMWPE NCs was interpreted with respect to considerable increase in abrasion resistance during wear process as a result of the reinforcing of CNTs which leads to increase some mechanical properties like surface hardness, toughness, shear strength, plowing, and cutting resistance [88]. However, considerable increase [158] or slightly increase [16,151], of the COF was observed in CNTs-UHMWPE NCs comparing with neat UHMWPE. This outcome could be due to the poor interaction between matrix and CNTs [146], or as a result of formation harder film with higher shear strength and rougher molecular profile brought by SWCNT on the UHMWPE film [159,160]. Comparing with conventional UHMWPE, some researchers have investigated the wear rate and wear effects on the bioactivity and biocompatibility of the UHMWPE/MWCNT NC material. They pointed out that the addition of MWCNTs leads to a considerable decrease in wear rate. Moreover, the results showed that the wear particles from the UHMWPE/MWCNT NC had a reduced osteolysis potential with equivalent cytotoxicity compared to those produced from the conventional PE. Depending on the obtained findings, the authors suggested that UHMWPE/MWCNT NCs show better alternative for joint replacements [1].

Numerous approaches can be used to develop the oxidation resistance of UHMWPE such as postirradiation thermal treatments and antioxidant systems. However, such processes have some major drawbacks. For instance, postirradiation thermal treatments may decrease the strength and the toughness of polymer [161], and possible physical loss by volatilization, migration, and water extraction with inadequate dispersion of the antioxidant in the host matrix may also take place [162,163]. Recently, use of compounds of natural sources has taken much attention in the field of stabilizing agents due to their high biocompatibility [164,165]. Arrigo et al. [166] have introduced innovative approach to prepare UHMWPE-based NC films from hybrid-nanoparticles coming from physical immobilization of a natural antioxidant molecule quercetin (Q) onto MWCNTs, in order to improve thermo-oxidation resistance. The results showed that the quercetin is an excellent antioxidant for PE as Q-functionalized CNTs disperse better in the host matrix. Moreover, a 10-fold increase of the onset of degradation occurrence in thermo-oxidative environment was found due to a synergic effect stemming from the physical interaction between Q and CNTs, which cannot provide a similar stabilizing action if used individually. The authors suggest that the process of physical immobilization of the Q molecules causes the formation of structural defects onto outer CNTs surfaces. Thus, remarkable improvements in the CNTs radical scavenging activity have been obtained which probably promote the Q regeneration. In addition, it seems that CNTs act as efficient nano-carriers for the quercetin molecules, improving the dispersion of the latter in the host matrix in spite of their poor solubility. On the other hand, MWCNTs offer better biocompatibility with high contact with bone for joint implants and other biomedical applications [167,168]. The contribution by Rama Sreekanth et al. [169] proved that the bone bonding ability of UHMWPE does not affect by the presence of MWCNTs due to the formation of a hydroxyapatite (HA) layer. This result is similar to that appearing in neat UHMWPE, after soaking the NC in simulated body fluid (SBS) solution for several days at 37°C. In addition, the results showed that in vivo inflammation of the surrounding tissues is analogous to that of the UHMWPE with a nonirritant biological level as a result of presence of various cells around the implantation [88].

10.8.2 Metal matrix NCs

Metals and alloys have been widely used as preferred biomaterials in many load-bearing orthopedic applications. They are inserted in large bones and designed as artificial joints. This is owing to their superior properties like strength and corrosion resistance. At present, metallic materials like 316L stainless steel (316LSS), Co-Cr alloys, as well as pure magnesium and Ti and their alloys are widely used for surgical implants [22]. However, metallic implants have some drawbacks such as corrosion, wear, and release of metallic ions, which support researchers to do further research with innovative technologies, leading to sufficient solutions for improving the properties of the metallic materials. The use of metal-based NCs with different metals and alloys reinforced by ceramic particles or fibers is one of these vital solutions to enhance the performance of metallic materials in hip surgery. In general, metal matrix NCs are suitable for load-bearing orthopedic implants since they present superior properties, particularly mechanical properties such as strength, hardness, and superplasticity, compared to unreinforced matrix. Various ceramic materials such as HA, Al_2O_3, and others have been used as reinforcements in metal based NCs for total hip applications.

10.8.2.1 Co-Cr based NCs

Co-Cr-based alloys are considered as the main representative of Co alloys and the most commonly used for different biomedical fields. The occurrence of Cr element in this alloy system can enhance the corrosion behavior and excellent high-temperature properties and abrasion resistance can be imparted using small amounts of other alloy elements such as molybdenum. Moreover, these materials cannot create a chemical bond with a bone, thus they are called "bioinert materials" [55].

As is well known, metallic materials like Co-Cr-Mo alloys have a main drawback related to their using in orthopedic surgery as their modulus (240 GPa) is higher than that of bone (10–30 GPa). The elastic modulus mismatch between the biomaterials and the surrounding bone causes loosening of the implant and bone resorption due to stress shielding effect [170]. The bone will force insufficient load under stress shielding effect owing to this mechanical mismatch [171]. The better balance between high strength, to withstand physiological loads, and low stiffness, to prevent stress shielding effect, is a necessary issue [172].

Manufacturing of NCs based on Co-Cr-Mo alloy with bioactive ceramic materials as reinforcements could be a successful technique to solve this problem of mechanical mismatch between the implant and the tissue. It was reported that bioceramic materials can be the potential solution to this essential challenge [173,174].

Use of calcium phosphate materials like HA as a secondary phase is a very common in bone tissue engineering [40]. These materials are attractive materials among all bioceramics as a result of their excellent biocompatibility [39] and promoted bone formation with the adjacent hard tissue [175,176]. HA ($Ca_{10}(PO_4)_6(OH)_2$) is the basic mineral component in hard tissues [177]. Indeed, nowadays, natural and synthetically HA can be produced with better chemical and biological affinity to bone tissue.

In comparison with Co-Cr-Mo alloys, HA has lower modulus, but its mechanical properties are scarce for load-bearing applications [178]. Therefore, using two constituents of Co-Cr-Mo alloy as matrix and nano-HA as reinforcement to form metal matrix NC is an advantageous method to sufficiently improve the mechanical and biological properties. Recently, Bahrami et al. [179] fabricated Co-Cr-Mo alloy NCs with diverse nano concentrations of HA (10, 15, and 20wt%), using ball milling to mix the starting powders, cold pressing, and sintering process. An enhancement of mechanical and biological properties is the main goal of this research. The results showed an obvious increase in microhardness with a considerable reduce in modulus (about 53%–63%) when HA nanoparticles were added to the Co-base alloy matrix. Moreover, the authors proved that using HA as reinforcement, it could successfully convert bioinert Co-Cr-Mo alloy to bioactive type as a result of formation nucleus of apatite on the surface of the NCs after 1, 14, and 28 days.

10.8.2.2 Titanium-based NCs

Titanium (Ti) and its alloys are widely used as an important class of engineering structural materials finding wide range of applications in biomedical area due to their low density, high mechanical strength at low temperatures, superior corrosion resistance, and excellent biocompatibility [180,181]. In short, Ti-based alloys are biocompatible and have ideal mechanical properties [182]. However, poor wear resistance, low mechanical strength at high temperatures, and relatively high cost are the main drawbacks of Ti which limit its use in load-bearing orthopedic applications. As the lifespan of artificial joints is significant, the problems of friction and wear in Ti implants have been studied by many authors [183,184] since these problems could negatively affect on the performance of designed devices.

In recent years, titanium-HA (Ti-HA) composites as developed biomaterials have found much research interest since they can offer excellent combination of mechanical properties of Ti and bioactivity of HA [185,186].

10.8.3 Ceramic matrix NCs

In general, crystalline and inorganic ceramics are hard and brittle materials with higher modulus compared to bone. They are signified by their high compressive strength and low tensile strength, providing high resistance to plastic deformation [187].

Ceramic materials were extensively applied in medicine because of their excellent biocompatibility and corrosion resistance, low friction and cost, as well as high wear resistance and strength [188]. Therefore, ceramics are used as alternative materials with essential characteristics to protect the hip joints from loosening.

Alumina (α-Al_2O_3), with hexagonal (corundum-type) crystallographic structure, was extensively used in fabrication of these prostheses owing to above-mentioned excellent properties. The surface of alumina has polar hydroxyl (–OH) groups which supports the contact with the aqueous body environments to deliver a lubricating layer. However, some of its drawbacks are low fracture toughness and insufficient wear resistance for longer duration in vivo, which cause large debris chips to form with a risk of infrequent and catastrophic in vivo fractures of femoral heads coupled with metallic stems [50,71,74,189]. It is believed that the existing ceramic bearings for hip joint replacements suffer from in vivo failures due to the formation of cracks propagation [32]. Thus, there is an imperative need to create new ceramic materials with higher mechanical properties compared to conventional ceramics like alumina and others. As a result, a great attention from researchers has been devoted to improve ceramic processing technology and ceramic matrix NCs for the clinical applications in THA [70,73,190]. These materials have been introduced in 1991 as novel candidate materials for joint implants [191]. As a result of introducing second nanophase, advantageous properties have been achieved using ceramic-based NCs such as strength, hardness, ductility, superplasticity, and wear resistance comparing with conventional counterparts. For example, ceria-stabilized tetragonal zirconia polycrystals (Ce-TZP) have provided very high toughness and high resistance to low-temperature aging degradation [192]. It was also reported that Ce-TZP/Al_2O_3 NC has higher toughness and equivalent strength of conventional Y-TZP [193]. Tanaka et al. [53] developed Ce-TZP/Al_2O_3 NC to be used as a bearing material in total joint prostheses. The results showed that this NC has significantly higher toughness and completed resistance to aging degradation compared to conventional ceramic materials (Y-TZP and Al_2O_3). Furthermore, to reduce the wear effect of PE, yttria-stabilized tetragonal zirconia polycrystals (Y-TZP) has been used as bearing components of total joint prosthesis in articulation with PE [194].

It was reported that alumina/zirconia composites have been developed to overcome the drawbacks of conventional ceramics [195,196]. The fracture toughness and other required characteristics of alumina are enhanced with dispersion of nanometer sized-second-phase particles of some ceramics such as TiO_2, Al_2O_3, HA, or Al_2O_3/TiO_2 into the matrix [197]. Therefore, ceramic matrix NCs can be considered as superior materials for the femoral head of total hip joint replacement due to their extraordinary fracture toughness and wear resistance.

Several studied have been made in last decade on zirconia/alumina NCs to develop their performance and properties for joint implants [53,198]. It was found out that the mechanical strength and toughness of 12Ce-TZP/Al_2O_3 NCs can be increased with adding a suitable amount of Al_2O_3 and TiO_2 [199,200].

Lee et al. [188] have suggested a new material for femoral head of hip joint replacement, Al_2O_3/TiO_2 NC with various amounts of TiO_2 (0–25 mol%). The authors pointed out that the addition of 10 mol% of TiO_2 to the matrix improves considerably the toughness, hardness, and wear resistance due to dispersion of $AlTiO_5$ and TiO_2 phases into alumina matrix. Benzaid et al. [86] studied the slow crack growth (SCG) behavior of an artificial NC involved zirconia and alumina with 10mol% of ceria (10Ce-TZP/Al_2O_3), with aim to develop its fracture toughness and strength for THR. The authors have found out that the tetragonal to monoclinic phase transformation of the ceria-stabilized zirconia (Ce-TZP) phase at the surface of ground specimens leads to surface compressive induced stresses with increasing in strength. Hence, the fracture toughness of this NC material is significantly higher than those of bioceramics (i.e., alumina and zirconia), with better cyclic fatigue threshold. Morilloa et al. [50] have fabricated alumina NCs by mixing and hot pressing different elements and ceramic nanopowders (Co, TiO_2, and Y-TZP) to Al_2O_3, in order to enhance the mechanical properties (hardness and fracture toughness), friction and wear resistance for hip prostheses under the environments of distilled water and fetal bovine serum solution (FBSS). The authors found that the addition of other elements and nanometer ceramic phases generally results in a reduction of the coefficients of friction and specific wear rate of alumina NCs, while the mechanical properties were not increased in this study. Irrespective of respectively low mechanical properties of alumina NCs, the low wear rate and the coefficients of friction may be related to the process by which the NCs did not change from mild wear to a fracture dominated.

10.8.3.1 New ceramics NCs with nanocarbon reinforcements

Many developments have been recorded over the last close years for fabrication new ceramic matrix NCs reinforced with GNPs. It was found that these composites enable to develop fracture toughness [201–205], flexural strength [205,206], mechanical damping [203], wear resistance [207–215], COF [208,210,211,213,214], and biocompatibility [216,217]. It was reported that the significant influence of GNPs on wear resistance was due to the formation of a thin GNP-rich lubricating tribofilm as a result of the exfoliation of graphene layers from GNPs [208] or owing to intrinsic lubrication provide by the GNPs distributed at the grain boundaries of the alumina matrix [213].

Nieto et al. [218] investigated the addition of GNPs as reinforcement material on the microscale tribological behavior of alumina ceramics for CoC hip implant applications. The results showed that the manufacturing of Al_2O_3-GNP NCs leads to enhance the fracture toughness and wear resistance by up to 21% and 39%, respectively. The authors attributed the improved wear resistance to GNP-induced toughening, which generates fine micro-cracks on the scratch surface. In addition, the Al_2O_3-GNP NCs demonstrated retention of biocompatibility which could develop next generation of hip implants with better lifespans and fewer catastrophic failure.

10.9 Conclusion

The number of total joint arthroplasty revisions increases annually owing to implant failure attributed to the type of material used and its function. The advance of the materials used for hip joint replacements becomes one of the most challenging issues of implant technology in the present time. Therefore, several great efforts from researchers, scientist, and engineers have been made for developing suitable biocompatible materials for this objective. As they can exhibit desirable properties, composite materials are widely used for specific biomedical applications. At present, NCs are being used quite frequently all over the world as alternative materials in hip joints due to their exceptional characteristics. The lower grain size of these materials compared to micro materials may induce higher performance in hip joint implants with various biological effects.

This chapter has provided a comprehensive overview of NCs with an equal focus on their application in total hip joint replacements. The fabrication techniques, characterizations, and performance of different kinds of NCs have been addressed in this chapter. The contemporary-based NCs represent a potential development in the field of biomaterials. Nowadays, there is still room for novel findings that may result in new NCs with lower production costs for the application of hip implants.

Acknowledgments

The author would like to sincerely thank Faculty of Engineering, University of Kufa, Najaf, Iraq for supporting his scientific works.

References

[1] S. Suner, C.L. Bladen, N. Gowland, J.L. Tipper, N. Emami, Investigation of wear and wear particles from a UHMWPE/multi-walled carbon nanotube nanocomposite for total joint replacements, Wear 317 (2014) 163–169.

[2] I.D. Learmonth, C. Young, C. Rorabeck, The operation of the century: total hip replacement, Lancet 370 (9597) (2007) 1508–1519, https://doi.org/10.1016/S0140-6736(07)60457-7.

[3] R. Pivec, A.J. Johnson, S.C. Mears, M.A. Mont, Hip arthroplasty, Lancet 380 (9855) (2012) 1768–1777.

[4] S. Kurtz, K. Ong, E. Lau, F. Mowat, M. Halpern, Projections of primary and revision hip and knee arthroplasty in the United States from 2005 to 2030, J. Bone Joint Surg. Am. 89 (2007) 780–785.

[5] M.S. Prime, J. Palmer, W.S. Khan, The National Joint Registry of England and Wales, Orthopedics 34 (2011) 107–110.

[6] G.H. Smith, S. Johnson, J.A. Ballantyne, E. Dunstan, I.J. Brenkel, Predictors of excellent early outcome after total hip arthroplasty, J. Orthop. Surg. Res. 7 (2012) 13–16, https://doi.org/10.1186/1749-799X-7-13.

[7] L.E. de Araújo, I.C. Leite, Analysis on quality of life of patients with osteoarthrosis undergoing total hip arthroplasty, Rev. Bras. Ortop. 47 (2015) 498–504.

[8] A. Chiba, K. Kumagai, N. Nomura, S. Miyakawa, Pin-on-disk wear behavior in a like on-like configuration in a biological environment of high carbon cast and low carbon forged Co–29Cr–6Mo alloys, Acta Mater. 55 (2007) 1309–1318.

[9] M. Gradzka, J. Dabrowski, B. Dabrowsky, Modification of mechanical properties of sintered implant materials on the base of Co–Cr–Mo alloy, J. Mater. Process. Technol. 204 (2008) 199–205.

[10] G. Pezzottia, K. Yamamoto, Artificial hip joints: the biomaterials challenge, J. Mech. Behav. Biomed. Mater. 31 (2014) 3–20.
[11] J.N. Katz, Total joint replacement in osteoarthritis, Best Pract. Res. Clin. Rheumatol. 20 (2006) 145–153.
[12] K. Plumlee, C.J. Schwartz, Improved wear resistance of orthopaedic UHMWPE by reinforcement with zirconium particles, Wear 267 (2009) 710–717.
[13] P.K. Rohatgi, M. Tabandeh-Khorshid, E. Omrani, M.R. Lovell, P.L. Menezes, Tribology of metal matrix composites, in: P.L. Menezes, M. Nosonovsky, S.P. Ingole, S.V. Kailas, M.R. Lovell (Eds.), Tribology for Scientists and Engineers, Springer, New York, 2013, pp. 233–268.
[14] L. Fang, Y. Leng, P. Gao, Processing and mechanical properties of HA/UHMWPE nanocomposites, Biomaterials 27 (20) (2006) 3701–3707.
[15] D.S. Xiong, J.M. Lin, D.L. Fan, Wear properties of nano-Al_2O_3/UHMWPE composites irradiated by gamma ray against a CoCrMo alloy, Biomed. Mater. 1 (2006) 175–179.
[16] Y. Xue, W. Wu, O. Jacobs, B. Schadel, Tribological behaviour of UHMWPE/HDPE blends reinforced with multi-wall carbon nanotubes, Polym. Test. 25 (2006) 221–229.
[17] H. Gleiter, Nanostructured materials: basic concepts and microstructure, Acta Mater. 48 (2000) 1–29.
[18] K. Dee, D. Puleo, R. Buzios, An Introduction to Tissue-Biomaterial Interactions, John Wiley & Sons Inc., Hoboken, 2002.
[19] S. Guelcher, J. Hollinger, An Introduction to Biomaterials, second ed., CRC Press, New York, 2011.
[20] M. Niinomi, Recent metallic materials for biomedical applications, Metall. Mater. Trans. A 33 (2002) 477–486.
[21] J. Park, R. Lakes, Biomaterials, Springer, New York, 2007.
[22] M. Geetha, A.K. Singh, R. Asokamani, A.K. Gogia, Ti based biomaterials, the ultimate choice for orthopaedic implants—a review, Prog. Mater. Sci. 54 (2009) 397–425.
[23] B.M. Holzapfel, J.C. Reichert, J.T. Schantz, U. Gbureck, L. Rackwitz, U. Noth, F. Jakob, M. Rudert, J. Groll, D.W. Hutmacher, How smart do biomaterials need to be? a translational science and clinical point of view, Adv. Drug Deliv. Rev. 65 (2013) 581–603.
[24] S. Affatato, A. Ruggiero, M. Merola, Advanced biomaterials in hip joint arthroplasty. a review on polymer and ceramics composites as alternative bearings, Compos. Part B 83 (2015) 276–283.
[25] I. Carvalho, T. Polcar, M. Henriques, S. Carvalho, Materials and incompatibility as a major cause of hip prosthesises rejection, Rev. Adv. Mater. Sci. 42 (2015) 36–49.
[26] S.R. Knight, R. Aujla, S.P. Biswas, Total hip arthroplasty—over 100 years of operative history, Orthop. Rev. 3 (2011) 72–74.
[27] K. Lozano, S. Yang, R. Jones, Nanofiber toughened polyethylene composites, Carbon 42 (2004) 2329–2331.
[28] H. Vasquez, K. Lozano, V. Soto, A. Rocha, Design of a wear tester for nano-reinforced polymer composites, Wear 41 (2008) 870–877.
[29] J.J. Jacobs, Implant wear in total joint replacement: clinical and biologic issues, in: T.M. Wright, S.B. Goodman (Eds.), Material and Design Considerations, American Academy of Orthopedic Surgeons, Rosemont, IL, 2001, pp. 31–42.
[30] W. Osterle, D. Klaffke, M. Griepentrog, U. Gross, I. Kranz, C. Knabe, Potential of wear resistant coatings on Ti–6Al–4V for artificial hip joint bearing surfaces, Wear 264 (2008) 505–517.

[31] A.H. Hosman, H.C. van der Mei, S.K. Bulstra, H.J. Busscher, D. Neut, Effects of metal-on-metal wear on the host immune system and infection in hip arthroplasty, Acta Orthop. 81 (2010) 526–534.
[32] M.N. Rahaman, A. Yao, Ceramics for prosthetic hip and knee joint replacement, J. Am. Ceram. Soc. 90 (2007) 1965–1988.
[33] E. Ingham, J. Fisher, The role of macrophages in osteolysis of total joint replacement, Biomaterials 26 (2005) 1271–1286.
[34] E.M. Branch del Prever, A. Bistolfi, P. Bracco, L. Costa, UHMWPE for arthroplasty: past or future? J. Orthop. Traumatol. 10 (2009) 1–8.
[35] J. Cawley, J. Metcalf, A. Jones, T. Band, D. Skupien, A tribological study of cobalt chromium molybdenum alloys used in metal-on-metal resurfacing hip arthroplasty, Wear 225 (2003) 999–1006.
[36] S. Affatato, Perspectives in Total Hip Arthroplasty, Advances in Biomaterials and their Tribological Interactions, Woodhead Publishing, United Kingdom, 2014.
[37] S. Pramanik, A.K. Agarwal, K.N. Rai, Chronology of total hip joint replacement and materials development, Trends Biomater. Artif. Organs 19 (2005) 15–26.
[38] C. Qizhi, A.T. George, Metallic implant biomaterials, Mater. Sci. Eng. R 87 (2015) 1–57.
[39] T.J. Webster, Nanophase ceramics as improved bone tissue engineering materials, Am. Ceram. Soc. Bull. 82 (2003) 23–28.
[40] R. Murugan, S. Ramakrishna, Development of nanocomposites for bone grafting, Compos. Sci. Technol. 65 (2005) 2385–2406.
[41] H. Liang, B. Shi, A. Fairchild, T. Cale, Applications of plasma coatings in artificial joints: an overview, Vacuum 73 (2004) 317–326.
[42] A.H. De Aza, J. Chevalier, G. Fantozzi, M. Schehl, R. Torrecillas, Crack growth resistance of alumina, zirconia and zirconia toughened alumina ceramics for joint prostheses, Biomaterials 23 (2002) 937–945.
[43] M.G. Creighton, J.J. Callaghan, J.P. Olejniczak, R.C. Johnston, J. Bone Joint Surg. Am. 80 (1998) 1439–1446.
[44] A. Sargeant, T. Goswami, Hip implants—paper VI—ion concentrations, Mater. Des. 28 (2007) 155–171.
[45] R. Murali, S.F. Bonar, G. Kirsh, W.K. Walter, W.L. Walter, Osteolysis in third-generation alumina ceramic-on-ceramic hip bearings with severe impingement and titanium metallosis, J. Arthroplasty 23 (2008) 1240.e13–1240.e19.
[46] M.S. Uddin, L.C. Zhang, Predicting the wear of hard-on-hard hip joint prostheses, Wear 301 (2013) 192–200.
[47] R.A. Brand, M.A. Mont, M.M. Manring, Biographical sketch: Themistocles Gluck (1853-1942), Clin. Orthop. 469 (2011) 1525–1527.
[48] M.G. Zywiel, S.A. Sayeed, A.J. Johnson, T.P. Schmalzried, M.A. Mont, State of the art in hard-on-hard bearings: how did we get here and what have we achieved? Expert Rev. Med. Dev. 8 (2011) 187–207.
[49] J. Charnley, Low Friction Arthroplasty of the Hip: Theory and practice, Springer, New York, 1979.
[50] C. Morilloa, Y. Sawaeb, T. Murakami, Effect of bovine serum constituents on the surface of the tribological pair alumina/alumina nanocomposites for total hip replacement, Tribol. Int. 43 (2010) 1158–1162.
[51] W.A. Jiranek, M. Machado, M.J. Jasty, D. Jevsevar, H.J. Wolfe, S.R. Goldring, M.J. Goldberg, W.J. Harris, Production of cytokines around loosened cemented acetabular components. Analysis with immunohistochemical techniques and in situ hybridization, J. Bone Joint Surg. 75A (1993) 863–879.

[52] L.P. Zichner, H.G. Willert, Comparison of alumina–polyethylene and metal–polyethylene in clinical trials, Clin. Orthop. (282) (1992) 86–94.
[53] K. Tanaka, J. Tamura, K. Kawanabe, M. Nawa, M. Oka, M. Uchida, T. Kokubo, T. Nakamura, Ce-TZP/Al_2O_3 nanocomposite as a bearing material in total joint replacement, J. Biomed. Mater. Res. 63 (2002) 262–270.
[54] J.H. Dumbleton, M.T. Manley, Metal-on-metal total hip replacement: what does the literature say? J. Arthoplasty 20 (2005) 174–188.
[55] J.R. Davis, Handbook of Materials for Medical Devices, ASM International, Materials Park, OH, 2003.
[56] J.J. Jacobs, A.K. Skipor, P.F. Doorn, Cobalt and chromium concentrations in patients with metal on metal total hip replacements, Clin. Orthop. 329 (1996) S256–S263.
[57] H.G. Willert, G.H. Buchhorn, A. Fayyazi, R. Flury, M. Windler, G. Koester, C.H. Lohmann, Metal-on-metal bearings and hyper-sensitivity in patients with artificial hip joints. A clinical and histomorphological study, J. Bone Joint Surg. 87 (2005) 28–36.
[58] J. Daniel, H. Ziaee, C. Pradhan, P.B. Pynsent, D.J. McMinn, Blood and urine metal ion levels in young and active patients after Birmingham hip resurfacing arthroplasty: four-year results of a prospective longitudinal study, J. Bone Joint Surg. 89 (2007) 169–173.
[59] M.A. Mont, T.P. Schmalzried, Modern metal-on-metal hip resurfacing: important observations from the first ten years, J. Bone Joint Surg. 90 (2008) 3–11.
[60] A.J. Hart, J.A. Skinner, P. Winship, N. Faria, E. Kulinskaya, D. Webster, S. Muirhead-Allwood, C.H. Aldam, H. Anwar, J.J. Powell, Circulating levels of cobalt and chromium from metal-on-metal hip replacement are associated with CD8+ T-cell lymphopenia, J. Bone Joint Surg. 91 (2009) 835–842.
[61] J. Yang, B. Shen, Z. Zhou, F. Pei, P. Kang, Changes in cobalt and chromium levels after metal-on-metal hip resurfacing in young, active Chinese patients, J. Arthroplasty 26 (2011) 65–70.
[62] M. Alvarez-Vera, J.A. Ortega-Saenz, M.A.L. Hernandez-Rodríguez, A study of the wear performance in a hip simulator of a metal–metal Co–Cr alloy with different boron additions, Wear 301 (2013) 175–181.
[63] H.P. Sieber, C.B. Rieker, P. Köttig, Analysis of 118 second generation metal-on-metal retrieved hip implants, J. Bone Joint Surg. 81 (1999) 46–50.
[64] B. Patel, F. Inam, M. Reece, M. Edirisinghe, W. Bonfield, J. Huang, A. Angadji, A novel route for processing cobalt–chromium–molybdenum orthopaedic alloys, J. R. Soc. Interface 7 (2010) 1641–1645.
[65] I. Clarke, T. Donaldson, J. Bowsher, S. Nasser, T. Takahashi, Current concepts of metal-on-metal hip resurfacing, Orthop. Clin. N. Am. 36 (2005) 143–162.
[66] P. Roberts, P. Gregoris, H. Bosch, N. Talwaker, Resurfacing arthroplasty of the hip, Curr. Orthop. 19 (2005) 263–279.
[67] A. Unsworth, Artificial joints, in: D. Dowson, V. Wright (Eds.), An Introduction to the Bio-mechanics of Joints and Joint Replacement, Mechanical Engineering Publications, London, 1981, pp. 134–139.
[68] I. Catelas, P. Cambell, J. Bobyn, J. Mendley, O. Huk, Wear particles from metal on-metal total hip replacements: effects of implant design and implantation time, J. Eng. Med. 220 (2006) 195–208.
[69] M. Alvarez-Vera, A. Juarez-Hernandez, C.E. Gonzalez-Rivera, R.D. Mercado-Solis, M.A.L. Hernandez-Rodriguez, Biotribological response of Co–Cr alloy with added boron under ball-on-disc tests, Wear 301 (2013) 243–249.

[70] R. Huet, A. Sakona, S.M. Kurtz, Strength and reliability of alumina ceramic femoral heads: review of design, testing, and retrieval analysis, J. Mech. Behav. Biomed. Mater. 4 (2011) 476–483.

[71] D. Hannouche, A. Zaoui, F. Zadegan, L. Sedel, R. Nizard, Thirty years of experience with alumina-on-alumina bearings in total hip arthroplasty, Int. Orthop. 35 (2011) 207–213.

[72] G. Willmann, Oxide ceramics for articulating components of total hip replacements, in: L. Sedel, C. Rey (Eds.), Bioceramics, Elsevier, Oxford, 1997, pp. 123–126.

[73] S.M.M. Tai, L. Parker, N.J. de Roeck, J.A. Skinner, Recurrent catastrophic ceramic femoral head failure in total hip arthroplasty, Case Rep. Orthop. (2014) https://doi.org/10.1155/2014/837954.

[74] F. Traina, M.D. Fine, A.D. Martino, C. Faldini, Fracture of ceramic bearing surfaces following total hip replacement: a systematic review. Biomed. Res. Int. (2013) https://doi.org/10.1155/2013/157247.

[75] L. Savarino, N. Baldini, G. Ciapetti, A. Pellacani, A. Giunti, Is wear debris responsible for failure in alumina-on-alumina implants? Acta Orthop. Scand. 80 (2009) 162–167.

[76] W.L. Walter, G.C. O'Toole, W.K. Walter, A. Ellis, B.A. Zicat, Squeaking in ceramic-on-ceramic hips: the importance of acetabular component orientation, J. Arthroplasty 22 (2007) 496–503.

[77] T.S. Huang, M.N. Rahaman, B.S. Bal, Alumina–tantalum composite for femoral head applications in total hip arthroplasty, Mater. Sci. Eng. C 29 (2009) 1935–1941.

[78] M.N. Rahaman, T. Huang, A. Yao, B.S. Bal, Y. Li, SiC nanoparticle-reinforced Al_2O_3–Nb composite as a potential femoral head material in total hip arthroplasty, Mater. Sci. Eng. C 30 (2010) 1197–1203.

[79] M.C. Roco, Nanotechnology: convergence with modern biology and medicine, Curr. Opin. Biotechnol. 14 (2003) 337–346.

[80] S.V. Dorozhkin, Calcium orthophosphate-based biocomposites and hybrid biomaterials, J. Mater. Sci. 44 (2009) 2343–2387.

[81] B. Bhushan (Ed.), Springer Handbook of Nanotechnology, Springer, Berlin, 2004.

[82] S. Fintova, L. Kunz, Fatigue properties of magnesium alloy AZ91 processed by severe plastic deformation, J. Mech. Behav. Biomed. Mater. 42 (2015) 219–228.

[83] A. Panigrahia, B. Sulkowskia, T. Waitza, K. Ozaltinc, W. Chrominskic, A. Pukenasd, J. Horkya, M. Lewandowskac, W. Skrotzkid, M. Zehetbauer, Mechanical properties, structural and texture evolution of biocompatible Ti–45Nb alloy processed by severe plastic deformation, J. Mech. Behav. Biomed. Mater. 62 (2016) 93–105.

[84] Y. Li, K. Sun, P. Liu, Y. Liu, P. Chui, Surface nanocrystallization induced by fast multiple rotation rolling on Ti-6Al-4V and its effect on microstructure and properties, Vacuum 101 (2014) 102–106.

[85] J. Lei, C. Wen-fang, S. Xiu, L. Gang, Z. Lian, Effects of surface nanocrystallization on corrosion resistance of β-type titanium alloy, Trans. Nonferrous Met. Soc. China 24 (2014) 2529–2535.

[86] R. Benzaid, J. Chevalier, M. Saâdaoui, G. Fantozzi, M. Nawa, L.A. Diaz, R. Torrecillas, Fracture toughness, strength and slow crack growth in a ceria stabilized zirconia–alumina nanocomposite for medical applications, Biomaterials 29 (2008) 3636–3641.

[87] J.R. Potts, D.R. Dreyer, C.W. Bielawski, R.S. Ruoff, Graphene-based polymer nanocomposites, Polymer 52 (2011) 5–25.

[88] J.A. Puértolas, S.M. Kurtz, Evaluation of carbon nanotubes and graphene as reinforcements for UHMWPE-based composites in arthroplastic applications: a review, J. Mech. Behav. Biomed. Mater. 39 (2014) 129–145.

[89] A.R. Fajardo, A.G.B. Pereira, E.C. Muniz, Hydrogels nanocomposites based on crystals, whiskers and fibrils derived from biopolymers, Adv. Struct. Mater. 74 (2015) 43–71.
[90] S. Weiner, H.D. Wagner, The material bone: structure-mechanical function relations, Annu. Rev. Mater. Sci. 28 (1998) 271–298.
[91] S. Ramakrishna, J. Mayer, Biomedical applications of polymer-composite materials: a review, Compos. Sci. Technol. 61 (2001) 1189–1224.
[92] M.J. Olszta, X.G. Cheng, S.S. Jee, B.R. Kumar, Y.Y. Kim, M.J. Kaufman, E.P. Douglas, L.B. Gower, Bone structure and formation: a new perspective, Mater. Sci. Eng. R 58 (2007) 77–116.
[93] S.M. Kurtz, O.K. Muratoglu, M. Evans, A.A. Edidin, Advances in the processing, sterilization, and crosslinking of ultra-high molecular weight polyethylene for total joint arthroplasty, Biomaterials 20 (1999) 1659–1688.
[94] H. McKellop, F.W. Shen, B. Lu, P. Campbell, R. Salovey, Development of an extremely wear-resistant ultra high molecular weight polyethylene for total hip replacements, J. Orthop. Res. 17 (1999) 157–167.
[95] T. Hanemann, D.V. Szabó, Polymer-nanoparticle composites: from synthesis to modern applications, Materials 3 (2010) 3468–3517.
[96] J. Berzen, Standardization of UHMW-PE, in: H.G. Willert, G.H. Buchorn, P. Eyerer (Eds.), Ultra-High Molecular Weight Polyethylene as Biomaterial in Orthopaedic Surgery, Hogrefeand Huber Publishers, Toronto, 1991, pp. 50–54.
[97] G. Lewis, Properties of crosslinked ultra-high-molecular-weight polyethylene, Biomaterials 22 (2001) 371–401.
[98] S. Kurtz, The UHMWPE Handbook, Principles and Clinical Applications in Total Joint Replacement, Elsevier Academic Press, San Diego, 2004.
[99] H. Ding, Y. Tian, L. Wang, B.J. Liu, Preparation of ultrahigh-molecular-weight polyethylene membranes via a thermally induced phase-separation method, J. Appl. Polym. Sci. 105 (2007) 3355–3362.
[100] J. Katta, Z. Jin, E. Ingham, J. Fisher, Biotribology of articular cartilage e a review of the recent advances, Med. Eng. Phys. 30 (2008) 1349–1363.
[101] T. Ogawa, H. Mukai, S. Osawa, Effects of functional groups and surface roughness on interfacial shear strength in ultrahigh molecular weight polyethylene fiber/polyethylene system, J. Appl. Polym. Sci. 71 (1999) 243–249.
[102] E. Oral, S.D. Christensen, A.S. Malhi, K.K. Wannomae, O.K. Muratoglu, Wear resistance and mechanical properties of highly cross-linked, ultrahigh-molecular weight polyethylene doped with vitamin-E, J. Arthroplasty 21 (2006) 580–591.
[103] E. García-Rey, E. García-Cimbrelo, A. Cruz-Pardos, J. OrtegaChamarro, New polyethylenes in total hip replacement: a prospective, comparative clinical study of two types of liner, J. Bone Joint Surg. (Br.) 90 ((2008) 149–153.
[104] M.P. Laurent, T.S. Johnson, R.D. Crowninshield, Characterization of a highly cross-linked ultrahigh molecular weight polyethylene in total hip arthroplasty, J. Arthroplasty 23 (2008) 751–761.
[105] S.M. Kurtz, UHMWPE, Biomaterials Handbook, Academic Press, Burlington/San Diego/London, 2009.
[106] J.L. Tipper, A.L. Galvin, E. Ingham, J. Fisher, Estimation of the osteolytic potential of noncrosslinked polyethylenes and ceramic-on-ceramic total hip prostheses, J. ASTM Int. 6 (2006) 1–16.
[107] ASTM F 2695-12, Standard Specification for Ultra-High Molecular Weight Polyethylene Powder Blended With Alphatocopherol (Vitamin E) and Fabricated Forms for Surgical Implant Applications, ASTM International, West Conshohocken, PA, 2012.

[108] P. Bracco, E. Oral, Vitamin E-stabilized UHMWPE for total joint implants, Clin. Orthop. Relat. Res. 469 (2011) 2286–2293.
[109] Y. Takahashi, T. Masaoka, K. Yamamoto, T. Shishido, T. Tateiwa, K. Kubo, G. Pezzotti, Vitamin-E blended and infused highly cross-linked polyethylene for total hip arthroplasty: a comparison of three-dimensional crystalline morphology and strain recovery behavior, J. Mech. Behav. Biomed. Mater. 36 (2014) 59–70.
[110] L. Puppulin, A. Leto, M. Hasegawa, G. Pezzotti, A comparative microstructural study of vitamin-E blended and infused highly crosslinked UHMWPE for total knee arthroplasty, J. Mech. Behav. Biomed. Mater. 39 (2014) 247–256.
[111] V. Premnath, W.H. Harris, M. Jasty, E.W. Merrill, Gamma sterilization of UHMWPE articular implants: an analysis of the oxidation problem, Biomaterials 17 (1996) 1741–1753.
[112] E. Gomez Barrena, J.A. Puertolas, L. Munuera, Y.T. Konttinen, Update on UHMWPE research from the bench to the bedside, Acta Orthop. 79 (2008) 832–840.
[113] A.D. Moghadam, E. Omrani, P.L. Menezes, P.K. Rohatgi, Mechanical and tribological properties of self-lubricating metal matrix nanocomposites reinforced by carbon nanotubes (CNTs) and graphene—a review, Compos. Part B Eng. 77 (2015) 402–420.
[114] J.H. Du, B. Jinbo, H.M. Cheng, The present status and key problems of carbon nanotube based polymer composites, Express Polym. Lett. 1 (2007) 253–273.
[115] S. Suñer, R. Joffe, J.L. Tipper, N. Emami, Ultra high molecular weight polyethylene/graphene oxide nanocomposites: thermal, mechanical and wettability characterization, Compos. Part B 78 (2015) 185–191.
[116] S.K. Smart, A.I. Cassady, G.Q. Lu, D.J. Martin, The biocompatibility of carbon nanotubes, Carbon 44 (2006) 1034–1047.
[117] H.C. Zhang, G. Gruner, Y.L. Zhao, Recent advancements of graphene in biomedicine, J. Mater. Chem. B 1 (2013) 2542–2567.
[118] T. Kuilla, S. Bhadra, D. Yao, N.H. Kim, S. Bose, J.H. Lee, Recent advances in graphene based polymer composites, Prog. Polym. Sci. 35 (2010) 1350–1375.
[119] C. Soldano, A. Mahmood, E. Dujardin, Production, properties and potential of graphene, Carbon 48 (2010) 2127–2150.
[120] D. Lahiri, R. Dua, C. Zhang, I. Socarraz-Novoa, A. Bhat, S. Ramaswamy, A. Agarwal, Graphene nanoplatelet-induced strengthening of UHMWPE and biocompatibility in vitro, ACS Appl. Mater. Interfaces 4 (2012) 2234–2241.
[121] P. Marques, G. Gonçalves, M.K. Singh, J. Gracio, Graphene oxide and hydroxyapatite as fillers of polylactic acid nanocomposites: preparation and characterization, J. Nanosci. Nanotechnol. 12 (2012) 6686–6692.
[122] Y.F. Chen, Y.Y. Qi, Z.X. Tai, X.B. Yan, F.L. Zhu, Q.J. Xue, Preparation, mechanical properties and biocompatibility of graphene oxide/ultrahigh molecular weight polyethylene composites, Eur. Polym. J. 48 (2012) 1026–1033.
[123] S. Suñer, N. Emami, Investigation of graphene oxide as reinforcement for orthopaedic applications, Tribol. Mater. Surf. Interfaces 8 (2014) 1–6.
[124] P.G. Ren, Y.Y. Di, Q. Zhang, L. Li, H. Pang, Z.M. Li, Composites of ultrahigh-molecular-weight polyethylene with graphene sheets and/or MWCNTs with segregated network structure: preparation and properties, Macromol. Mater. Eng. 297 (2012) 437–443.
[125] F.J. Pascual, D. Molina-Manso, J.A. Puertolas, E. Gómez-Barrena, J. Esteban, in: Bacterial adherence on UHMWPE/MWCNT composites: searching new clinical applications for multi-wall carbon nanotubes structures, Proceedings of EORS, Amsterdam, 2012.

[126] M. Stuerzel, F. Kempe, Y. Thomann, S. Mark, M. Enders, R. Muelhaupt, Novel graphene UHMWPE nanocomposites prepared by polymerization filling using single-site catalysts supported on functionalized graphene nanosheet dispersions, Macromolecules 45 (2012) 6878–6887.
[127] J.F. Ou, J.Q. Wang, S. Liu, B. Mu, J.F. Ren, H.G. Wang, S.R. Yang, Tribology study of reduced graphene oxide sheets on silicon substrate synthesized via covalent assembly, Langmuir 26 (2010) 15830–15836.
[128] D. Berman, A. Erdemir, A.V. Sumant, Reduced wear and friction enabled by graphene layers on sliding steel surfaces in dry nitrogen, Carbon 59 (2013) 167–175.
[129] D. Lahiri, F. Hec, M. Thiesse, A. Durygind, C. Zhang, A. Agarwal, Nanotribological behavior of graphene nanoplatelet reinforced ultra high molecular weight polyethylene composites, Tribol. Int. 70 (2014) 165–169.
[130] B. Pan, J. Zhao, Y. Zhang, Y. Zhang, Wear performance and mechanisms of polyphenylene sulfide/polytetrafluoroethylene wax composite coatings reinforced by graphene, J. Macromol. Sci. 51 (2012) 1218–1227.
[131] S.S. Kandanur, M.A. Rafiee, F. Yavari, M. Schrameyer, Z.Z. Yu, T.A. Blanchet, N. Koratkar, Suppression of wear in graphene polymer composites, Carbon 50 (2012) 3178–3183.
[132] G. Ren, Z. Zhang, X. Zhu, B. Ge, F. Guo, X. Men, W. Liu, Influence of functional graphene as filler on the tribological behaviors of Nomex fabric/phenolic composite, Composites A 49 (2013) 157–164.
[133] M.M. Sokrieh, M.R. Hosseinkhani, M.R. Naimi-Jamal, H. Tourani, Nanoindentation and nanoscratch investigations on graphene-based nanocomposites, Polym. Test. 32 (2013) 45–51.
[134] Z. Tai, Y. Chen, Y. An, X. Yan, Q. Xue, Tribological behaviour of UHMWPE reinforced with graphene oxide nanosheets, Tribol. Lett. 46 (2012) 55–63.
[135] M. Yang, J. Yao, Y.X. Duan, Graphene and its derivatives for cell biotechnology, Analyst 138 (2013) 72–86.
[136] C. Chung, Y.K. Kim, D. Shin, S.R. Ryoo, B.H. Hong, D.H. Min, Biomedical applications of graphene and graphene oxide, Acc. Chem. Res. 46 (2013) 2211–2224.
[137] K. Yang, L.Z. Feng, X.Z. Shi, Z. Liu, Nano-graphene in biomedicine: theranostic applications, Chem. Soc. Rev. 42 (2013) 530–547.
[138] S.L. Ruan, P. Gao, X.G. Yang, T.X. Yu, Toughening high performance ultrahigh molecular weight polyethylene using multiwalled carbon nanotubes, Polymer 44 (2003) 5643–5654.
[139] D.I. Chukov, A.A. Stepashkin, A.V. Maksimkin, V.V. Tcherdyntsev, S.D. Kaloshkin, K.V. Kuskov, V.I. Bugakov, Investigation of structure, mechanical and tribological properties of short carbon fiber reinforced UHMWPE-matrix composites, Compos. Part B Eng. 76 (2015) 79–88.
[140] P.S. Rama Sreekanth, N. Naresh Kumar, S. Kanagaraj, Improving post irradiation stability of high density polyethylene by multi walled carbon nanotubes, Compos. Sci. Technol. 72 (2012) 390–396.
[141] R. Andrews, D. Jaqques, D. Qian, T. Rantell, Multiwall carbon nanotubes: synthesis and application, J. Acc. Chem. Res. 35 (2002) 1008–1017.
[142] A. Mierczynska, M. Mayne-L'Hermite, G. Boiteux, J.K. Jeszka, Electrical and mechanical properties of carbon nanotube/ultrahigh-molecular-weight polyethylene composites prepared by a filler prelocalization method, J. Appl. Polym. Sci. 105 (2007) 158–168.
[143] N. Campo, A.M. Visco, Incorporation of carbon nanotubes into ultra high molecular weight polyethylene by high energy ball milling, Int. J. Polym. Anal. Charact. 15 (2010) 438–449.

[144] A. Fonseca, S. Kanagaraj, M.S.A. Oliveira, J.A.O. Simoes, in: A. Ochsner, G.E. Murch, J.M.P. Delgado (Eds.), Enhanced UHMWPE reinforced with MWCNT through mechanical ball-milling, Defect Diffusion Forum, 2011, pp. 1238–1243.
[145] M.J. Martinez, P. Castell, V. Martinez-Nogues, M.T. Martinez, P.J. Alonso, J.A. Puertolas, Effects of gamma-irradiation on UHMWPE/MWNT nanocomposites, Compos. Sci. Technol. 71 (2011) 282–288.
[146] A.V. Maksimkin, S.D. Kaloshkin, M.S. Kaloshkina, M.V. Gorshenkov, V.V. Tcherdyntsev, K.S. Ergin, I.V. Shchetinin, Ultra-high molecular weight polyethylene reinforced with multi-walled carbon nanotubes: fabrication method and properties, J. Alloys Compd. 536 (2012) S538–S540.
[147] P.S. Rama Sreekanth, S. Kanagaraj, Restricting the ageing degradation of the mechanical properties of gamma irradiated UHMWPE using MWCNTs, J. Mech. Behav. Biomed. Mater. 21 (2013) 57–66.
[148] P.S. Rama Sreekanth, S. Kanagaraj, Assessment of bulk and surface properties of medical grade UHMWPE based nanocomposites using nanoindentation and microtensile testing, J. Mech. Behav. Biomed. Mater. 18 (2013) 140–151.
[149] R. Ormsby, T. McNally, C. Mitchell, N. Dunne, Incorporation of multiwalled carbon nanotubes to acrylic based bone cements: effects on mechanical and thermal properties, J. Mech. Behav. Biomed. Mater. 3 (2010) 136–145.
[150] H. Pang, D.X. Yan, Y. Bao, J.B. Chen, C. Chen, Z.M. Li, Super-tough conducting carbon nanotube/ultrahighmolecular-weight polyethylene composites with segregated and double-percolated structure, J. Mater. Chem. 22 (2012) 23568–23575.
[151] V. Martínez-Nogués, M.J. Martínez-Morlanes, F.J. Pascual, S. Kurtz, J.A. Puértolas, in: Tribological behavior of carbon fibers and multi-wall carbon nanotubes reinforced UHMWPE, Proceedings World Tribology Congress. Torino, Italy, 2013.
[152] L.S. Schadler, S.C. Giannaris, P.M. Ajayan, Load transfer in carbon nanotube epoxy composites, Appl. Phys. Lett. 73 (1998) 3842–3844.
[153] B. Safibonab, A. Reyhani, A.N. Golikand, S.Z. Mortazavi, S. Mirershadi, M. Ghoranneviss, Improving the surface properties of multi-walled carbon nanotubes after irradiation with gamma rays, Appl. Surf. Sci. 258 (2011) 766–773.
[154] Z.W. Xu, L. Chen, B.M. Zhou, Y.L. Li, B.D. Li, J.R. Niu, M.J. Shan, Q.W. Guo, Z. Wang, X.M. Qian, Nano-structure and property transformations of carbon systems under gammaray irradiation: a review, RSC Adv. 3 (2013) 10579–10597.
[155] S. Bose, R.A. Khare, P. Moldenaers, Assessing the strengths and weaknesses of various types of pre-treatments of carbon nanotubes on the properties of polymer/carbon nanotubes composites: a critical review, Polymer 51 (2010) 975–993.
[156] H.J. Park, J. Kim, Y. Seo, J. Shim, M.Y. Sung, S. Kwak, Wear behavior of in situ polymerized carbon nanotube/ultra high molecular weight polyethylene composites, Macromol. Res. 21 (2013) 965–970.
[157] B.B. Johnson, M.H. Santare, J.E. Novotny, S.G. Advani, Wear behavior of carbon nanotube/high density polyethylene composites, Mech. Mater. 41 (2009) 1108–1115.
[158] Y.S. Zoo, J.W. An, D.P. Lim, D.S. Lim, Effect of carbon nanotube addition on tribological behavior of UHMWPE, Tribol. Lett. 16 (2004) 305–309.
[159] S. Kanagaraj, M.T. Mathew, A. Fonseca, M.S.A. Oliveira, J.A.O. Simoes, L.A. Rocha, Tribological characterisation of carbon nanotubes/ultrahigh molecular weight polyethylene composites: the effect of sliding distance, Int. J. Surf. Sci. Eng. 4 (2010) 305–321.
[160] M.A. Samad, S.K. Sinha, Mechanical, thermal and tribological characterization of a UHMWPE film reinforced with carbon nanotubes coated on steel, Tribol. Int. 44 (2011) 1932–1941.

[161] S. Atwood, D.W. Van Citters, E.W. Patten, J. Furmanski, M.D. Ries, L.A. Pruitt, Tradeoffs amongst fatigue, wear, and oxidation resistance of cross-linked ultra-high molecular weight polyethylene, J. Mech. Behav. Biomed. Mater. 4 (2011) 1033–1045.

[162] N.C. Billingham, The physical chemistry of polymer oxidation and stabilization, in: G. Scott (Ed.), Atmospheric Oxidation and Antioxidants Atmospheric Oxidation and Antioxidants, Elsevier, Amsterdam, 1993, pp. 219–277.

[163] M. Lundbaeck, C. Strandberg, A. Albertsson, M.S. Hedenqvist, U.W. Gedde, Loss of stability by migration and chemical reaction of Santonox R in branched polyethylene under anaerobic and aerobic conditions, Polym. Degrad. Stab. 91 (2006) 1071–1078.

[164] V. Ambrogi, P. Cerruti, C. Carfagna, M. Malinconico, V. Marturano, M. Perrotti, P. Persico, Natural antioxidants for polypropylene stabilization, Polym. Degrad. Stab. 96 (2011) 2152–2158.

[165] D. Tatraaljai, L. Major, E. Foldes, B. Pukanszky, Study of the effect of natural antioxidants in polyethylene: performance of b-carotene, Polym. Degrad. Stab. 102 (2014) 33–40.

[166] R. Arrigo, N. Dintcheva, M. Guenzi, C. Gambarotti, G. Filippone, S. Coiai, Thermo-oxidative resistant nanocomposites containing novel hybrid-nanoparticles based on natural polyphenol and carbon nanotubes, Polym. Degrad. Stab. 115 (2015) 129–137.

[167] N. Saito, Y. Usui, K. Aoki, N. Narita, M. Shimizu, N. Ogiwara, K. Nakamura, N. Ishigaki, H. Kato, S. Taruta, M. Endo, Carbon nanotubes for biomaterials in contact with bone, Curr. Med. Chem. 15 (2008) 523–527.

[168] N. Narita, Y. Kobayashi, H. Nakamura, K. Maeda, A. Ishihara, T. Mizoguchi, Y. Usui, K. Aoki, M. Simizu, H. Kato, H. Ozawa, N. Udagawa, M. Endo, N. Takahashi, N. Saito, Multiwalled carbon nanotubes specifically inhibit osteoclast differentiation and function, Nano Lett. 9 (2009) 1406–1413.

[169] P.S. Rama Sreekanth, N. Ravindra Reddy, M. Lahkar, S. Kanagaraj, Biocompatibility studies on MWCNTs reinforced ultra high molecular weight polyethylene nanocomposites, Trends Biomater. Artif. Organs 27 (2013) 1–9.

[170] W.C. Head, D.J. Bauk, R.H. Emerson, Titanium as the material of choice for cementless femoral components in total hip arthroplasty, Clin. Orthop. Relat. Res. 311 (1995) 85–90.

[171] H. Kröger, P. Venesmaa, J. Jurvelin, H. Miettinen, O. Suomalainen, E. Alhava, Bone density at the proximal femur after total hip arthroplasty, Clin. Orthop. Relat. Res. 352 (1998) 66–74.

[172] F. Simancik, Introduction: the strange world of cellular metals, in: H.P. Degischer, B. Kriszt (Eds.), Handbook of Cellular Metals, Wiley-VCH Verlag, Weinheim, 2002, pp. 1–4.

[173] R.H. Doremus, Bioceramics, J. Mater. Sci. 27 (1992) 285–297.

[174] C. Ergun, T.J. Webster, R. Bizios, R.H. Doremus, Hydroxylapatite with substituted magnesium, zinc, cadmium, and yttrium. I. Structure and microstructure, J. Biomed. Mater. Res. 59 (2002) 305–311.

[175] B.M. Tracy, R.H. Doremus, Direct electron microscopy studies of the bone—hydroxylapatite interface, J. Biomed. Mater. Res. 18 (1984) 719–726.

[176] C. Du, F.Z. Cui, Q.L. Feng, X.D. Zhu, K. De Groot, Tissue response to nano-hydroxyapatite/collagen composite implants in marrow cavity, J. Biomed. Mater. Res. 42 (1998) 540–548.

[177] L. Peru, G. Daculsi, Synthetic calcium phosphates: models for biological crystals? Clin. Mater. 15 (1994) 267–272.

[178] K. de Groot, R. Geesink, P. Serekian, Plasma sprayed coatings of hydroxylapatite, J. Biomed. Mater. Res. 21 (1987) 1375–1381.

[179] M. Bahrami, M.H. Fathi, M. Ahmadian, The effect of nanobioceramic reinforcement on mechanical and biological properties of Co-base alloy/hydroxyapatite nanocomposite, Mater. Sci. Eng. C 48 (2015) 572–578.
[180] W.D. Brewer, R.K. Bird, T.A. Wallace, Titanium alloys and processing for high speed aircraft, Mater. Sci. Eng. A 243 (1998) 299–304.
[181] M.T. Mohammed, Z.A. Khan, A.N. Siddiquee, Beta titanium alloys: the lowest elastic modulus for biomedical applications: a review, Int. J. Chem. Nucl. Metall. Mater. Eng. 8 (2014) 726–731.
[182] S. Tada, E. Timucin, T. Kitajima, O.U. Sezerman, Y. Ito, Direct in vitro selection of titanium-binding epidermal growth factor, Biomaterials 35 (2014) 3497–3503.
[183] M. Fellah, M. Labaïz, O. Assala, L. Dekhil, A. Taleb, H. Rezzag, Tribological behavior of Ti-6Al-4Vand Ti-6Al-7Nb alloys for total hip prosthesis, Adv. Tribol. 2014 (2014) 1–13.
[184] M. Fellah, M. Labaïz, O. Assala, L. Dekhil, N. Zerniz, A. Iost, Tribological behavior of biomaterial for total hip prosthesis, Matér. Tech. 102 (2014) 601.
[185] C. Ning, Y. Zhou, Correlations between the in vitro and in vivo bioactivity of the Ti/HA composites fabricated by a powder metallurgy method, Acta Biomater. 4 (2008) 1944–1952.
[186] Q. Chang, H.Q. Ru, D.L. Chen, X.Y. Yue, L. Yu, C.P. Zhang, An in-vitro investigation of iron-containing hydroxyapatite/titanium composites, J. Mater. Sci. Technol. 27 (2011) 546–552.
[187] B.D. Ratner, Biomaterials Science: An Introduction to Materials in Medicine, Elsevier Academic Press, Amsterdam, 2004.
[188] S.W. Lee, C. Morillo, J. Lira-Olivares, S.H. Kim, T. Sekino, K. Niihara, B.J. Hockey, Tribological and microstructural analysis of Al_2O_3/TiO_2 nanocomposites to use in the femoral head of hip replacement, Wear 255 (2003) 1040–1044.
[189] R.L. Barrack, C. Burak, H.B. Skinner, Concerns about ceramics in THA, Clin. Orthop. Relat. Res. 429 (2004) 73–79.
[190] B. Masson, M. Kuntz, Total hip arthroplasty: wear behaviour of different articulations, in: K. Knahr (Ed.), Tribology in Total Hip and Knee Arthroplasty, Springer, New York, 2012, pp. 145–152.
[191] K. Niihara, New design concept of structural ceramics—ceramic nanocomposites, J. Ceram. Soc. Jpn. 99 (1991) 974–982.
[192] K. Tsukuma, Mechanical properties and thermal stability of CeO_2 containing tetragonal zirconia polycrystals, Am. Ceram. Soc. Bull. 65 (1986) 1386–1389.
[193] M. Nawa, S. Nakamoto, T. Sekino, K. Niihara, Tough and strong Ce-TZP/alumina nanocomposites doped with titania, Ceram. Int. 18 (1992) 11–17.
[194] P. Kumar, M. Oka, K. Ikeuchi, K. Shimizu, T. Yamamuro, H. Okumura, Y. Kotoura, Low wear rate of UHMWPE against zirconia ceramic (Y-PSZ) in comparison to alumina ceramic and SUS 316L alloy, J. Biomed. Mater. Res. 25 (1991) 813–828.
[195] W. Burger, H.G. Richter, High strength and toughness alumina matrix composites by transformation toughening and 'in situ' platelet reinforcement (ZPTA), in: S. Giannini, A. Moroni (Eds.), The New Generation of Bioceramics, Trans Tech, Bologna, 2000, pp. 545–548.
[196] S. Affatato, M. Goldoni, M. Testoni, A. Toni, Mixed-oxides prosthetic ceramic ball heads, part 3: effect of the ZrO_2 fraction on the wear of ceramic on ceramic hip joint prostheses, a long-term in vitro wear study, Biomaterials 22 (2001) 717–723.
[197] J. Chevalier, S. Deville, G. Fantozzi, Nanostructured ceramic oxides with a slow crack growth resistance close to covalent materials, Nano Lett. 4 (2005) 1297–1301.

[198] M. Uchida, H.M. Kim, T. Kokubo, M. Nawa, T. Asano, K. Tanaka, T. Nakamura, Apatite forming ability of a zirconia/alumina nano-composite induced by chemical treatment, J. Biomed. Mater. Res. 60 (2002) 277–282.
[199] M. Nawa, S. Nakamoto, T. Sekino, K. Niihara, Tough and strong Ce-TZP/alumina nanocomposites doped with titania, Ceram. Int. 24 (1997) 497–506.
[200] M. Nawa, N. Bamba, T. Sekino, K. Niihara, The effect of TiO_2 addition of strengthening and toughening in intragranular type of 12 Ce-TZP/Al_2O_3 nanocomposites, J. Eur. Ceram. Soc. 18 (1998) 209–219.
[201] L.S. Walker, V.R. Marotto, M.A. Rafiee, N. Koratkar, E.L. Corral, Toughening in graphene ceramic composites, ACS Nano 5 (2011) 3182–3190.
[202] G.B. Yadhukulakrishnan, S. Karumuri, A. Rahman, R.P. Singh, A.K. Kalkan, S.P. Harimkar, Spark plasma sintering of graphene reinforced zirconium diboride ultra-high temperature ceramic composites, Ceram. Int. 39 (2013) 6637–6646.
[203] A. Nieto, D. Lahiri, A. Agarwal, Graphene nanoplatelets reinforced tantalum carbide consolidated by spark plasma sintering, Mater. Sci. Eng. A 582 (2013) 338–346.
[204] J.H. Lee, J. Marroquin, K.Y. Rhee, S.J. Park, D. Hui, Cryomilling application of graphene to improve material properties of graphene/chitosan nanocomposites, Compos. Part B Eng. 45 (2013) 682–687.
[205] C. Ramirez, P. Miranzo, M. Belmonte, M.I. Osendi, P. Poza, S.M. Vega Diaz, M. Terrones, Extraordinary toughening enhancement and flexural strength in Si_3N_4 composites using graphene sheets, J. Eur. Ceram. Soc. 34 (2014) 161–169.
[206] Y. Liu, S.K. Sinha, Wear performances and wear mechanism study of bulk UHMWPE composites with nacre and CNT fillers and PFPE overcoat, Wear 300 (2013) 44–54.
[207] A. Nieto, D. Lahiri, A. Agarwal, Synthesis and properties of bulk graphene nanoplatelets consolidated by spark plasma sintering, Carbon 50 (2012) 4068–4077.
[208] M. Belmonte, C. Ramirez, J. Gonzalez-Julian, J. Schneider, P. Miranzo, M.I. Osendi, The beneficial effect of graphene nanofillers on the tribological performance of ceramics, Carbon 61 (2013) 431–435.
[209] P. Hvizdos, J. Dusza, C. Balazsi, Tribological properties of Si_3N_4–graphene nanocomposites, J. Eur. Ceram. Soc. 33 (2013) 2359–2364.
[210] H. Porwal, P. Tatarko, R. Saggar, S. Grasso, M.K. Mani, I. Dlouhy, J. Dusza, M.J. Reece, Tribological properties of silica–graphene nanoplatelet composites, Ceram. Int. 40 (2014) 12067–12074.
[211] H. Li, Y. Xie, K. Li, L. Huang, S. Huang, B. Zhao, X. Zheng, Microstructure and wear behavior of graphene nanosheets reinforced zirconia coating, Ceram. Int. 40 (2014) 12821–12829.
[212] J. Balko, P. Hvizdos, J. Dusza, C. Balazsi, J. Gamcova, Wear damage of Si_3N_4–graphene nanocomposites at room and elevated temperatures, J. Eur. Ceram. Soc. 34 (2014) 3309–3317.
[213] H.J. Kim, S.-M. Lee, Y.-S. Oh, Y.-H. Yang, Y.S. Lim, D.H. Yoon, C. Lee, J.-Y. Kim, R.S. Ruoff, Unoxidized graphene/alumina nanocomposite: fracture- and wear-resistance effects of graphene on alumina matrix, Sci. Rep. 4 (2014) 5176.
[214] C.F. Gutierrez-Gonzalez, A. Smirnov, A. Centeno, B. Fernandez, B. Alonso, V.G. Rocha, R. Torrecillas, A. Zurutuza, J.F. Bartolome, Wear behavior of graphene/alumina composite, Ceram. Int. 41 (2015) 7434–7438.
[215] P. Rutkowski, L. Stobierski, D. Zientara, L. Jaworska, P. Klimczyk, M. Urbanik, The influence of the graphene additive on mechanical properties and wear of hot-pressed Si_3N_4 matrix composites, J. Eur. Ceram. Soc. 35 (2015) 87–94.

[216] Y.W. Zhang, Z.H. Gao, H.L. Ma, S.J. Wang, J. Peng, J.Q. Li, M.L. Zhai, A facile synthesis of platinum nanoparticle decorated graphene by one-step gamma-ray induced reduction for high rate supercapacitors, J. Mater. Chem. C 1 (2013) 321–328.

[217] S.B. Liu, T.H. Zeng, M. Hofmann, E. Burcombe, J. Wei, R.R. Jiang, J. Kong, Y. Chen, Antibacterial activity of graphite, graphite oxide, graphene oxide, and reduced graphene oxide: membrane and oxidative stress, ACS Nano 5 (2011) 6971–6980.

[218] A. Nieto, J.M. Zhao, Y. Han, K.H. Hwang, J.M. Schoenung, Microscale tribological behavior and in vitro biocompatibility of graphene nanoplatelet reinforced alumina, J. Mech. Behav. Biomed. Mater. 61 (2016) 122–134.

Further reading

[219] M.J. Martinez-Morlanes, P. Castell, P.J. Alonso, M.T. Martinez, J.A. Puertolas, Multiwalled carbon nanotubes acting as free radical scavengers in gammairradiated ultrahigh molecular weight polyethylene composites, Carbon 50 (2012) 2442–2452.

Chitosan-based nanocomposites for cardiac, liver, and wound healing applications

11

Mudasir Ahmad, Kaiser Manzoor*, Suhail Ahmad*, Nadia Akram*[†], *Saiqa Ikram**

*Department of Chemistry, Faculty of Natural Science, Jamia Millia Islamia, New Delhi, India, [†]Department of Chemistry, Government College University, Faisalabad, Pakistan

11.1 Introduction

Chitosan is a polyaminosaccharide typically obtained from chitin in basic medium involving deproteination and deacetylation (Fig. 11.1). It is the second largest biopolymer present on the earth obtained from the biological systems such as, insects, crustaceans, and fungi [1]. Chitosan comprises (1-4)-2-amino-2-deoxy-β-D-glucan and (1-4)-2-acetamido-2-deoxy-β-D-glucan. It is semisynthetic natural organic biomaterial with special functionality and carries wide applications in biomedical and other fields [2–4]. It is widely used in research work not only because it carries special functionality, but it is also obtained from the natural sources [5,6]. Chitin, the precursor of the chitosan has limited applications due to its least solubility in aqueous and nonaqueous solvents whereas the chitosan has solubility in most of the organic solvents which enhances the functionality and proves better than the chitin. The most important aspect of the chitosan is the nontoxicity at all concentrations and hence widely used in the medical field due to high biocompatibility, biodegradability, flexibility, and microbial activity.

Recently, there has been growing interest in the chitosan to increase the functionality via chemical cross-linking or by introducing the large or small groups in the chitosan backbone to increase its solubility and application in various fields [7–9]. Introduction of small groups like alkyl or carboxymethyl groups increase the solubility without affecting the cationic property at neutral and basic pH [10,11]. Chitosan with physical, chemical, and biological properties has played a wide role in various applications such as, adsorption of metals from water, agriculture, cosmetics and toiletries, medical sciences, fibers and textiles, food and beverages, optical and paper technology [12,13]. Among all these modifications, grafting polymerization is important that has been widely used. Grafting polymerization allows the formation of derivates by covalent bonding of the chemical grafts onto the chitosan backbone. Chitosan has two types of functionality that can be grafted. Firstly, the free amino group on the deacetylated group and secondly the hydroxyl groups at C3 and C6 on acetylated or deacetylated unit. Recently, the researchers have shown that after the primary derivatization of the chitosan followed by graft polymerization, better solubility, antibacterial and antioxidant activity can be achieved [14,15]. In common way, the grafting

Applications of Nanocomposite Materials in Orthopedics. https://doi.org/10.1016/B978-0-12-813740-6.00013-2

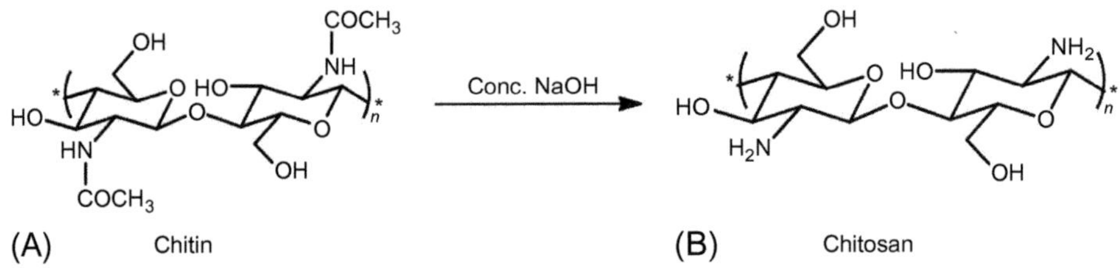

Fig. 11.1 Structure of (A) Chitin and (B) Chitosan.

polymerization increases the chelation [16], bacteriostatic effect [17], and complexation properties [18]. Although the modification of chitosan improves the properties via the grafting polymerization, it is easy to maintain some properties such as mucoadhasive [19], biocompatibility [20–22], and biodegradability [23,24]. The researchers take the advantage of chemical grafting of chitosan and modified into various forms such as chitosan-based nanocomposites. These composites have been used in various applications with the increasing functionality in various fields such as wound healing and tissue engineering [25].

11.2 Tissue engineering

Tissue engineering is the application of engineering, chemistry, biology, and medicine for the fabrication and replacement of damaged tissues to restore, maintain, improve structurally, and functionally damaged organs [21]. The present tissue engineering is the seedling of cells onto the biodegradable biopolymer. It involves the in vitro cultivation and proliferation of cells which needs special support and temperature dependence. A primary factor is the availability of good biomaterials to serve as the ordinary material. For example, Cindy et al. have prepared injectable-type thermoreponsive hydrogel scaffold for tissue engineering. It was injected as liquid at room temperature and gelled at the target site due to change in the body temperature [24]. Being biodegradable and nontoxic, its presence in various forms such as hydrogel, powder, and films can be widely used in tissue engineering such as liver and cardiac tissue engineering and wound healing [26]. In the present scenario, chitosan and its derivatives are considered as potential scaffold in tissue engineering as these are degraded when the new tissue starts to grow without any inflammation or toxic degradations [27,28]. In tissue engineering application, the cationic nature of chitosan is responsible for the electrostatic interaction with the anionic molecules, glycosaminoglycans, and proteoglycans [29].

11.2.1 Chitosan nanocomposites in liver tissue engineering

Liver is the largest gland and almost carries about 500 functions within the body such as detoxification of endogenous and exogenous compounds, lipids, carbohydrates and protein metabolism, secretion of much serum protein (i.e., albumin), and production of bile juice for secretion. In case of the liver damage, orthotropic liver transplantation is the only solution for liver failure. The treatment of hepatic failure is currently the replacement of damaged liver by bio artificial liver (BAL). It is an important invention of tissue engineering. In BAL technology, primary hepatocytes, hepatic cell lines, and liver stem cells are significant sources for the development of proper BAL devices [30]. Chitosan nanofiber mesh is a culture substrate for hepatocyte and its properties such as adhesion, morphology, and activity of cultured hepatocytes showed that the chitosan mesh has the potential for tissue engineering for liver regeneration as it is biocompatible with hepatocyte [31]. Regenerated silk fibroin and galactosylated chitosan in the presence of polyethylene oxide have been used to prepare biomimetic artificial extracellular matrix (ECM), for hepatic tissue engineering applications through electrospinning method [32]. The oxidized alginate covalently cross-linked with galactosylated chitosan via Schiff base reaction, without employing any chemical cross-linking agent has been used as ECM for hepatocytes culture to maintain high level of liver-specific functions [33]. Frohbergh et al. have proposed a unique method for fabrication of chitosan-based microfiber device without the use of chemical additives for liver tissue engineering applications [33]. The multilayer radial-flow bioreactor based on galactosylated chitosan nanofiber scaffolds may afford the short-term support of patients with hepatic failure [34]. The natural nanofibrous scaffold from galactosylated chitosan was fabricated by electrospinning to be used as BAL-assist devices and tissue engineering for liver regenerations [35].

11.2.2 Chitosan nanocomposites in cardiac tissue engineering

Cardiac muscle is an electroactive tissue capable of transferring electrical signals and allowing the heart to beat. Adult heart muscle has poor capability to repair itself due to insignificant regeneration potential of cardiomyocytes [36]. Recently great interest has arisen from the possibility to regenerate lost tissue by implanting of therapeutic cells, biomaterials, and cardiac patches. A lot of work has been carried out on natural biopolymers. For example, chitosan-based carbon scaffolds were prepared by doping carbon nanotubes through precipitation method. These scaffolds played a major role in cardiac tissue engineering [37]. Chitosan nanoparticles were synthesized at different concentrations of chitosan and tripolyphosphate through ionic gelation method to deliver drugs to particular organs, like heart [38]. Electrospinning technique was used for the fabrication of bioactive three-dimensional (3D) chitosan nanofiber scaffolds and the resultant structure showed its potential for long-term cardiac function [39]. In case of porous chitosan scaffolds, several other methods such as solution extrusion and neutralization methods were used for the fabrication of porous chitosan matrix. Several factors beneficial for the tissue engineering such as fiber scaffold mass ratio, fiber mechanical properties, and fiber length of scaffold on porous chitosan frameworks makes chitosan nanocomposites as good material with enhanced mechanical properties for heart valve supports [40]. Polycaprolactone sandwiched gelatin-chitosan hydrogels prepared in a 3D scaffold with different molecular weights (Fig. 11.2), were used as biodegradable matrix with a potential for surgical reconstruction of congenital heart defects [41].

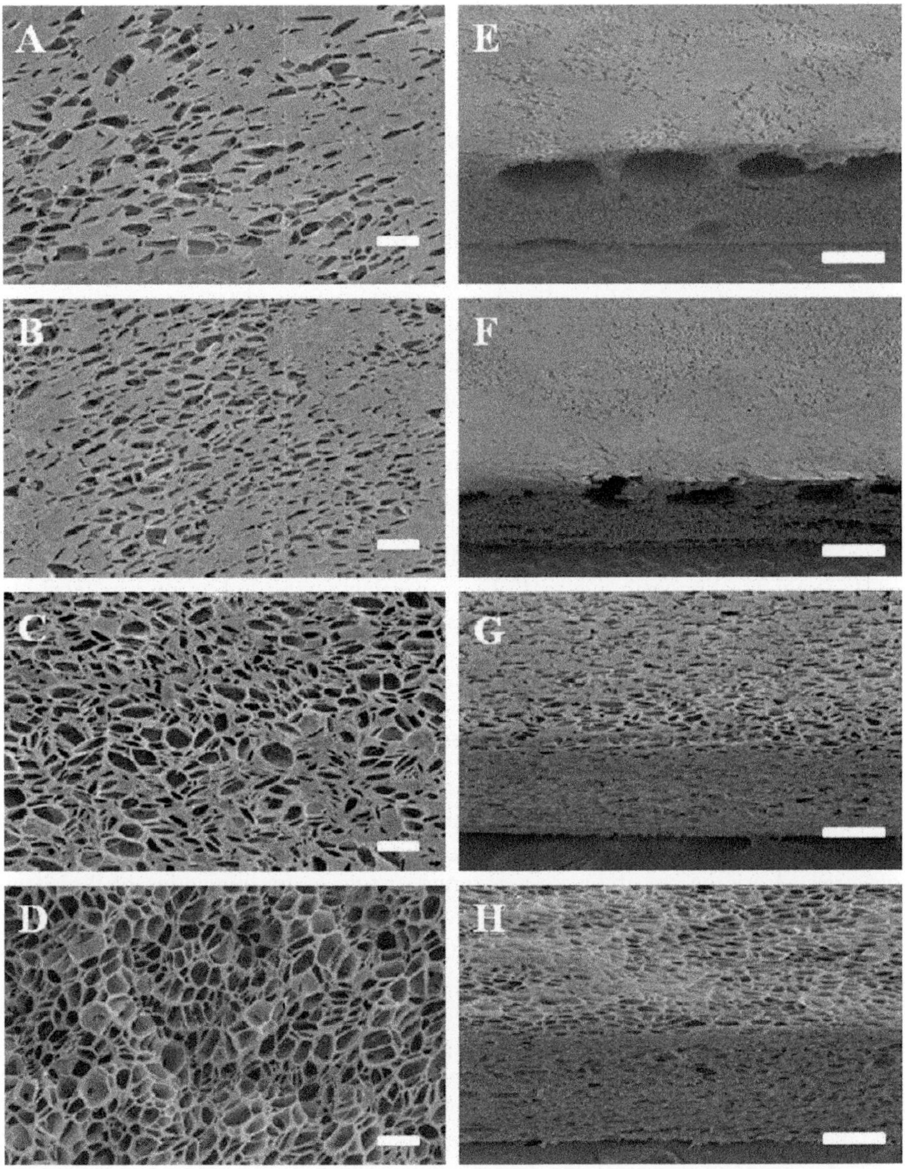

Fig. 11.2 Microscopic structures of multilayered gelatin-chitosan hydrogel. (A–D) SEM micrographs of multilayered scaffolds with different gelatin:chitosan blending ratios (vol%): (A) 0:100; (B) 25:75; (C) 50:50; and (D) 75:25). (E–H) 3D view of multilayered scaffolds: (E) 0:100; (F) 25:75; (G) 50:50; and (H) 75:25. All samples formed a 3D porous structure after lyophilization.
Modified from S. Pok, J.D. Myers, S.V. Madihally, J.G. Jacot, A multilayered scaffold of a chitosan and gelatin hydrogel supported by a PCL core for cardiac tissue engineering, Acta Biomater. 9 (2013) 5630–5642, copyright (August 2017), with permission from Elsevier.

11.2.3 Chitosan nanocomposite in wound healing applications

Wound healing, a typical biological progression in the human body, is worldwide medicinal alarms with numerous challenges attained through five precisely and extremely automated phases: hemostasis, migration, inflammation, proliferation, and remodeling [42]. Covering materials used for skin injuries to avoid bleeding, protect the wound from environmental infections, preserved the wound moist, quicken the healing way, and take away excess exudates. The main aim of wound healing is a quick recovery, nontoxic and antiseptic, oxygen permeability, permeable edifice with minimal damaging, and highest function, meanwhile of the direct interaction with the wound. The choice of materials is very significant for wound healing application. In case of autologous skin grafts are not available, biopolymers may be used to endorse and initiate the usual epidermal and dermal wound healing process [43]. Bioactive polymers are generally used because of their biodegradability and biocompatibility, but more notably, they have a vigorous therapeutic effect on one or further stages of wound healing. Bioactive polymer dressing constituents chitosan, collagen and gelatin, sodium alginate, and hyaluronic acid [44]. Chitosan has no mutagenic property that makes it an entrant for biomedical application as per the chemical carcinogenesis research information system [45]. The *N*-acetyl glucosamine (NAG) in chitin and chitosan is a chief constituent of dermal tissue, which is crucial for repair of scar tissues [46]. Chitosan has numerous advantages over other type of antiseptics since it possesses a broader spectrum of activity, higher antibacterial activity, sophisticated killing rate, and minimal toxicity to mammalian cells. Electrospun chitosan nanocomposite fibers for cytocompatible chitin nanocrystals as moisture steadiness enhanced mechanical properties, and water vapor porousness of the electrospun nanocomposite mats. The electrospun fiber mats in wound dressing showed compatibility toward adipose-derived stem cells; moreover, authorizing their possible use as wound dressing materials [47]. Chitosan and halloysite nanocomposite have been of deliberate therapeutic choice for problematic wound healing burns and skin lesions [48]. Chitosan hybrid materials including metal and metal oxide nanoparticles have shown outstanding properties arising from synergistic effects [49]. Kumar et al. proposed chitosan hydrogel/nano zinc oxide composite bandages for wound dressing. These bandages have shown antimicrobial assessment, strong effects on *Staphylococcus aureus* and *Escherichia coli*, with an advanced antimicrobial impact on *E. coli*. Moreover, in vivo wound healing evaluations exposed that the microporous chitosan hydrogel/nano zinc oxide composite bandages as equated with Kaltosat and bare wound with high wound-dressing efficacy without any toxicity [50]. Chitosan/silk sericin scaffolds combined with lauric acid (LA) and zinc oxide nanoparticles (ZnO NPs) were used for the synthesis of wound dressing materials and found that there was no effect on pore dimensions and porosity on introducing ZnO NPs or LA [51]. The involvement of silver nanoparticles (Ag NPs) with chitosan for the improvement of antimicrobial effectiveness in the matrix was examined. Chitosan has capability to alleviate the profile of Ag NPs [52]. It has been shown that the combined chitosan or modified chitosan with Ag, carrying higher antimicrobial activity, enlarged tensile strength but reduced water vapor permeability. Nanocomposites with nanoparticle stacking ≤ 5.0 wt% displayed very virtuous in vitro cytocompatibility and aided to quicker wound healing than neat castor oil and regulate gauze, hence illustrated excessive potential as antibacterial wound dressings [53]. The method used for the fabrication of Ag NPs, in most cases, have been electrospinning based, due to the resourceful, simple, speedy, and low-cost nanofiber manufacturing [54]. Chitosan cannot be twisted directly by an electrospinning method, due to its polycationic nature in solution. To overcome this problem, several studies were carried out to modify the chitosan with other polymers. Polyvinyl alcohol/chitosan blend solutions activated a decrease in the diameters of the electrospinning fibers and an improvement in the tensile strength, but with a reduction in the elongation [55]. There has been continued attention to produce inorganic-organic hydrogels nanocomposite, which concluded their sophisticated biomedical significance [56]. Chitosan-hyaluronan/chondroitin sulfate nanoparticle sponge is used for wound dressing applications. The nanocomposite sponge displayed porosity in the range of 60%–70%, which is satisfactory for a wound dressing. It has also enhanced blood clotting, exhibited platelet activation, and controlled swelling and biodegradation. Cytotoxicity revisions onto human dermal fibroblast cells are the strong evidence for the nontoxic nature of composite sponges. Therefore, the incorporation of chondroitin sulfate nanoparticles into chitosan-hyaluronan sponge improved the overall efficacy of the sponge [57]. The bacterial infection at the wound site is common; to stop the infection, the wound site needs bioadhesive materials with inherent antibacterial properties. The prepared hydrogel from *N*-(2-hydroxypropyl)-3-trimethylammonium chitosan chloride and polydextran aldehyde showed negligible toxicity toward human red blood cells (only 2%–3% hemolysis) and no inflammation to the neighboring tissue upon dermal implantation in mice, thus verifying it as a safe and operative antibacterial sealant [58]. The polyvinyl alcohol/chitosan/montmorillonite nanocomposite hydrogels have better mechanical properties due to the incorporation of montmorillonite clay and its various characteristics such as antibacterial activity, good swelling behavior, and biocompatibility. All these factors have made it a necessary entrant for wound dressing applications [59]. Chitosan and poly(ethylene oxide) have been fabricated into CS/polyethylene oxide nanofibrous-type ECM, which releases dual growth factor for wound healing in such a way as the wound healing occurs through the natural process. The design and release of the dual growth factor occurs at the wound site (Fig. 11.3) [60].

11.3 Conclusion

Chitosan, a copolymer of glucosamine and *N*-acetylglucosamine units, is linked by β-1,4-glycosidic linkages. It is nontoxic, biocompatible, antifungal, and biodegradability and its derivatives have been broadly used in the fields of tissue engineering, medicine, biochemical separation systems, and wound healing. This chapter enlightens the biomedical applications of chitosan-based nanocomposites in tissue engineering and wound dressing. It also unlocks the innovative applications of natural biopolymers that can be used in a variety of nanostructural forms. Multifunctional chitosan nanostructured scaffolds are advanced gift for tissue engineering that has been proved to aid for healing of wounds.

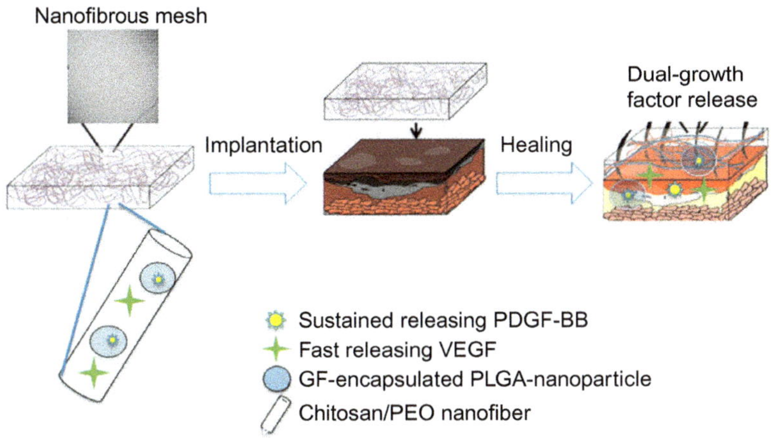

Fig. 11.3 The preparation of chitosan-based composite loaded with two growth factors for enhancing the wound healing by electrospun fibers through a simple method.
Modified from S. Chen, Y. Wang, Study on β-cyclodextrin grafting with chitosan and slow release of its inclusion complex with radioactive iodine, J. Polym. Sci. 82 (2001) 2414–2421, copyright (2013), with permission from Elsevier.

Acknowledgments

The author Mudasir Ahmad acknowledges University Grants Commission (UGC), New Delhi, India, for financial support as Basic Science Researcher (SRF) Fellow. The author further acknowledges Mr Gulzar Rafiqi Department of Chemistry (IUST) for support.

References

[1] M. Ahmad, S. Ahmed, B.L. Swami, S. Ikram, Preparation & characterization of antibacterial thiosemicarbazide chitosan as efficient cu(II) adsorbent, Carbohydr. Polym. 132 (2015) 164–172.

[2] M. Ahmad, K. Manzoor, P. Venkatachalam, S. Ikram, Kinetic and thermodynamic evaluation of adsorption of cu(II) by thiosemicarbazide chitosan, Int. J. Biol. Macromol. 92 (2016) 910–919.

[3] M. Ahmad, K. Manzoor, R.R. Chaudhuri, S. Ikram, Thiocarbohydrazide cross-linked oxidized chitosan and poly(vinyl alcohol): a green framework as efficient cu(II), Pb(II), and hg(II) adsorbent, J. Chem. Eng. Data 62 (2017) 2044–2055.

[4] M. Ahmad, S. Ahmed, B.L. Swami, S. Ikram, Adsorption of heavy metal ions: role of chitosan & cellulose for water treatment, Int. J. Pharmacogn. 2 (2015) 280–289.

[5] M. Ahmad, K. Manzoor, S. Singh, S. Ikram, Chitosan centered Bionanocomposites for medical specialty and curative applications: a review, Int. J. Pharm. 529 (2017) 200–217.

[6] S. Ahmed, M. Ahmad, S. Ikram, Chitosan: a natural antimicrobial agent—a review, J. Appl. Chem. 3 (2014) 493–503.

[7] K. Kurita, T. Kojima, T. Munakata, H. Akao, T. Mori, Y. Nishiyama, M. Shimojoh, Preparation of Nonnatural branched chitin and chitosan, Chem. Lett. 27 (1998) 317–318.

[8] H. Sashiwa, Y. Shigemasa, Chemical modification of chitin and chitosan 2: preparation and water soluble property of N-acylated or N-alkylated partially deacetylated chitins, Carbohydr. Polym. 39 (1999) 127–138.
[9] A. Heras, N.M. Rodriguez, V.M. Ramos, N-methylene phosphonic chitosan: a novel soluble derivative, Carbohydr. Polym. 44 (2001) 1–8.
[10] R. Jayakumar, R.L. Reis, J.F. Mano, Synthesis of N-carboxymethyl chitosan beads for controlled drug delivery applications, Mater. Sci. Forum 514-516 (2006) 1015–1019.
[11] G. Lu, L. Kong, B. Sheng, G. Wang, Y. Gong, X. Zhang, Degradation of covalently crosslinked carboxymethyl chitosan and its potential application for peripheral nerve regeneration, Eur. Polym. J. 43 (2007) 3807–3818.
[12] M. Rinaudo, Chitin and chitosan: properties and applications, Prog. Polym. Sci. 31 (2006) 603–632.
[13] M.N. Ravi Kumar, A review of chitin and chitosan applications, React. Funct. Polym. 46 (2000) 1–27.
[14] R.A.A. Muzzarelli, M. Emmanueli, S. Mariotti, N-(carboxymethylidene)chitosans and N-(carboxymethyl)chitosans: novel chelating polyampholytes obtained from chitosan glyoxylate, Carbohydr. Res. 107 (1982) 199–214.
[15] W.M. Xie, P.X. Xu, W. Wang, Q. Lu, Antioxidant activity of water-soluble chitosan derivatives, Bioorg. Med. Chem. Lett. 11 (2001) 1699–1703.
[16] Z.K. Yang, Y. Yuan, Studies on the synthesis and properties of hydroxyl azacrown ether-grafted chitosan, J. Appl. Polym. Sci. 82 (2001) 1838–1843.
[17] B.O. Jung, C.H. Kim, K.S. Choi, Y.M. Lee, J.J. Kim, Preparation of amphiphilic chitosan and their antimicrobial activities, J. Appl. Polym. Sci. 72 (1999) 1713–1719.
[18] S. Chen, Y. Wang, Study on β-cyclodextrin grafting with chitosan and slow release of its inclusion complex with radioactive iodine, J. Polym. Sci. 82 (2001) 2414–2421.
[19] T.K. Giri, A. Thakur, A. Alexander, H.B. Ajazuddin, D.K. Tripathi, Modified chitosan hydrogels as drug delivery and tissue engineering systems: present status and applications, Acta Pharm. Sin. B 2 (2012) 439–449.
[20] R.A. Tasker, B.J. Connell, S.J. Ross, C.M. Elson, Development of an injectable sustained-release formulation of morphine: antinociceptive properties in rats, Lab. Anim. 32 (1998) 270–275.
[21] E. Travers, B. Mecheri, C. Mandoli, S. Soliman, A. Rinaldi, S. Licoccia, G. Forte, F. Pagliari, S. Pagliari, F. Carotenuto, M. Minieri, P. Di Nardo, Tuning hierarchical architecture of 3D polymeric scaffolds for cardiac tissue engineering, J. Exp. Nanosci. 3 (2008) 97–110.
[22] K. Ono, Y. Saito, H. Yura, K. Ishikawa, Photocrosslinkable chitosan as a biological adhesive, J. Biomed. Mater. Res. 49 (2000) 289–295.
[23] D.K. Singh, A.R. Ray, Characterization of grafted chitosan films, Carbohydr. Polym. 36 (1998) 251–255.
[24] I.-Y. Kim, S.-J. Seo, H.-S. Moon, M.-K. Yoo, I.-Y. Park, B.-C. Kim, C.-S. Cho, Chitosan and its derivatives for tissue engineering applications, Biotechnol. Adv. 26 (2008) 1–21.
[25] V.-M. Le, J.-J. Wang, M. Yuan, T.-L. Nguyen, G.-F. Yin, Y.-H. Zheng, W.-B. Shi, M.-D. Lang, L.-M. Xu, J.-W. Liu, An investigation of antitumor efficiency of novel sustained and targeted 5-fluorouracil nanoparticles, Eur. J. Med. Chem. 92 (2015) 882–889.
[26] C.D. Ren, M. Kurisawa, J.E. Chunga, J.Y. Ying, Liposomal delivery of horseradish peroxidase for thermally triggered injectable hyaluronic acid–tyramine hydrogel scaffolds, J. Mater. Chem. B 3 (2015) 4663–4670.
[27] Y.H. Kim, S.H. Kim, C.R. Park, Structural characteristics of size-controlled self-aggregates of Deoxycholic acid-modified chitosan and their application as a DNA delivery carrier, Bioconjug. Chem. 12 (2001) 932–938.

[28] K. Tuzlakoglu, C.M. Alves, J.F. Mano, R.L. Reis, Production and characterization of chitosan fibers and 3-D Fiber mesh scaffolds for tissue engineering applications, Macromol. Biosci. 4 (2004) 811–819.
[29] N.M. Alves, J.F. Mano, Chitosan derivatives obtained by chemical modifications for biomedical and environmental applications, Int. J. Biol. Macromol. 43 (2008) 401–414.
[30] A. Mishra, Y. Loo, R. Deng, Y.J. Chuah, H.T. Hee, J.Y. Ying, C.A.E. Hauser, Ultrasmall natural peptides self-assemble to strong temperature-resistant helical fibers in scaffolds suitable for tissue engineering, Nano Today 6 (2011) 232–239.
[31] Z.-Q. Feng, M.K. Leach, X.-H. Chu, Y.-C. Wang, T. Tian, X.-L. Shi, Y.-T. Ding, Z.-Z. Gu, Electrospun chitosan nanofibers for hepatocyte culture, J. Biomed. Nanotechnol. 6 (2010) 658–666.
[32] N. Kasoju, U. Bora, Silk fibroin based biomimetic artificial extracellular matrix for hepatic tissue engineering applications, Biomed. Mater. 8 (2013) 049501.
[33] F. Chen, M. Tian, D. Zhang, J. Wang, Q. Wang, X. Yu, X. Zhang, C. Wan, Preparation and characterization of oxidized alginate covalently cross-linked galactosylated chitosan scaffold for liver tissue engineering, Mater. Sci. Eng. C 32 (2012) 310–320.
[34] K.H. Lee, S.J. Shin, C.-B. Kim, J.K. Kim, Y.W. Cho, B.G. Chung, S.-H. Lee, Microfluidic synthesis of pure chitosan microfibers for bio-artificial liver chip, Lab Chip 10 (2010) 1328–1334.
[35] X. Chu, X. Shi, Z. Feng, J. Gu, H. Xu, Y. Zhang, Z. Gu, Y. Ding, In vitro evaluation of a multi-layer radial-flow bioreactor based on galactosylated chitosan nanofiber scaffolds, Biomaterials 30 (2009) 4533–4538.
[36] M. Shin, O. Ishii, T. Sueda, J.P. Vacanti, Contractile cardiac grafts using a novel nanofibrous mesh, Biomaterials 25 (2004) 3717–3723.
[37] A. Martins, G. Eng, S. Caridade, J.F. Mano, R.L. Reis, G.V. Novakovic, Electrically conductive chitosan/carbon scaffolds for cardiac tissue engineering, Biomacromolecules 15 (2014) 635–643.
[38] G. Narasimhan, D.R. de Alba-Aguayo, R. Mondragón-Flores, S. González-Pozos, M.J. Miranda-Saturnino, M. Sridharan, A. Rueda, Acute administration of chitosan nanoparticles increases Ca2+ leak in rat cardiomyocytes, J. Nano Res. (2014) 29–38.
[39] A. Hussain, G. Collins, D. Yip, C.H. Cho, You have full text access to this content-Functional 3-D cardiac co-culture model using bioactive chitosan nanofiber scaffolds, Biotechnol. Bioeng. 110 (2013) 637–647.
[40] M.Z. Albanna, T.H. Bou-Akl, H.L. Walters, H.W.T. Matthew, Improving the mechanical properties of chitosan-based heart valve scaffolds using chitosan fibers, J. Mech. Behav. Biomed. Mater. 5 (2012) 171–180.
[41] S. Pok, J.D. Myers, S.V. Madihally, J.G. Jacot, A multilayered scaffold of a chitosan and gelatin hydrogel supported by a PCL core for cardiac tissue engineering, Acta Biomater. 9 (2013) 5630–5642.
[42] J.S. Boateng, K.H. Matthews, H.N.E. Stevens, G.M. Eccleston, Wound healing dressings and drug delivery systems: a review, J. Pharm. Sci. 97 (8) (2008) 2892–2923.
[43] S. Barrientos, O. Stojadinovic, M.S. Golinko, H. Brem, M. Tomic-Canic, Growth factors and cytokines in wound healing, Wound Repair Regen. 16 (2008) 585–601.
[44] J. Boateng, O. Catanzano, Advanced therapeutic dressings for effective wound healing-a review, J. Pharm. Sci. 104 (2015) 3653–3680.
[45] Chemical Carcinogenesis Research Information System, (CCRIS), <http://toxnet.nlm.nih.gov>
[46] D.K. Singh, A. Ray, Biomedical applications of chitin, chitosan and their derivatives, J. Macromol. Sci. Polym. Rev. 40 (2000) 69–83.

[47] N. Naseri, C. Algan, V. Jacobs, M. John, K. Oksman, A.P. Mathew, Electrospun chitosan-based nanocomposite mats reinforced with chitin nanocrystals for wound dressing, Carbohydr. Polymers 109 (2014) 7–15.

[48] G. Sandri, C. Aguzzi, S. Rossi, M.C. Bonferoni, G. Bruni, C. Boselli, A.I. Cornaglia, F. Riva, C. Viseras, C. Caramella, F. Ferrari, Halloysite and chitosan oligosaccharide nanocomposite for wound healing, Acta Biomater. 15 (57) (2017) 216–224.

[49] D. Archana, B.K. Singh, J. Dutta, P.K. Dutta, In vivo evaluation of chitosan-PVP-titanium dioxide nanocomposite as wound dressing material, Carbohydr. Polym. 95 (2013) 530–539.

[50] P.T.S. Kumar, V.K. Laskmanan, T.V. Anilkumar, C. Ramya, P. Reshmi, A.G. Unnikrishnan, S.V. Nair, R. Jayakumar, Flexible and microporous chitosan hydrogel/nano ZnO composite bandages for wound dressing: in vitro and in vivo evaluation, ACS Appl. Mater. Interfaces 4 (2012) 2618–2629.

[51] Z. Karahaliloglu, E. Kilicay, E.B. Denkbas, Antibacterial chitosan/silk sericin 3D porous scaffolds as a wound dressing material, Artif. Cells Nanomed. Biotechnol. (2016) 1–14.

[52] M.B. Ahmad, J.J. Lim, K. Shameli, N.A. Ibrahim, M.Y. Tay, Synthesis of silver nanoparticles in chitosan, gelatin and chitosan/gelatin bionanocomposites by a chemical reducing agent and their characterization, Molecules 16 (2011) 7237–7248.

[53] M. Ana, D. Pascual, A.L. Díez-Vicente, Wound healing bionanocomposites based on castor oil polymeric films reinforced with chitosan-modified ZnO nanoparticles, Biomacromolecules 16 (9) (2015) 2631–2644.

[54] W.E. Teo, S. Ramakrishna, Electrospun nanofibers as a platform for multifunctional, hierarchically organized nanocomposite, Compos. Sci. Technol. 69 (2009) 1804–1817.

[55] A.T. Hang, B. Tae, J.S. Park, Non-woven mats of poly(vinyl alcohol)/chitosan blends containing silver nanopartiocles: fabrication and characterization, Carbohydr. Polym. 82 (2010) 472–479.

[56] M. Yadollahi, I. Gholamali, H. Namazi, M. Aghazadeh Synthesis, Characterization of antibacterial carboxymethylcellulose/CuO bio-nanocomposite hydrogels, Int. J. Biol. Macromol. 73 (2014) 109–114.

[57] B.S. Anisha, D. Sankar, A. Mohandas, K.P. Chennazhi, S.V. Nair, R. Jayakumar, Chitosan-hyaluronan/nano chondroitin sulfate ternary composite sponges for medical use, Carbohydr. Polym. 92 (2013) 1470–1476.

[58] J. Hoque, R.G. Prakash, K. Paramanandham, B.R. Shome, J. Haldar, Biocompatible injectable hydrogel with potent wound healing and antibacterial properties, Mol. Pharm. 14 (4) (2017) 1218–1230.

[59] S. Nooria, M. Kokabia, Z.M. Hassan, Nanoclay enhanced the mechanical properties of poly(vinyl alcohol)/chitosan/montmorillonite nanocomposite hydrogel as wound dressing, Proc. Mater. Sci. 11 (2015) 152–156.

[60] Z. Xie, C.B. Paras, H. Weng, P. Punnakitikashem, L.-C. Su, K. Vu, L. Tang, J. Yang, K.T. Nguyen, Dual growth factor releasing multi-functional nanofibers for wound healing, Acta Biomater. 12 (2013) 9351–9359.

Extracellular matrix: The ideal natural fibrous nanocomposite products

12

Mehdi Ebrahimi
Oral Rehabilitation, Prince Philip Dental Hospital, The University of Hong Kong, Sai Ying Pun, Hong Kong

12.1 Introduction

Extracellular matrix (ECM) is a complex and dynamic network of extracellular macromolecules secreted by cells unique to each tissue. ECM provides a specific three-dimensional (3D) environment for biophysical and biochemical support of surrounding cells and can regulate cell behavior through direct or indirect pathways [1]. In fact, the ECM is a dynamic matrix that can control and modify the deposition, degradation, and remodeling of its component to support the tissue/organ physiological functions [2]. Fig. 12.1 illustrates basic functions and properties of ECM.

It has been reported that physical properties of matrix such as nanotopography, elasticity, rigidity, porosity, and solubility can regulate the cell behavior and anchorage-related cell functions (e.g., cell division and migration) [3]. With regard to biochemical properties, the ECM can play a role in both direct and indirect cellular signaling pathways; directly by binding to various types of cell surface receptors or co-receptors, and indirectly by presenting noncanonical growth factors or by enzymatic remodeling and release of active ingredients [1]. Both of direct and indirect pathways result in the initiation of different intracellular signaling and influence the cell behavior in their environment. Furthermore, the biomechanical properties of ECM, such as stiffness, can influence the process of mechanotransduction which is defined as the cell's perception of the external physical stimuli and their subsequent response to the surrounding environment [4]. It is important to mention that ECM and cells have mutual and dynamic interaction, where they influence each other in a closely related reciprocal way. Live cells actively and continuously secret and modify the composite nature of ECM of their environment based on different stimuli received from them, while this remodeling consequently regulates the cell behavior [1,5].

"Niche" was first defined for hematopoietic microenvironment in 1978 [6] as a specialized dynamic microenvironment of ECM that hosts the stem cells. It is unique to each type of stem cell, determines the identity, and regulates the behavior by controlling stem cells self-renewal and differentiation profile [7,8]. It is well known that stem cells possess high proliferation capacity, while the niche plays a significant role in the maintenance of essential balance between active and quiescence cell states by keeping them at a reasonable economic metabolic state [9]. These cell states are

Applications of Nanocomposite Materials in Orthopedics. https://doi.org/10.1016/B978-0-12-813740-6.00014-4

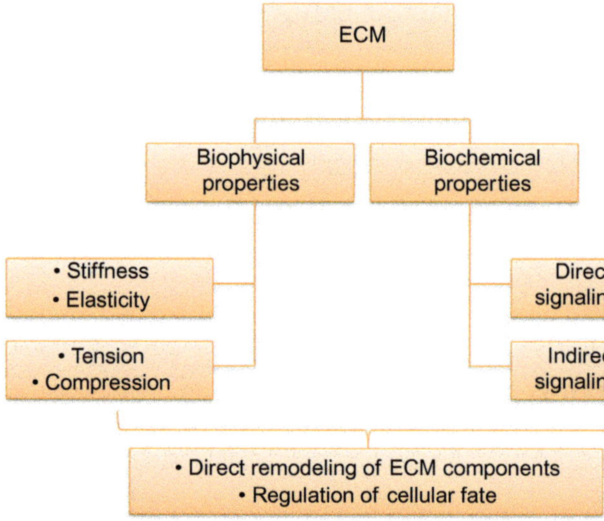

Fig. 12.1 ECM functions and properties. ECM provides a dynamic network that host and supports various cell functions. It can convey two main signals of biochemical and biophysical nature. ECM can mediate ECM-cell interaction by presenting different signaling molecules to the cell receptors through direct or indirect pathways. ECM can also regulate cell behavior through alteration of its physical matrix properties, such as stiffness and elasticity.

reported to be present in several tissues in defined anatomical locations [10,11]. A precise and balanced interaction between stem cells and their niche is critical to guaranty the long-term maintenance of stem cells in an undifferentiated state and also maintain the proper tissue functioning by recruitment of required cells into cell cycle for differentiation [12,13]. Furthermore, the niche may have a protective anticancer role by inhibiting the accumulation of gene mutations [11,14].

Niche provides highly complex microenvironment where many factors can interact with each other. The niche composed of stem cells and supportive stromal cells which communicate with each other via cell surface receptors, soluble factors, gap junctions, tight junctions, and adhesion junctions [15]. Furthermore, the presence of both vascular and neural inputs can modify the niche microenvironment by carrying various systemic signals that can trigger different physiological responses by recruitments of other cells and signaling molecules. In addition to the dynamic stem cell-niche interaction and the interaction between stem cells themselves, other secreted factors, oxygen tension gradients, temperature, and shear stress can also modify stem cells behaviors [12,13].

The ECM is also considered as a reservoir and carrier of different factors that present them to the cell surface. The interplay between these factors and interaction between cells and ECM initiate different signaling mechanism within ECM. This, in turn, results in 3D self-assembly and organization of constituted macromolecules within ECM that provides an instructive power to supply the microenvironment for controlling of stem cell homeostasis [16,17]. The crucial role of ECM in stem cell niche and the control of stem cell behavior have been shown in altered or aged niche with reduced function [5]. Other studies reported the role of de-cellularized ECM tissue in directing differentiation of stem cells into original residing cell type within the ECM [18–20].

12.2 ECM-cell interaction: Cell receptors and biochemical cues

The interaction between ECM and stem cells is mediated through different types of cell receptors. Integrins are considered the main class of cell surface receptors involved in ECM-stem cell interaction. The integrins are a superfamily of heterodimeric transmembrane glycoproteins as adhesion receptors that attach to ECM ligands, cell surface ligands, and soluble ligands. The integrins proteins consist of a large (α) and a small (β) subunits of sizes 120–170 kDa and 90–100 kDa, respectively [21,22], which form transmembrane heterodimers.

The family of human integrins consists of at least 18 α-subtypes and 8 β-subtypes which together generate 24 known binding pairs for the integrin heterodimers, each of which is composed of one α-subunit and one β-subunit [23]. Integrins play a key role in the organization of cytoskeleton and function bidirectionally by transducing the intracellular biochemical signals and transmitting the information from the cells to outside ECM. They mediate different processes such as cell adhesion, homing of stem cells, proliferation, migration, differentiation, and apoptosis [24]. For example, in response to growth factors and cytokines (i.e., IL-3 and TGF-β), the integrins regulate the signaling pathways, which in turn influence the integrins expression [24–27]. Fig. 12.2 shows the basic cell-ECM interaction mediated by biochemical cues on exposure of the cell to nanotopography of ECM substrate.

Different subtypes of integrins bind to various types of ECM protein components, such as collagen, fibronectin, vitronectin, fibrinogen, laminin, and glycoproteins (i.e., osteopontin). The RGD motif (the sequence of arginine-glycine-aspartic amino acid)

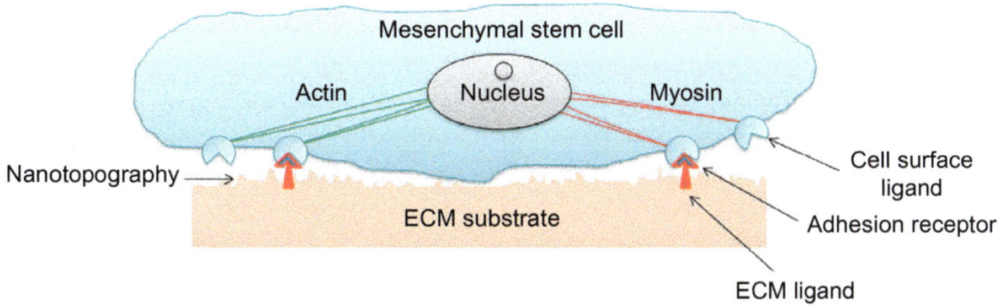

Fig. 12.2 Cell-ECM interaction. The topography of the ECM substrate can induce differentiation of MSC through mechanotransduction pathway. This is modulated by integrin activated focal adhesion. This results in alteration of cell morphology due to contractility of actomyosin fibers that produce cytoskeleton changes.

or LDV motif (leucine-aspartate-valine) are considered the most common integrins-binding motif on target ligands [28]. The RGD motif also acts as recognition sequence of fibronectin, vitronectin, and fibrinogen [22,28,29]. β1 integrins are reported to play an important role in controlling the symmetric and asymmetric division of stem cells, thereby maintaining the balance and preserving the reservoir of stem cells [30,31]. The type of ligands and the biological background after integrins activation determine the signaling pathway and the subsequent short-term or long-term biological responses [32].

However, the role of other nonintegrins receptors in ECM-cell interaction is minimally investigated. For example, adhesion receptor CD44 and its major ligand hyaluronic acid (HA) play a key role in homing of hematopoietic stem cells (HSC) and their migration and anchorage within the specific niche of bone marrow [33]. It is also demonstrated that the guidance molecule "Robo4" functions competitively to specifically anchor HSCs to BM niches [34].

Another unique feature of ECM is its ability of presentation of noncanonical growth factors that also influence stem cell responses. In addition to offering binding sites to adhesion receptors such as integrins, the ECM proteins also allow binding of soluble growth factors, thereby regulating their availability, distribution, and activation. In this way, ECM plays an important role by presenting these soluble factors to cells via an established biochemical gradient and triggering complex signaling pathway in an organized pattern [1].

Heparin – a component of ECM proteoglycans – allows binding of various growth factors such as FGF (fibroblast growth factor) and VEGF (vascular endothelial growth factor). Therefore, it can function as a localized pool of growth factors in solid phase and help in the maintenance of stable growth factors gradient within ECM. Further examples of this action are the binding of fibronectin, vitronectin, collagens, and proteoglycans to soluble ligands, such as FGFs, VEGFs, BMPs, and TGF-β [35]. Furthermore, ECM can act as a dispenser of growth factors upon enzymatic degradation (i.e., collagenase and matrix metalloproteinases) of proteins (proteolysis) or glycosaminoglycan components. The degradation process results in the breakdown of ECM-bound growth factors and release of growth factors in the bioactive soluble state to bind their receptors and induces remodeling of ECM components [1].

Therefore, ECM proteins and growth factors can be presented in complex form as solid-phase ligands bound to ECM or be released locally in soluble form. Furthermore, both ECM proteins and growth factors can act synergistically in influencing cellular responses, considering the possibility of specific direct binding between them without the involvement of glycosaminoglycans. It has been shown that integrins communicate with receptors of various growth factors to control cellular responses. In fact, there is a reciprocal communication where integrins are regulating activation and specificity of growth factors receptors, while their functions also being modulated by growth factors [36]. This reciprocal modulatory mechanism determines the form and presentation of different ligands within ECM depending on physiological requirements. Rather than biochemical cues, ECM biophysical properties have also been shown to affect the type of adhesion structure formed following integrins activation [37]. Therefore, both biochemical and biophysical aspects of ECM should be considered when evaluating cellular behaviors.

12.3 ECM-cell interaction: Cell fate and biophysical cues

Both intrinsic and extrinsic mechanisms are involved in the determination of stem cell fate. Intrinsic transcriptional pathways interact with extrinsic signaling from local microenvironment to modulate cell response in its niche. Extrinsic signals are received by cells from different sources, such as: ECM, growth factors, proteins, and other neighboring cells [38].

ECM provides a unique extrinsic microenvironment not only specific in chemical composition but also highly specialized in physical parameters such as elasticity, stiffness, and topography (Fig. 12.3). Therefore, a particular cell in its local microenvironment is subjected to different biomechanical forces from the surrounding ECM and neighboring cells. These external mechanical forces from the outer environment are conveyed to cells via ECM. Subsequently, the cells modulate their actin cytoskeleton by the generation of internal forces that are transmitted to their local ECM [39,40]. Consequently, these internal forces may result in alterations of cell morphology and size. The ECM-cell interplay allows fine tuning of cell behavior and maintenance of specific cell morphology through the generation of isometric tension within the cytoskeleton [39,41].

Many studies have investigated the mechanism of interactions between cell morphology and its ECM that ultimately influences the stem cell lineage performances.

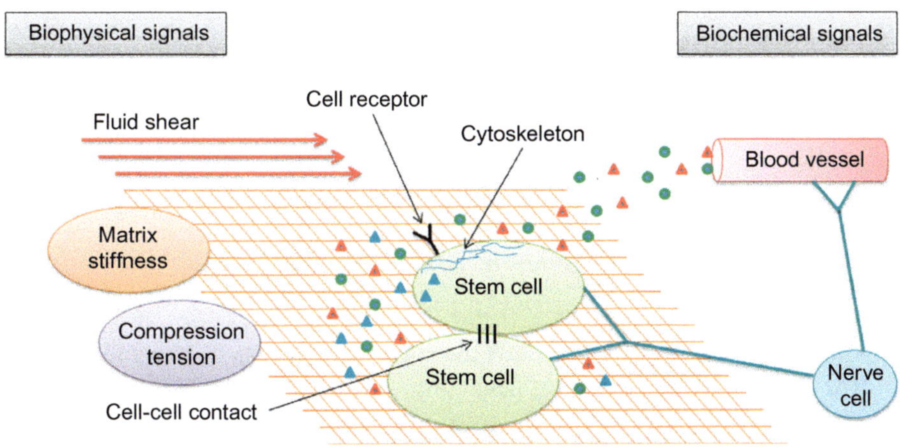

Fig. 12.3 Stem cell microenvironment. Different elements contribute to the regulation of stem cell behavior within its microenvironment that can be categorized under biophysical and biochemical signals. Biophysical signals are received by a cell from fluid shear, matrix stiffness or elasticity, compression and tension forces, and cell-cell contacts. Biochemical signals are received from blood supply as soluble factors, nervous system stimulation, and cell-released factors. *Orange mesh*: ECM; *green circle*: oxygen molecules; *red triangle*: signaling factors from the blood vessel (humoral factors; i.e., hormones); *blue triangle*: signaling factors excreted by cells (autocrine and paracrine factors; i.e., growth and differentiation factors); *black lines*: cell-cell contact.

Several approaches have been developed to study the in vivo stem cell-ECM interactions at single cell level by using specialized custom-made substrates such as micropatterned islands and microfluidic chips. These procedures allow single cell accommodation without contact with other cells, and they direct each cell to attain the specific morphologies. The substrate pattern and the spatial distribution of ECM can influence the cell orientation and distribution pattern of ECM. Furthermore, the island area or the substrates with specific morphologies can also determine the differentiation profile of MSC. It has been shown that the substrates which permit adhesion, flattening, and spreading of cells, allow osteogenic differentiation of MSC, while those that prevent cell spreading and limit them to round morphology are adipogenic [42,43].

The cell morphology itself, along with other cues may act as a signal to regulate the differentiation profile of MSC [44]. The cell morphology provides mechanical cues within ECM that can regulate the lineage commitment of MSC. Cell morphology can modulate intrinsic RhoA/Rho-associated protein kinase (ROCK) activity, where active and dominant-negative RhoA result in osteogenic and adipogenic lineage commitments of MSC, respectively. However, the RhoA/ROCK-mediated osteogenesis was shown to be dependent on flat shape as well. Furthermore, it has been reported that cell morphology and cytoskeletal mechanics can regulate BMP-2 induced signaling and osteogenesis commitment of MSC. Using micro-patterned substrate, it has been demonstrated that BMP induced osteogenesis is inhibited by restricting cell spreading, reducing ROCK signaling, and blocking cytoskeletal tension. Only in spread cells, BMP could induce rapid and constant RhoA/Rho-associated protein kinase (ROCK) activity and contractile cytoskeletal tension that is required for BMP-induced osteogenesis [45].

Biomechanical signaling also plays crucial roles in embryogenesis and development. Mechanical forces provide instructive information for normal tissue homeostasis including stem cell maintenance, proliferation, and differentiation. It has been shown that only stiff substrates (10 kPa hydrogel compared to the soft substrate) can maintain long-term human embryonic stem cells proliferation and pluripotency where the cells can bind to the substrate via glycosaminoglycans and respond to mechanical stimuli transmitted via glycosaminoglycans engagement. Furthermore, stiff matrix involved in activation of transcriptional co-activators proteins YAP/TAZ which are required in mechanosensing and stem cell pluripotency [46].

The mechanical properties of ECM play important roles in pathological processes as well. The rigidity of ECM affects the proliferation and metastasis of cancer cells. Indeed, it has been shown that proliferation rate of cancer cells can increase two- to fivefolds on stiffer substrates [47]. The hallmark for cancer development within the soft tissue is an increased rigidity, for example, breast tumors develop an elastic modulus of about 4000 Pa compared to a healthy mammary gland which is only about 160 Pa [48].

In summary, it is important to highlight the impact of both biochemical and biophysical cues interplay during stem cell differentiation and functions. With regard to the biophysical cues, the cells respond to overall intrinsic and extrinsic mechanical cues by remodeling of actin-myosin and generation of cytoskeletal tension that influence cell morphology which in turn modify cell fate [42,49]. It should be mentioned that the interactions between cell and ECM is dynamic and reciprocal. This is because the cells response to different extrinsic stimuli by continuous remodeling of their microenvironment which subsequently influence their overall fate. Therefore, it is crucial to study and understand different types of biophysical cues from ECM or substrate that have a role in modulating the cell behaviors. Here, we explain briefly the importance of some biophysical factors such as stiffness, elasticity, tension, compression, and fluid shear stress.

12.3.1 Stiffness and matrix elasticity

It has been demonstrated that stem cell behavior is influenced by ECM composition and organization that in turn determine the stiffness [4,50]. The Young's modulus represents the stiffness (tissue elasticity) of ECM microenvironment which is also affected by aging, disease and after injury [51,52].

The matrix elasticity can regulate the stem cell lineage specification (Fig. 12.3). The relationship between substrate stiffness and differentiation profile of stem cells has also been investigated in detail. It is reported that the substrate stiffness cannot influence embryonic stem cells attachment, however the increased substrate stiffness can encourage cell spreading, proliferation, and differentiation. For example, a substrate with similar rigidity and stiffness to that of collagenous natural bone marrow could stimulate osteogenic differentiation while a relatively soft matrix identical to brain tissue elasticity proved to be neurogenic [53–55]. It has been shown through the study of electrospun nanofibers with similar microstructures and surface chemistries but of different modulus that stiffer fibers with higher modulus can promote enhanced osteogenesis by increasing osteogenic Runx2, ALP activity, and osteocalcin gene expression, while nanofibers of lower modulus support chondrogenesis [56]. Furthermore, mineralization of MSC has also been linked to substrate stiffness [37]. Highlighting the impact of elasticity and stiffness, it is reported that although the reprogramming of stem cells during the initial culture time is possible, ultimately, the stem cells undergo differentiation pattern determined by matrix elasticity and stiffness [54].

Furthermore, it has been shown that elasticity of ECM can regulate self-renewal and multipotency beside differentiation of MSC. When human bone marrow-derived MSC are cultured on the soft hydrogel, they become quiescent, but these nonproliferative cells could reenter the cell cycle once cultured on a stiff substrate. In fact, the microenvironment of bone marrow hMSC can mediate the biomechanical signals from ECM elasticity that enable long-term maintenance of MSC in bone marrow niche [57].

The stiffness of ECM has also been found to modulate durotaxis—migration of cells within ECM toward the area of higher stiffness [58]. However, it is reported that the initial cell attachment to the substrate is independent of substrate rigidity. Moreover, the formation and spreading of focal adhesion are linked directly to the stiffness of ECM [59,60].

There is emerging evidence that modification of substrate stiffness is an efficient and promising way in modulating cellular responses. Therefore, many studies attempted to investigate and analyze cell response toward a various range of biophysical cues such as stiffness. Furthermore, in addition to substrate stiffness that influences the stem cell behaviors to various concentration of ECM protein attached to the substrate, the stem cells can also perceive the ECM tethering, independently [61,62]. The responses of MSC can be modulated by controlling the concentration of cross-linkers used to tether collagen fibers [61]. As a result, the idea of gradients stiffness has also been put forward demonstrating that substrate platform exhibiting continuously graded stiffness is superior as they allow systematic exploration and modification of cell response [37].

Other important topics to consider is the density and stiffness of ECM proteins (collagen) themselves [63]. A new insight regarding the mechanical properties of ECM has revealed the importance of collagen fibers network in regulating the stem cell behaviors through their stiffness characteristics. The stem cells determine their fate by exerting a mechanical force on local collagen network (substrate-bound ECM) and assessing the mechanical feedback from them. For example, hydrogel substrates of a variety of stiffness can alter this mechanical feedback from collagen that results in different anchoring densities. It is reported that the nature of cell-ECM interface and its mechanical properties vary depending on the type of substrate. Loose bonding of ECM to the substrate cannot initiate the required mechanical feedback for integrins complex to function appropriately [61]. It has been shown that the ability of cells to spread is limited to the availability of collagen binding sites, where initially increasing the collagen density resulted in increased cell spreading, while at high collagen concentration this relationship was reversed. This could be related to limited availability of free integrins on the cell surface at higher ECM protein concentration [63].

An interesting topic is how deeply the cells can feel into a substrate matrix. The adhesive spreading and the responses of MSC have been evaluated on an elastic gel of various thicknesses that were affixed onto a rigid substrate. The results indicated that MSC can respond to the rigidity of underlying "hidden" substrate at an average of 15-μm-gel thickness with a characteristic tactile length of less than about 5 μm [64].

In general, it is accepted that pliability and biophysical stimuli such as microstructural stiffness of substrate or a dynamic scaffold could play a critical role in controlling stem cell morphology, proliferation, apoptosis, and differentiation fate by modulating their cytoskeletal organization [56,65]. In fact, the cell gathers information about its environment by converting the received data from local forces or geometry into biochemical signals that ultimately results in modulation of cell response. Therefore, the fact that the relationship between mechanical environments such as substrate stiffness and cell behaviors is translated through the mechanical feedback should be considered in the fabrication of biomaterials as it plays an important role in the early and late cell differentiation [55]. Application of these fact during biomaterials synthesis would allow better control of stem cell fate in tissue engineering [61]. However, it should be noted that the final cell commitment is influenced by the interplay of all other factors, such as substrate physicochemical properties, tension, compression, shear stress of interstitial fluids, cell morphology, and cell spreading.

12.3.2 Tension and compression

The maintenance of the normal tissues such as bone physiologic functions, requires the mechanical loading in the form of both traction and compression stimuli. The absence of mechanical forces influences the stem cell functions and fate which subsequently results in disuse atrophy or bone volume shrinkage. For example, it has been demonstrated in animal studies that upon bone immobilization and reduction of mechanical loading, there was a 50% reduction in the number of MSC isolated from unloaded bone marrow. Furthermore, these MSC showed a reduced osteogenic differentiation potential by exhibiting a decreased ALP activity and osteopontin gene expression [66,67]. Therefore, lack of proper mechanical loading negatively affects both proliferation and differentiation profile of MSC which results in disuse atrophy of the bone.

From the clinical aspect, the idea of tension/compression is commonly applied in orthopedic or orthodontic fields. For example, in distraction osteogenesis, the bone and periosteum are subjected to tension forces that stimulate MSC to differentiate into bone forming cells and lay down bone. The same concept is true in orthodontic tooth movement in dental field where the area of tooth subjected to compression stimulate osteoclastic differentiation and bone resorption while new bone will be generated at the area of tension.

Cells subjected to tension forces can exhibit different behaviors based on type and magnitude of cell traction forces. It has been shown that the applied tension can induce osteogenic lineage commitment of stem cells. However, the magnitude, duration, and frequency of application are different among studies. For example, when human mesenchymal stem cells (hMSCs) were confined to within various shapes of convex and concave edges by micropatterning technique, cells at convex edges that subjected to higher traction forces exhibited a higher osteogenic profile with a greater ALP activity [68].

Moreover, the tensile strength range of 8%–12% has been shown to result in reduced proliferation rate and increased expression of various bone markers such as Runx2, ALP, collagen type 1, and BMP-2 [69–71]. Similarly, a lower level of tension (0.2%–5%) inhibited the expression of adipogenic, chondrogenic, and neurogenic markers but promoted expression of osteogenic gene markers, higher ALP activity, and matrix mineralization [72–75].

Furthermore, the applied compression forces have also been investigated in the literature showing their ability to induce chondrogenic and osteogenic lineage differentiation of hMSC. For example, application of 10% cyclic compression along with continuous perfusion promoted osteogenic lineage commitment by increasing expression of bone markers in MSC, such as Runx2 and osteocalcin [76]. However, it is noted that both tension and compression forces and their interplay may be more important for MSC residing in the periosteum [77].

Application of external physical force can also affect the reorganization of focal adhesions and increase the bindings between ECM and integrins receptors [78]. For instance, it has been shown that mechanical stretch results in reorientation of stress fibers, therefore, many cell types reorient their actin cytoskeleton and main cell axis almost perpendicular to the direction of stretch [79].

Critical analysis of the literature reveals that the role of biomechanical interactions on large-scale cellular and matrix organization is mostly unknown. Focal adhesion sites mediate the attachment of cells to ECM and allow cells to transmit contractile forces to substrate via cytoskeleton actin fibers. This biomechanical contraction produces traction stresses within ECM and substrate. Traction force microscopy (TFM) allows mapping of traction fields of a single cell [80]. In TFM, the magnitude of substrate displacement caused by cells contractility is tracked and measured as traction stress field. The measurement of traction forces provides quantitative information about the nature of ECM-cell interaction. Different studies attempted to evaluate and quantify ECM-cell interaction in two-dimensional (2D) environments. Recently, new approaches like three-dimensional traction force microscopy (3D TFM) technique allowed 3D imaging with the ability to compute cellular traction forces dynamically during cell migration and locomotion [81]. In addition to traction forces, other biomechanical cues may affect the traction stress, such as cell stiffness, adhesion energy density, and cell morphology [82]. The alignment of cells in the direction of maximum stiffness has been observed as a consequence of ECM-cell re-arrangement due to the accumulation of tension forces. This indicates the close link between the cellular organization and biomechanical feedbacks [83]. Understanding of these mechanisms can be useful for design and development of optimized scaffolds for better tissue engineering applications.

12.3.3 Fluid shear stress

The fluid shear stress or tension is also important factor that can modify stem cell fate (Fig. 12.3). The technique of application, direction pattern, magnitude, duration, and frequency can influence cell fate and produce different proliferation and differentiation pattern. It has been reported that the fluid shear can produce mechanical stimuli at the cellular level with potential induction of osteogenic lineage [84]. Furthermore, exposure of cells to fluid flow can reduce the proliferation rate but enhance the expression of collagen type 1 and matrix mineralization. However, other bone markers (e.g., osteopontin and sialoprotein) were more dependent on the type of substrate [85]. Studies have shown that both static and dynamic hydrostatic pressures can induce osteogenic lineage commitment of stem cells [86]. Both fluid shear and hydrostatic pressure may be the dominant mechanical stimuli pathways for MSC residing in bone marrow [77].

12.4 Cell perception of biophysical cues from the ECM microenvironment

Cells try to cope with their ECM through the physicochemical interactions by modulating their survival and proliferation and adapting the cytoskeletal structure. This is ultimately crucial in normal physiological processes, such as homeostasis, remodeling, and morphogenesis. Furthermore, as long as this interaction is bidirectional, the cells also actively modify their local environment in response to the perceived stimuli from them. Therefore, it is important to mention that how a stem cell is able to feel the mechanical and physical changes in their environment? Explanation of this mechanism helps the researchers in better designing of the substrate microenvironment structure with instructive clues that guide the stem cell into the required pathways. The cell is considered an intelligent basic structural unit that continuously evaluates the properties and nature of their microenvironment to determine the required modification and conformational changes. The term "mechanotransduction" can refer to overall cellular behaviors in response to biomechanical stimuli. All types of mechanical stimuli from local ECM, substrate and neighboring cells are primarily converted to internal biochemical signals by cellular mechanosensors. In addition, the extracellular mechanical stimuli are able to induce nucleus distortion by modification of nucleus architecture or by activation of various cytoplasmic signaling cascades through mechanotransduction pathways. These processes thereby alter gene expression through changes of transcription factors. Moreover, nuclear stiffness can also be changed through this nucleus remodeling by mechanoregulation, which has a potential influence on stem cell fate and differentiation profile [87,88]. In this context, different mechanotransduction pathways have been postulated among which "YAP/TAZ" transcriptional factors found to be the key mediators of the biological effects of ECM elasticity, cell geometry, and cytoskeletal organization [39,89].

In order to perceive the environmental mechanical stimuli and establish physicochemical interaction with ECM, the stem cells possess different mechanosensors including cytoskeleton, nonmotile primary cilia, integrins-based focal adhesions, gap junctions, membrane channels, and podosomes (Fig. 12.4) [90]. Here, three main cellular mechanosensors are discussed within the limit of this chapter.

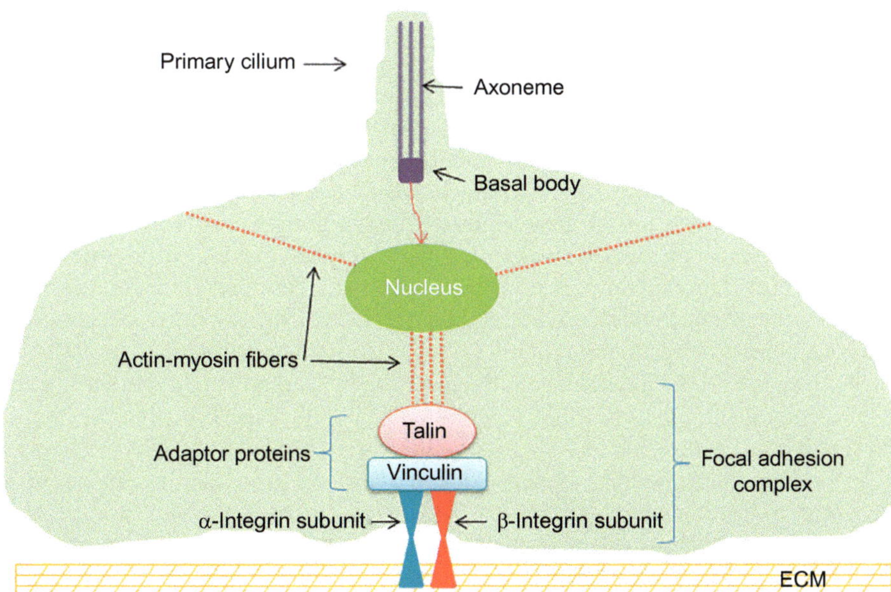

Fig. 12.4 Basic cellular mechanosensors. Cytoskeletal forces generated through the interaction of actin and myosin fibers are important mean of cellular mechanosensing. The primary cilium consists of parallel microtubules called "Axoneme." The "basal body" is responsible for anchoring the axoneme to the cell. Focal adhesion is another type of cellular mechanosensor which is a macromolecular protein complex consists of cell surface receptors (i.e., integrins), and adaptor proteins that provide a mean of connection between focal adhesion and cytoskeleton.

12.4.1 Focal adhesions

They are macromolecular proteins complex acting as one of the key cellular mechanosensors. They link the intracellular cytoskeleton to ECM through active anchorage through integrins family of cell surface receptors which is the basic component of focal adhesion. Adapter proteins such as *talin* and *vinculin* mediate the link between focal adhesions and cytoskeleton [90]. Due to their unique structural components, the focal adhesions are able to transmit both physical and chemical cues from and to ECM. The focal adhesions are also responsible for attachment of cells to the substrate involving integrins and other signaling molecules which transmit the physical stimuli to the substrate as well. It has been reported that an intact focal adhesion is important for mechanically induced osteogenesis [91].

12.4.2 The cytoskeletal

The cytoskeletal tension that is generated through actin and myosin fibers interactions is considered another important mean of cellular mechanosensing. This is mediated via focal adhesions as adaptor and signaling molecules that provide a link between the intracellular actin microfilaments and integrins receptors. This binds the cells to ECM, which is crucial for mechanical induction of stem cells to undergo osteogenic differentiation pathway. Maintenance of the integrity of focal adhesions is dependent on their connection to the actin cytoskeleton and the tensions applied to them via internal actin-mediated mechanism (rho-induced actomyosin contractility) or by external forces to cell surface receptors [92]. It has been reported that increase in intracellular tension produces biomechanical forces that result in conformational changes by unfolding of some proteins or modifying protein's functions, thereby, generating biochemical signaling [65].

Furthermore, the ability of cells to apply cytoskeletal forces to their ECM via integrins receptors is important for their adherence and migration process. The recruitment of *vinculin* as a marker for a focal complex that are precursors of focal adhesion allows exertion of force by initial integrins-ECM adhesions. Therefore, cells use mechanical stimuli as the means to reinforce initial integrins-ECM adhesion into focal complexes and control the number of migration forces applied to focal adhesion sites [92].

Studies have attempted to explain the mechanism for direct and rapid transduction of mechanical stimuli from ECM to inner cells and nucleus. For example, this has been proposed to be mediated by a specialized *peri-nuclear cap* that is composed of actomyosin microfilament bundles. These actin cap fibers use the linkers of nucleoskeleton, cytoskeleton, and the *LINC complex* (linker of the nucleoskeleton and cytoskeleton) to bind to nucleus from one side and then connect the nucleus to specific focal adhesion sites at other ends [88].

12.4.3 The primary cilium

The nonmotile primary cilia are slender microscopic protuberances or organelles that project from the surface of nearly all mammalian cells into ECM. Primary cilia appear typically as the appendages microtubules on the apical surface of cells. They particularly serve as sensory organelles receiving signals from ECM and neighboring cells and play a crucial role in human and animal development. Some genetic diseases may result in dysfunction of primary cilia that cause serious health burden conditions termed *ciliopathies* [93]. The primary cilia have been shown to be the important signaling center acting as the means for mechanosensation in multiple cell types. They also act as a microdomain for housing a various concentrations of proteins and enzymes [94]. The cells (osteocytes) respond to the mechanical stimuli by release of soluble factors that induce and enhance osteogenic gene expression in MSC. The primary cilia also play an important role in the normal osteogenesis process by providing a sensory mechanism for various extracellular biochemical and biomechanical signals including fluid shear stress. This has been proved by blocking the formation of primary cilia in MSC where the osteogenic differentiation did not occur due to lack of upregulation of osteopontin as osteogenic gene expression factors [95].

It is well known that the maintenance of normal bone homeostasis requires a continuous and appropriate mechanical stimulus which supports the bone remodeling by osteogenic differentiation of MSC. The role of primary cilia in the cellular mechanotransduction and lineage commitment of MSC has been recently emphasized. A higher hMSC proliferation rate and an enhanced osteogenic gene expression were reported in response to a short period of mechanical stimuli (oscillatory fluid flow) due to an increase in cyclooxygenase-2 and BMP-2 gene expression. Thereby, primary cilium has been postulated as an important mechanosensor that directly regulates these mechanically induced signaling in hMSC and their osteogenic differentiation. The primary cilia could be a potential target for controlling stem cell fate for enhanced bone regeneration and therapeutic tissue engineering [95,96].

Disruption of action of a potential mechanosensors results in the inhibition of mechanotransduction pathways that interfere with mechanically induced osteogenesis [97]. There are multiple proposed pathways which play essential roles in osteogenic lineage commitment. Several mechanotransduction pathways have been postulated to be important in this process, such as calcium signaling pathways, mitogen activated protein kinase (MAPK)/ERK, Wnt, Hippo, and RhoA/ROCK pathways. However, the detailed mechanism and importance of these pathways are largely unknown and require further investigations.

12.5 Strategies for investigation of ECM and stem cell interaction

There are many reports in the literature about the responsiveness of MSC to biophysical stimuli. However, it is not clear yet how these mechanical factors should be applied and controlled to induce osteogenic lineage commitment. For example, different levels of tension have been postulated to be osteogenic [70–72,75], and many profiles of fluid flow shear stress are reported to promote osteogenic differentiation of MSC [84,85,98]. Furthermore, a relatively few studies investigated the mechanobiology nature and mechanotransduction pathways of stem cells. Thereby, the ideal methods of mechanical stimulation for induction of osteogenesis pathways are to be explored further. In addition, the mechanical signals are able to induce epigenetic changes of DNA without alteration of its sequence by reducing DNA methylation that can be passed to daughter cells as well. These changes may activate or deactivate different genes which in turn influence the cellular behaviors by controlling their osteogenic fate [84]. Finally, one should consider the possible differences in cellular responses toward biomechanical cues within the body when compared to in vitro studies.

The ideal model for investigation of stem cells-ECM interaction within the stem cell niche is in vivo study. This allows analysis of a particular component of ECM and its importance in modulating the stem cell functions [99,100]. However, the disadvantages of this approach are the complexity of niche microenvironment, the interaction of various factors and cofactors, and the difficulty of controlling them, the presence of compensatory mechanisms, and dynamic interactions between stem cells and ECM. Furthermore, improper function of a component within ECM may not influence the stem cells directly, but could generate sequences of events that produce indirect effects on stem cells.

In order to overcome these obstacles, in vitro mimicking of stem cell niche through engineering of specialized tools has been investigated by several research groups. Stem cell therapeutic approach requires proper techniques to re-create local stem cell microenvironment that can serve as a platform for cells transplantation for regenerative therapies. Synthesis of novel multiphasic biomaterials, fabrication of 3D scaffolds with nano-topographical modifications, development of bioactive hydrogel mimicking ECM, ECM micro-patterning and microarrays are among the major interesting efforts made in this field [101,102].

Development and study of bioactive hydrogels [103] allowed a better understanding of stem cell biology as these hydrogels can better mimic ECM by retaining a remarkable high water volume. Furthermore, the stiffness and other physicochemical properties [104] of these hydrogels can be easily controlled reproducing healthy and pathological conditions [105–107]. Moreover, gradients of stiffness have also been re-created in the hybrid hydrogel to study the influences of various stiffness on stem cell fate simultaneously [108].

The high-throughput ECM microarray as a micro-scale engineering strategy is another useful approach that allows reconstruction and screening of individual niche components and their role in stem cell-ECM interactions [101,109–111]. Another approach is the use of micro-patterned islands that allow patterning of ECM onto a substrate with predefined shape, size, and topography for stem cell adherence. This approach can be used to investigate the influence of different factors on stem cell fate and behaviors [112,113]. It is known that special topographical features can direct MSC toward osteogenic lineage and differentiation; however, the nature of the interplay between surface topography and cell behavior is not yet understood [114]. This opens a new research window for development and optimization of novel bioactive surfaces with possible smart or instructive features to guide the biological performances of cells.

Rather than cell-ECM interaction, the impact of cell-cell interaction should also be considered as it plays an essential role in differentiation and fate determination of stem cells [115]. This is true, since the stem cells work in a group rather than individually, and isolated cells have shown less intention to differentiate. Furthermore, to better simulate the exposure of stem cells to various growth factors, researchers have developed microfluidic systems to evaluate cell response to spatiotemporal exposure of various factors [116]. Bio-mimicking approaches also allow incorporation of various factors into biomaterials allowing their controlled release during biodegradation [101].

Despite the obvious advancements in the in vitro study of ECM-stem cell interplay, the major challenges still exist in the reconstruction of 3D microenvironment similar to in vivo condition. It is well known that the cells respond differently once exposed to the 3D environment [117,118]. With this aim, bio-encapsulation of stem cells inside various hydrogels has been investigated [119,120]. Alternatively, recent advances in 3D bio-printing techniques may provide promising alternative approaches. This is because they allow high-throughput and spatially defined immobilization of cells, ECM components, and other biomolecules on different surfaces to study stem cell behavior [119].

Recently, the idea of 4D biology has been postulated which allows a full spatiotemporal control of ECM functional molecules for dynamic evaluation of ECM signals and their influence on cell behavior over time [121]. Therefore, the future perspectives will be focused on controlling the ECM-stem cell interaction, guiding stem cells proliferation and differentiation profile, and manipulating the pattern of ECM deposition to allow synthesis and regeneration of different tissue and organs [122,123].

12.6 Conclusion

Cells require structural 3D support from their ECM, and the dynamic ECM-cell and cell-cell interaction influence their fate. The dynamic interaction between biochemical processes and biomechanical signaling is crucial during various physiological and biological processes, such as embryogenesis, morphogenesis, wound healing, and cellular behaviors. It also plays important rule in pathological processes, such as tumor development and metastasis. The cells continuously interact with surrounding ECM and substrate during these processes, and the feedback signals play a crucial role to define cellular behaviors [81]. It has been demonstrated by various studies that the ECM provides an instructive environment for stem cells, as well as other local cells. This is achieved by ECM unique biomechanical and biochemical properties and the dynamic interplay between cells and ECM. The importance of ECM nanotopography and rigidity in enhancing the proliferation and inducing the differentiation of stem cells has also been highlighted. All of these aspects are dynamically orchestrated to provide instructive cues for physiological cell behavior. The studies involving the interaction of stem cells and ECM will open new horizons as they allow more accurate mimicking of essential properties of in vivo niches by synthesizing and designing instructive and smart in vitro systems. Alternatively, these tissue-specific extracellular networks can act as a nanocomposite scaffold for hosting, maintaining, and directing various stem cells activities [124–126]. Further in vitro studies together with in vivo applications of these engineered biomaterials can open new scope to improve the knowledge of stem cell biology and develop new areas in cell-based therapy and tissue engineering.

References

[1] R.O. Hynes, The extracellular matrix: not just pretty fibrils. Science 326 (2009) 1216–1219, https://doi.org/10.1126/science.1176009.

[2] P. Lu, K. Takai, V.M. Weaver, Z. Werb, Extracellular matrix degradation and remodeling in development and disease. Cold Spring Harb. Perspect. Biol. 3 (2011), https://doi.org/10.1101/cshperspect.a005058.

[3] P. Lu, V.M. Weaver, Z. Werb, The extracellular matrix: a dynamic niche in cancer progression. J. Cell Biol. 196 (2012) 395–406, https://doi.org/10.1083/jcb.201102147.

[4] C.C. DuFort, M.J. Paszek, V.M. Weaver, Balancing forces: architectural control of mechanotransduction. Nat. Rev. Mol. Cell Biol. 12 (2011) 308–319, https://doi.org/10.1038/nrm3112.

[5] A. Kurtz, S.-J. Oh, Age related changes of the extracellular matrix and stem cell maintenance. Prev. Med. (Baltim) 54 (Suppl) (2012) S50–S56, https://doi.org/10.1016/j.ypmed.2012.01.003.

[6] R. Schofield, The relationship between the spleen colony-forming cell and the haemopoietic stem cell, Blood Cells 4 (1978) 7–25, http://www.ncbi.nlm.nih.gov/pubmed/747780.

[7] E. Fuchs, T. Tumbar, G. Guasch, Socializing with the neighbors: stem cells and their niche, Cell 116 (2004) 769–778, http://www.ncbi.nlm.nih.gov/pubmed/15035980.

[8] S.J. Morrison, A.C. Spradling, Stem cells and niches: mechanisms that promote stem cell maintenance throughout life. Cell 132 (2008) 598–611, https://doi.org/10.1016/j.cell.2008.01.038.

[9] K.W. Orford, D.T. Scadden, Deconstructing stem cell self-renewal: genetic insights into cell-cycle regulation. Nat. Rev. Genet. 9 (2008) 115–128, https://doi.org/10.1038/nrg2269.

[10] L. Li, H. Clevers, Coexistence of quiescent and active adult stem cells in mammals. Science 327 (2010) 542–545, https://doi.org/10.1126/science.1180794.

[11] V. Greco, S. Guo, Compartmentalized organization: a common and required feature of stem cell niches? Development 137 (2010) 1586–1594, https://doi.org/10.1242/dev.041103.

[12] A.D. Lander, J. Kimble, H. Clevers, E. Fuchs, D. Montarras, M. Buckingham, A.L. Calof, A. Trumpp, T. Oskarsson, What does the concept of the stem cell niche really mean today? BMC Biol. 10 (2012) 19, https://doi.org/10.1186/1741-7007-10-19.

[13] A.J. Wagers, The stem cell niche in regenerative medicine. Cell Stem Cell 10 (2012) 362–369, https://doi.org/10.1016/j.stem.2012.02.018.

[14] D.L. Jones, A.J. Wagers, No place like home: anatomy and function of the stem cell niche. Nat. Rev. Mol. Cell Biol. 9 (2008) 11–21, https://doi.org/10.1038/nrm2319.

[15] M. Walker, K. Patel, T. Stappenbeck, The stem cell niche. J. Pathol. 217 (2009) 169–180, https://doi.org/10.1002/path.2474.

[16] F.M. Watt, H. Fujiwara, Cell-extracellular matrix interactions in normal and diseased skin. Cold Spring Harb. Perspect. Biol. 3 (2011) 1–14, https://doi.org/10.1101/cshperspect.a005124.

[17] D.E. Discher, D.J. Mooney, P.W. Zandstra, Growth factors, matrices, and forces combine and control stem cells. Science 324 (2009) 1673–1677, https://doi.org/10.1126/science.1171643.

[18] J.J. Song, H.C. Ott, Organ engineering based on decellularized matrix scaffolds. Trends Mol. Med. 17 (2011) 424–432, https://doi.org/10.1016/j.molmed.2011.03.005.

[19] D.C. Sullivan, S.-H. Mirmalek-Sani, D.B. Deegan, P.M. Baptista, T. Aboushwareb, A. Atala, J.J. Yoo, Decellularization methods of porcine kidneys for whole organ engineering using a high-throughput system. Biomaterials 33 (2012) 7756–7764, https://doi.org/10.1016/j.biomaterials.2012.07.023.

[20] K.H. Nakayama, C.A. Batchelder, C.I. Lee, A.F. Tarantal, Decellularized rhesus monkey kidney as a three-dimensional scaffold for renal tissue engineering. Tissue Eng. Part A 16 (2010) 2207–2216, https://doi.org/10.1089/ten.tea.2009.0602.

[21] I.D. Campbell, M.J. Humphries, Integrin structure, activation, and interactions. Cold Spring Harb. Perspect. Biol. 3 (2011) a004994, https://doi.org/10.1101/cshperspect.a004994.

[22] Y. Takada, X. Ye, S. Simon, The integrins. Genome Biol. 8 (2007) 215, https://doi.org/10.1186/gb-2007-8-5-215.

[23] A.E. Berman, N.I. Kozlova, G.E. Morozevich, Integrins: structure and signaling, Biochemistry (Mosc) 68 (2003) 1284–1299, http://www.ncbi.nlm.nih.gov/pubmed/14756624.

[24] K.R. Legate, S.A. Wickström, R. Fässler, Genetic and cell biological analysis of integrin outside-in signaling. Genes Dev. 23 (2009) 397–418, https://doi.org/10.1101/gad.1758709.

[25] M.F. Brizzi, G. Tarone, P. Defilippi, Extracellular matrix, integrins, and growth factors as tailors of the stem cell niche. Curr. Opin. Cell Biol. 24 (2012) 645–651, https://doi.org/10.1016/j.ceb.2012.07.001.

[26] C. Margadant, A. Sonnenberg, Integrin-TGF-beta crosstalk in fibrosis, cancer and wound healing. EMBO Rep. 11 (2010) 97–105, https://doi.org/10.1038/embor.2009.276.

[27] J. Ivaska, J. Heino, Cooperation between integrins and growth factor receptors in signaling and endocytosis. Annu. Rev. Cell Dev. Biol. 27 (2011) 291–320, https://doi.org/10.1146/annurev-cellbio-092910-154017.

[28] J.D. Humphries, A. Byron, M.J. Humphries, Integrin ligands at a glance. J. Cell Sci. 119 (2006) 3901–3903, https://doi.org/10.1242/jcs.03098.

[29] M. Barczyk, S. Carracedo, D. Gullberg, Integrins. Cell Tissue Res. 339 (2010) 269–280, https://doi.org/10.1007/s00441-009-0834-6.

[30] P. Defilippi, A. Rosso, P. Dentelli, C. Calvi, G. Garbarino, G. Tarone, L. Pegoraro, M.F. Brizzi, {beta}1 Integrin and IL-3R coordinately regulate STAT5 activation and anchorage-dependent proliferation. J. Cell Biol. 168 (2005) 1099–1108, https://doi.org/10.1083/jcb.200405116.

[31] H.N. Suh, H.J. Han, Collagen I regulates the self-renewal of mouse embryonic stem cells through $\alpha2\beta1$ integrin- and DDR1-dependent Bmi-1. J. Cell. Physiol. 226 (2011) 3422–3432, https://doi.org/10.1002/jcp.22697.

[32] J.A. Askari, C.J. Tynan, S.E.D. Webb, M.L. Martin-Fernandez, C. Ballestrem, M.J. Humphries, Focal adhesions are sites of integrin extension. J. Cell Biol. 188 (2010) 891–903, https://doi.org/10.1083/jcb.200907174.

[33] A. Avigdor, P. Goichberg, S. Shivtiel, A. Dar, A. Peled, S. Samira, O. Kollet, R. Hershkoviz, R. Alon, I. Hardan, H. Ben-Hur, D. Naor, A. Nagler, T. Lapidot, CD44 and hyaluronic acid cooperate with SDF-1 in the trafficking of human CD34+ stem/progenitor cells to bone marrow. Blood 103 (2004) 2981–2989, https://doi.org/10.1182/blood-2003-10-3611.

[34] S. Smith-Berdan, A. Nguyen, D. Hassanein, M. Zimmer, F. Ugarte, J. Ciriza, D. Li, M.E. García-Ojeda, L. Hinck, E.C. Forsberg, Robo4 cooperates with CXCR4 to specify hematopoietic stem cell localization to bone marrow niches. Cell Stem Cell 8 (2011) 72–83, https://doi.org/10.1016/j.stem.2010.11.030.

[35] F. Gattazzo, A. Urciuolo, P. Bonaldo, Extracellular matrix: a dynamic microenvironment for stem cell niche. Biochim. Biophys. Acta 1840 (2014) 2506–2519, https://doi.org/10.1016/j.bbagen.2014.01.010.

[36] N. Alam, H.L. Goel, M.J. Zarif, J.E. Butterfield, H.M. Perkins, B.G. Sansoucy, T.K. Sawyer, L.R. Languino, The integrin-growth factor receptor duet. J. Cell. Physiol. 213 (2007) 649–653, https://doi.org/10.1002/jcp.21278.

[37] P.-Y. Wang, W.-B. Tsai, N.H. Voelcker, Screening of rat mesenchymal stem cell behaviour on polydimethylsiloxane stiffness gradients. Acta Biomater. 8 (2012) 519–530, https://doi.org/10.1016/j.actbio.2011.09.030.

[38] F.M. Watt, R.R. Driskell, The therapeutic potential of stem cells. Philos. Trans. R. Soc. B Biol. Sci. 365 (2010) 155–163, https://doi.org/10.1098/rstb.2009.0149.

[39] G. Halder, S. Dupont, S. Piccolo, Transduction of mechanical and cytoskeletal cues by YAP and TAZ. Nat. Rev. Mol. Cell Biol. 13 (2012) 591–600, https://doi.org/10.1038/nrm3416.

[40] D.E. Discher, Tissue cells feel and respond to the stiffness of their substrate. Science 310 (2005) 1139–1143, https://doi.org/10.1126/science.1116995.

[41] A. Mammoto, T. Mammoto, D.E. Ingber, Mechanosensitive mechanisms in transcriptional regulation. J. Cell Sci. 125 (2012) 3061–3073, https://doi.org/10.1242/jcs.093005.

[42] R. McBeath, D.M. Pirone, C.M. Nelson, K. Bhadriraju, C.S. Chen, Cell shape, cytoskeletal tension, and RhoA regulate stem cell lineage commitment, Dev. Cell 6 (2004) 483–495, http://www.ncbi.nlm.nih.gov/pubmed/15068789.

[43] S. Watari, K. Hayashi, J.A. Wood, P. Russell, P.F. Nealey, C.J. Murphy, D.C. Genetos, Modulation of osteogenic differentiation in hMSCs cells by submicron topographically-patterned ridges and grooves. Biomaterials 33 (2012) 128–136, https://doi.org/10.1016/j.biomaterials.2011.09.058.

[44] F.M. Watt, P.W. Jordan, C.H. O'Neill, Cell shape controls terminal differentiation of human epidermal keratinocytes, Proc. Natl. Acad. Sci. U. S. A. 85 (1988) 5576–5580, http://www.ncbi.nlm.nih.gov/pubmed/2456572.

[45] Y.-K. Wang, X. Yu, D.M. Cohen, M.A. Wozniak, M.T. Yang, L. Gao, J. Eyckmans, C.S. Chen, Bone morphogenetic protein-2-induced signaling and osteogenesis is regulated by cell shape, RhoA/ROCK, and cytoskeletal tension. Stem Cells Dev. 21 (2012) 1176–1186, https://doi.org/10.1089/scd.2011.0293.

[46] S. Musah, S.A. Morin, P.J. Wrighton, D.B. Zwick, S. Jin, L.L. Kiessling, Glycosaminoglycan-binding hydrogels enable mechanical control of human pluripotent stem cell self-renewal. ACS Nano 6 (2012) 10168–10177, https://doi.org/10.1021/nn3039148.

[47] R.W. Tilghman, E.M. Blais, C.R. Cowan, N.E. Sherman, P.R. Grigera, E.D. Jeffery, J.W. Fox, B.R. Blackman, D.J. Tschumperlin, J.A. Papin, J.T. Parsons, Matrix rigidity regulates cancer cell growth by modulating cellular metabolism and protein synthesis. PLoS ONE 7 (2012) e37231, https://doi.org/10.1371/journal.pone.0037231.

[48] M.J. Paszek, N. Zahir, K.R. Johnson, J.N. Lakins, G.I. Rozenberg, A. Gefen, C.A. Reinhart-King, S.S. Margulies, M. Dembo, D. Boettiger, D.A. Hammer, V.M. Weaver, Tensional homeostasis and the malignant phenotype. Cancer Cell 8 (2005) 241–254, https://doi.org/10.1016/j.ccr.2005.08.010.

[49] M. Théry, V. Racine, A. Pépin, M. Piel, Y. Chen, J.-B. Sibarita, M. Bornens, The extracellular matrix guides the orientation of the cell division axis. Nat. Cell Biol. 7 (2005) 947–953, https://doi.org/10.1038/ncb1307.

[50] T. Mammoto, D.E. Ingber, Mechanical control of tissue and organ development. Development 137 (2010) 1407–1420, https://doi.org/10.1242/dev.024166.

[51] B.D. Cosgrove, A. Sacco, P.M. Gilbert, H.M. Blau, A home away from home: challenges and opportunities in engineering in vitro muscle satellite cell niches. Differentiation 78 (2009) 185–194, https://doi.org/10.1016/j.diff.2009.08.004.

[52] A.J. Keung, E.M. de Juan-Pardo, D.V. Schaffer, S. Kumar, Rho GTPases mediate the mechanosensitive lineage commitment of neural stem cells. Stem Cells 29 (2011) 1886–1897, https://doi.org/10.1002/stem.746.

[53] P.C. Georges, W.J. Miller, D.F. Meaney, E.S. Sawyer, P.A. Janmey, Matrices with compliance comparable to that of brain tissue select neuronal over glial growth in mixed cortical cultures. Biophys. J. 90 (2006) 3012–3018, https://doi.org/10.1529/biophysj.105.073114.

[54] A.J. Engler, S. Sen, H.L. Sweeney, D.E. Discher, Matrix elasticity directs stem cell lineage specification. Cell 126 (2006) 677–689, https://doi.org/10.1016/j.cell.2006.06.044.

[55] N.D. Evans, C. Minelli, E. Gentleman, V. LaPointe, S.N. Patankar, M. Kallivretaki, X. Chen, C.J. Roberts, M.M. Stevens, Substrate stiffness affects early differentiation events in embryonic stem cells, Eur. Cell. Mater. 18 (2009) 1-13–4, http://www.ncbi.nlm.nih.gov/pubmed/19768669.

[56] J. Nam, J. Johnson, J.J. Lannutti, S. Agarwal, Modulation of embryonic mesenchymal progenitor cell differentiation via control over pure mechanical modulus in electrospun nanofibers. Acta Biomater. 7 (2011) 1516–1524, https://doi.org/10.1016/j.actbio.2010.11.022.

[57] J.P. Winer, P.A. Janmey, M.E. McCormick, M. Funaki, Bone marrow-derived human mesenchymal stem cells become quiescent on soft substrates but remain responsive to chemical or mechanical stimuli. Tissue Eng. Part A 15 (2009) 147–154, https://doi.org/10.1089/ten.tea.2007.0388.

[58] S.V. Plotnikov, A.M. Pasapera, B. Sabass, C.M. Waterman, Force fluctuations within focal adhesions mediate ECM-rigidity sensing to guide directed cell migration. Cell 151 (2012) 1513–1527, https://doi.org/10.1016/j.cell.2012.11.034.

[59] X. Zhang, G. Jiang, Y. Cai, S.J. Monkley, D.R. Critchley, M.P. Sheetz, Talin depletion reveals independence of initial cell spreading from integrin activation and traction. Nat. Cell Biol. 10 (2008) 1062–1068, https://doi.org/10.1038/ncb1765.

[60] B. Zimerman, T. Volberg, B. Geiger, Early molecular events in the assembly of the focal adhesion-stress fiber complex during fibroblast spreading. Cell Motil. Cytoskeleton 58 (2004) 143–159, https://doi.org/10.1002/cm.20005.

[61] B. Trappmann, J.E. Gautrot, J.T. Connelly, D.G.T. Strange, Y. Li, M.L. Oyen, M.A. Cohen Stuart, H. Boehm, B. Li, V. Vogel, J.P. Spatz, F.M. Watt, W.T.S. Huck, Extracellular-matrix tethering regulates stem-cell fate. Nat. Mater. 11 (2012) 642–649, https://doi.org/10.1038/nmat3339.

[62] A. Engler, L. Bacakova, C. Newman, A. Hategan, M. Griffin, D. Discher, Substrate compliance versus ligand density in cell on gel responses. Biophys. J. 86 (2004) 617–628, https://doi.org/10.1016/S0006-3495(04)74140-5.

[63] C. Gaudet, W.A. Marganski, S. Kim, C.T. Brown, V. Gunderia, M. Dembo, J.Y. Wong, Influence of type I collagen surface density on fibroblast spreading, motility, and contractility. Biophys. J. 85 (2003) 3329–3335, https://doi.org/10.1016/S0006-3495(03)74752-3.

[64] A. Buxboim, K. Rajagopal, A.E.X. Brown, D.E. Discher, How deeply cells feel: methods for thin gels. J. Phys. Condens. Matter 22 (2010) 194116, https://doi.org/10.1088/0953-8984/22/19/194116.

[65] V. Vogel, M. Sheetz, Local force and geometry sensing regulate cell functions. Nat. Rev. Mol. Cell Biol. 7 (2006) 265–275, https://doi.org/10.1038/nrm1890.

[66] S. Keila, S. Pitaru, A. Grosskopf, M. Weinreb, Bone marrow from mechanically unloaded rat bones expresses reduced osteogenic capacity in vitro. J. Bone Miner. Res. 9 (1994) 321–327, https://doi.org/10.1002/jbmr.5650090306.

[67] R. Zhang, S.C. Supowit, G.L. Klein, Z. Lu, M.D. Christensen, R. Lozano, D.J. Simmons, Rat tail suspension reduces messenger RNA level for growth factors and osteopontin and decreases the osteoblastic differentiation of bone marrow stromal cells. J. Bone Miner. Res. 10 (1995) 415–423, https://doi.org/10.1002/jbmr.5650100312.

[68] S.A. Ruiz, C.S. Chen, Emergence of patterned stem cell differentiation within multicellular structures. Stem Cells 26 (2008) 2921–2927, https://doi.org/10.1634/stemcells.2008-0432.

[69] M. Jagodzinski, M. Drescher, J. Zeichen, S. Hankemeier, C. Krettek, U. Bosch, M. van Griensven, Effects of cyclic longitudinal mechanical strain and dexamethasone on osteogenic differentiation of human bone marrow stromal cells, Eur. Cell. Mater. 7 (2004) 35–41, http://www.ncbi.nlm.nih.gov/pubmed/15095254.

[70] P. Zhang, Y. Wu, Z. Jiang, L. Jiang, B. Fang, Osteogenic response of mesenchymal stem cells to continuous mechanical strain is dependent on ERK1/2-Runx2 signaling. Int. J. Mol. Med. 29 (2012) 1083–1089, https://doi.org/10.3892/ijmm.2012.934.

[71] R.D. Sumanasinghe, S.H. Bernacki, E.G. Loboa, Osteogenic differentiation of human mesenchymal stem cells in collagen matrices: effect of uniaxial cyclic tensile strain on bone morphogenetic protein (BMP-2) mRNA expression. Tissue Eng. 12 (2006) 3459–3465, https://doi.org/10.1089/ten.2006.12.3459.

[72] E.M. Kearney, E. Farrell, P.J. Prendergast, V.A. Campbell, Tensile strain as a regulator of mesenchymal stem cell osteogenesis. Ann. Biomed. Eng. 38 (2010) 1767–1779, https://doi.org/10.1007/s10439-010-9979-4.

[73] M.-C. Qi, J. Hu, S.-J. Zou, H.-Q. Chen, H.-X. Zhou, L.-C. Han, Mechanical strain induces osteogenic differentiation: Cbfa1 and Ets-1 expression in stretched rat mesenchymal stem cells. Int. J. Oral Maxillofac. Surg. 37 (2008) 453–458, https://doi.org/10.1016/j.ijom.2007.12.008.

[74] C.-H. Huang, M.-H. Chen, T.-H. Young, J.-H. Jeng, Y.-J. Chen, Interactive effects of mechanical stretching and extracellular matrix proteins on initiating osteogenic differentiation of human mesenchymal stem cells. J. Cell. Biochem. 108 (2009) 1263–1273, https://doi.org/10.1002/jcb.22356.

[75] J.R. Mauney, S. Sjostorm, J. Blumberg, R. Horan, J.P. O'Leary, G. Vunjak-Novakovic, V. Volloch, D.L. Kaplan, Mechanical stimulation promotes osteogenic differentiation of human bone marrow stromal cells on 3-D partially demineralized bone scaffolds in vitro. Calcif. Tissue Int. 74 (2004) 458–468, https://doi.org/10.1007/s00223-003-0104-7.

[76] M. Jagodzinski, A. Breitbart, M. Wehmeier, E. Hesse, C. Haasper, C. Krettek, J. Zeichen, S. Hankemeier, Influence of perfusion and cyclic compression on proliferation and differentiation of bone marrow stromal cells in 3-dimensional culture. J. Biomech. 41 (2008) 1885–1891, https://doi.org/10.1016/j.jbiomech.2008.04.001.

[77] U.A. Gurkan, O. Akkus, The mechanical environment of bone marrow: a review. Ann. Biomed. Eng. 36 (2008) 1978–1991, https://doi.org/10.1007/s10439-008-9577-x.

[78] F. Kong, Z. Li, W.M. Parks, D.W. Dumbauld, A.J. García, A.P. Mould, M.J. Humphries, C. Zhu, Cyclic mechanical reinforcement of integrin-ligand interactions. Mol. Cell 49 (2013) 1060–1068, https://doi.org/10.1016/j.molcel.2013.01.015.

[79] V. Niediek, S. Born, N. Hampe, N. Kirchgessner, R. Merkel, B. Hoffmann, Cyclic stretch induces reorientation of cells in a Src family kinase- and p130Cas-dependent manner. Eur. J. Cell Biol. 91 (2012) 118–128, https://doi.org/10.1016/j.ejcb.2011.10.003.
[80] M.S. Hall, R. Long, X. Feng, Y. Huang, C.-Y. Hui, M. Wu, Toward single cell traction microscopy within 3D collagen matrices. Exp. Cell Res. 319 (2013) 2396–2408, https://doi.org/10.1016/j.yexcr.2013.06.009.
[81] C. Franck, S.A. Maskarinec, D.A. Tirrell, G. Ravichandran, Three-dimensional traction force microscopy: a new tool for quantifying cell-matrix interactions. PLoS ONE 6 (2011) e17833, https://doi.org/10.1371/journal.pone.0017833.
[82] R. Zielinski, C. Mihai, D. Kniss, S.N. Ghadiali, Finite element analysis of traction force microscopy: influence of cell mechanics, adhesion, and morphology. J. Biomech. Eng. 135 (2013) 71009, https://doi.org/10.1115/1.4024467.
[83] S. Checa, M.K. Rausch, A. Petersen, E. Kuhl, G.N. Duda, The emergence of extracellular matrix mechanics and cell traction forces as important regulators of cellular self-organization. Biomech. Model. Mechanobiol. 14 (2015) 1–13, https://doi.org/10.1007/s10237-014-0581-9.
[84] E.J. Arnsdorf, P. Tummala, A.B. Castillo, F. Zhang, C.R. Jacobs, The epigenetic mechanism of mechanically induced osteogenic differentiation. J. Biomech. 43 (2010) 2881–2886, https://doi.org/10.1016/j.jbiomech.2010.07.033.
[85] S. Scaglione, D. Wendt, S. Miggino, A. Papadimitropoulos, M. Fato, R. Quarto, I. Martin, Effects of fluid flow and calcium phosphate coating on human bone marrow stromal cells cultured in a defined 2D model system. J. Biomed. Mater. Res. A 86 (2008) 411–419, https://doi.org/10.1002/jbm.a.31607.
[86] J. Liu, Z. Zhao, J. Li, L. Zou, C. Shuler, Y. Zou, X. Huang, M. Li, J. Wang, Hydrostatic pressures promote initial osteodifferentiation with ERK1/2 not p38 MAPK signaling involved. J. Cell. Biochem. 107 (2009) 224–232, https://doi.org/10.1002/jcb.22118.
[87] D.-H. Kim, S.B. Khatau, Y. Feng, S. Walcott, S.X. Sun, G.D. Longmore, D. Wirtz, Actin cap associated focal adhesions and their distinct role in cellular mechanosensing. Sci. Rep. 2 (2012) 1–13, https://doi.org/10.1038/srep00555.
[88] R.P. Martins, J.D. Finan, F. Guilak, D.A. Lee, Mechanical regulation of nuclear structure and function. Annu. Rev. Biomed. Eng. 14 (2012) 431–455, https://doi.org/10.1146/annurev-bioeng-071910-124638.
[89] S. Dupont, L. Morsut, M. Aragona, E. Enzo, S. Giulitti, M. Cordenonsi, F. Zanconato, J. Le Digabel, M. Forcato, S. Bicciato, N. Elvassore, S. Piccolo, Role of YAP/TAZ in mechanotransduction. Nature 474 (2011) 179–183, https://doi.org/10.1038/nature10137.
[90] R. Janoštiak, A.C. Pataki, J. Brábek, D. Rösel, Mechanosensors in integrin signaling: the emerging role of p130Cas. Eur. J. Cell Biol. 93 (2014) 445–454, https://doi.org/10.1016/j.ejcb.2014.07.002.
[91] K. Burridge, K. Fath, T. Kelly, G. Nuckolls, C. Turner, Focal adhesions: transmembrane junctions between the extracellular matrix and the cytoskeleton. Annu. Rev. Cell Biol. 4 (1988) 487–525, https://doi.org/10.1146/annurev.cb.04.110188.002415.
[92] C.G. Galbraith, K.M. Yamada, M.P. Sheetz, The relationship between force and focal complex development. J. Cell Biol. 159 (2002) 695–705, https://doi.org/10.1083/jcb.200204153.
[93] L.B. Pedersen, J.L. Rosenbaum, Intraflagellar transport (IFT) role in ciliary assembly, resorption and signalling. Curr. Top. Dev. Biol. 85 (2008) 23–61, https://doi.org/10.1016/S0070-2153(08)00802-8.
[94] R.Y. Kwon, S. Temiyasathit, P. Tummala, C.C. Quah, C.R. Jacobs, Primary cilium-dependent mechanosensing is mediated by adenylyl cyclase 6 and cyclic AMP in bone cells. FASEB J. 24 (2010) 2859–2868, https://doi.org/10.1096/fj.09-148007.

[95] D.A. Hoey, D.J. Kelly, C.R. Jacobs, A role for the primary cilium in paracrine signaling between mechanically stimulated osteocytes and mesenchymal stem cells. Biochem. Biophys. Res. Commun. 412 (2011) 182–187, https://doi.org/10.1016/j.bbrc.2011.07.072.
[96] D.A. Hoey, S. Tormey, S. Ramcharan, F.J. O'Brien, C.R. Jacobs, Primary cilia-mediated mechanotransduction in human mesenchymal stem cells. Stem Cells 30 (2012) 2561–2570, https://doi.org/10.1002/stem.1235.
[97] J.C. Chen, C.R. Jacobs, Mechanically induced osteogenic lineage commitment of stem cells. Stem Cell Res. Ther. 4 (2013) 107, https://doi.org/10.1186/scrt318.
[98] K.D. Kavlock, A.S. Goldstein, Effect of pulse frequency on the osteogenic differentiation of mesenchymal stem cells in a pulsatile perfusion bioreactor. J. Biomech. Eng. 133 (2011) 91005, https://doi.org/10.1115/1.4004919.
[99] C.F. Bentzinger, Y.X. Wang, J. von Maltzahn, V.D. Soleimani, H. Yin, M.A. Rudnicki, Fibronectin regulates Wnt7a signaling and satellite cell expansion. Cell Stem Cell 12 (2013) 75–87, https://doi.org/10.1016/j.stem.2012.09.015.
[100] A. Urciuolo, M. Quarta, V. Morbidoni, F. Gattazzo, S. Molon, P. Grumati, F. Montemurro, F.S. Tedesco, B. Blaauw, G. Cossu, G. Vozzi, T.A. Rando, P. Bonaldo, Collagen VI regulates satellite cell self-renewal and muscle regeneration. Nat. Commun. 4 (2013) 1–13, https://doi.org/10.1038/ncomms2964.
[101] R. Peerani, P.W. Zandstra, Enabling stem cell therapies through synthetic stem cell-niche engineering. J. Clin. Invest. 120 (2010) 60–70, https://doi.org/10.1172/JCI41158.
[102] J. Wu, E.S. Tzanakakis, Deconstructing stem cell population heterogeneity: single-cell analysis and modeling approaches. Biotechnol. Adv. 31 (2013) 1047–1062, https://doi.org/10.1016/j.biotechadv.2013.09.001.
[103] D. Seliktar, Designing cell-compatible hydrogels for biomedical applications. Science 336 (2012) 1124–1128, https://doi.org/10.1126/science.1214804.
[104] F. Guilak, D.M. Cohen, B.T. Estes, J.M. Gimble, W. Liedtke, C.S. Chen, Control of stem cell fate by physical interactions with the extracellular matrix. Cell Stem Cell 5 (2009) 17–26, https://doi.org/10.1016/j.stem.2009.06.016.
[105] P.M. Gilbert, S. Corbel, R. Doyonnas, K. Havenstrite, K.E.G. Magnusson, H.M. Blau, A single cell bioengineering approach to elucidate mechanisms of adult stem cell self-renewal. Integr. Biol. (Camb) 4 (2012) 360–367, https://doi.org/10.1039/c2ib00148a.
[106] K. Saha, A.J. Keung, E.F. Irwin, Y. Li, L. Little, D.V. Schaffer, K.E. Healy, Substrate modulus directs neural stem cell behavior. Biophys. J. 95 (2008) 4426–4438, https://doi.org/10.1529/biophysj.108.132217.
[107] P.M. Gilbert, K.L. Havenstrite, K.E.G. Magnusson, A. Sacco, N.A. Leonardi, P. Kraft, N.K. Nguyen, S. Thrun, M.P. Lutolf, H.M. Blau, Substrate elasticity regulates skeletal muscle stem cell self-renewal in culture. Science 329 (2010) 1078–1081, https://doi.org/10.1126/science.1191035.
[108] J.R. Tse, A.J. Engler, Stiffness gradients mimicking in vivo tissue variation regulate mesenchymal stem cell fate. PLoS ONE 6 (2011) e15978, https://doi.org/10.1371/journal.pone.0015978.
[109] S. Gobaa, S. Hoehnel, M. Roccio, A. Negro, S. Kobel, M.P. Lutolf, Artificial niche microarrays for probing single stem cell fate in high throughput. Nat. Methods 8 (2011) 949–955, https://doi.org/10.1038/nmeth.1732.
[110] C.J. Flaim, D. Teng, S. Chien, S.N. Bhatia, Combinatorial signaling microenvironments for studying stem cell fate. Stem Cells Dev. 17 (2008) 29–39, https://doi.org/10.1089/scd.2007.0085.

[111] M.A. LaBarge, C.M. Nelson, R. Villadsen, A. Fridriksdottir, J.R. Ruth, M.R. Stampfer, O.W. Petersen, M.J. Bissell, Human mammary progenitor cell fate decisions are products of interactions with combinatorial microenvironments. Integr. Biol. (Camb) 1 (2009) 70–79, https://doi.org/10.1039/b816472j.

[112] K.A. Kilian, B. Bugarija, B.T. Lahn, M. Mrksich, Geometric cues for directing the differentiation of mesenchymal stem cells. Proc. Natl. Acad. Sci. 107 (2010) 4872–4877, https://doi.org/10.1073/pnas.0903269107.

[113] J.T. Connelly, J.E. Gautrot, B. Trappmann, D.W.-M. Tan, G. Donati, W.T.S. Huck, F.M. Watt, Actin and serum response factor transduce physical cues from the microenvironment to regulate epidermal stem cell fate decisions. Nat. Cell Biol. 12 (2010) 711–718, https://doi.org/10.1038/ncb2074.

[114] H.V. Unadkat, M. Hulsman, K. Cornelissen, B.J. Papenburg, R.K. Truckenmuller, A.E. Carpenter, M. Wessling, G.F. Post, M. Uetz, M.J.T. Reinders, D. Stamatialis, C.A. van Blitterswijk, J. de Boer, An algorithm-based topographical biomaterials library to instruct cell fate. Proc. Natl. Acad. Sci. 108 (2011) 16565–16570, https://doi.org/10.1073/pnas.1109861108.

[115] J. Tang, R. Peng, J. Ding, The regulation of stem cell differentiation by cell-cell contact on micropatterned material surfaces. Biomaterials 31 (2010) 2470–2476, https://doi.org/10.1016/j.biomaterials.2009.12.006.

[116] B.G. Chung, L.A. Flanagan, S.W. Rhee, P.H. Schwartz, A.P. Lee, E.S. Monuki, N.L. Jeon, Human neural stem cell growth and differentiation in a gradient-generating microfluidic device. Lab Chip 5 (2005) 401–406, https://doi.org/10.1039/b417651k.

[117] A.W. Lund, B. Yener, J.P. Stegemann, G.E. Plopper, The natural and engineered 3D microenvironment as a regulatory cue during stem cell fate determination. Tissue Eng. Part B Rev. 15 (2009) 371–380, https://doi.org/10.1089/ten.teb.2009.0270.

[118] S. Adler, D. Basketter, S. Creton, O. Pelkonen, J. van Benthem, V. Zuang, K.E. Andersen, A. Angers-Loustau, A. Aptula, A. Bal-Price, E. Benfenati, U. Bernauer, J. Bessems, F.Y. Bois, A. Boobis, E. Brandon, S. Bremer, T. Broschard, S. Casati, S. Coecke, R. Corvi, M. Cronin, G. Daston, W. Dekant, S. Felter, E. Grignard, U. Gundert-Remy, T. Heinonen, I. Kimber, J. Kleinjans, H. Komulainen, R. Kreiling, J. Kreysa, S.B. Leite, G. Loizou, G. Maxwell, P. Mazzatorta, S. Munn, S. Pfuhler, P. Phrakonkham, A. Piersma, A. Poth, P. Prieto, G. Repetto, V. Rogiers, G. Schoeters, M. Schwarz, R. Serafimova, H. Tähti, E. Testai, J. van Delft, H. van Loveren, M. Vinken, A. Worth, J.-M. Zaldivar, Alternative (non-animal) methods for cosmetics testing: current status and future prospects-2010. Arch. Toxicol. 85 (2011) 367–485, https://doi.org/10.1007/s00204-011-0693-2.

[119] S. Tasoglu, U. Demirci, Bioprinting for stem cell research. Trends Biotechnol. 31 (2013) 10–19, https://doi.org/10.1016/j.tibtech.2012.10.005.

[120] J. Kim, P. Sachdev, K. Sidhu, Alginate microcapsule as a 3D platform for the efficient differentiation of human embryonic stem cells to dopamine neurons. Stem Cell Res. 11 (2013) 978–989, https://doi.org/10.1016/j.scr.2013.06.005.

[121] M.W. Tibbitt, K.S. Anseth, Dynamic microenvironments: the fourth dimension. Sci. Transl. Med. 4 (2012) 24, https://doi.org/10.1126/scitranslmed.3004804.

[122] Y. Sasai, Next-generation regenerative medicine: organogenesis from stem cells in 3D culture. Cell Stem Cell 12 (2013) 520–530, https://doi.org/10.1016/j.stem.2013.04.009.

[123] M.A. Lancaster, M. Renner, C.-A. Martin, D. Wenzel, L.S. Bicknell, M.E. Hurles, T. Homfray, J.M. Penninger, A.P. Jackson, J.A. Knoblich, Cerebral organoids model human brain development and microcephaly. Nature 501 (2013) 373–379, https://doi.org/10.1038/nature12517.

[124] H. Naderi, M.M. Matin, A.R. Bahrami, Review paper: critical issues in tissue engineering: biomaterials, cell sources, angiogenesis, and drug delivery systems. J. Biomater. Appl. 26 (2011) 383–417, https://doi.org/10.1177/0885328211408946.

[125] F.-M. Chen, L.-A. Wu, M. Zhang, R. Zhang, H.-H. Sun, Homing of endogenous stem/progenitor cells for in situ tissue regeneration: promises, strategies, and translational perspectives. Biomaterials 32 (2011) 3189–3209, https://doi.org/10.1016/j.biomaterials.2010.12.032.

[126] G. Orlando, P. Baptista, M. Birchall, P. De Coppi, A. Farney, N.K. Guimaraes-Souza, E. Opara, J. Rogers, D. Seliktar, K. Shapira-Schweitzer, R.J. Stratta, A. Atala, K.J. Wood, S. Soker, Regenerative medicine as applied to solid organ transplantation: current status and future challenges. Transpl. Int. 24 (2011) 223–232, https://doi.org/10.1111/j.1432-2277.2010.01182.x.

Biomimetic principle for development of nanocomposite biomaterials in tissue engineering

13

Mehdi Ebrahimi
Oral Rehabilitation, Prince Philip Dental Hospital, The University of Hong Kong, Sai Ying Pun, Hong Kong

13.1 Introduction

The main goal of tissue engineering during production of the ideal substrate is the replication of extracellular matrix (ECM) of the tissue that is planned to be replaced. Human tissues have their basic structural units arranged at the nanoscale level. Therefore, the ideal biomaterials for tissue engineering are those produced at nanoscale level with close similarity to the three-dimensional (3D) structure of ECM. Different types of the natural scaffold (i.e., collagen) as tissue substitutes have been developed for applications in tissue engineering. In addition, several synthetic materials and scaffolds have been designed by surface modification with biomolecules or as hybrid composites to serve as a biomimetic template that mimics the native ECM [1]. Among them, the nanofibrous scaffold exhibits great potential in mimicking the architecture of native ECM at the nanoscale. Nanofibers exhibit close similarity to collagen fibrils that form a complex network in native ECM (50–500 nm diameter). Nanofibers are highly porous with the high surface area to volume ratio which facilitates cell attachment, proliferation, infiltration, and differentiation [2]. Both physical (elasticity and porosity) and chemical properties of nanofibers can be tailored to match the requirements of the substituted tissues [3]. It has been shown that fibers with dimensions less than the diameter of cells support cell adhesion and organization [4]. Nanofibers scaffolds can be fabricated using different methods such as phase separation, electrospinning, and self-assembly techniques [4]. Based on the general method adopted, fabrication strategies of biomaterial scaffold for tissue engineering fall into one of the two main approaches; namely top-down and bottom-up approach (Fig. 13.1).

13.2 Top-down approach in tissue engineering

This is the traditional and most commonly used approach in designing biomaterials in tissue engineering [5]. Here, the prefabricated porous tissue constructs of desired size and shape are produced which act as temporary scaffolds during tissue regeneration. The scaffold should replicate structure and function of natural ECM and possess essential basic features such as adequate pore size and interconnected porosity, mechanical properties, and biodegradability. The scaffold is also expected to support attachment

Applications of Nanocomposite Materials in Orthopedics. https://doi.org/10.1016/B978-0-12-813740-6.00015-6

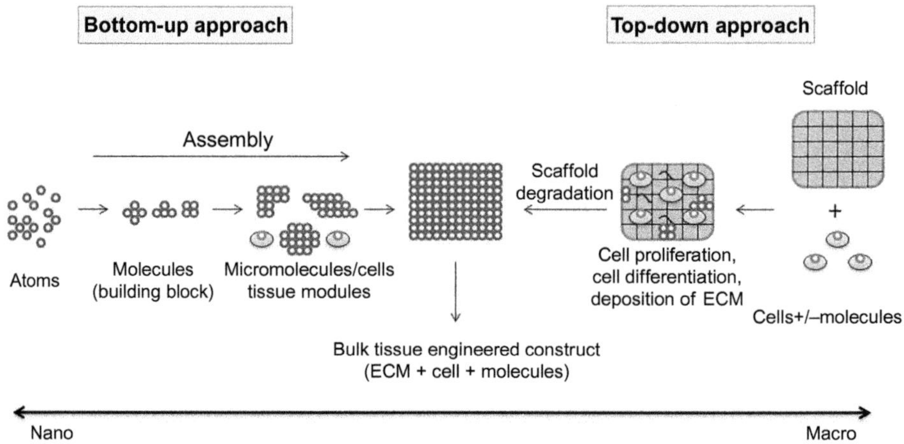

Fig. 13.1 Two main strategies in tissue engineering. The top-down approach is the conventional strategy that aims to develop bulk constitutes as a temporary scaffold for cell function and tissue regeneration. The scaffold can be enriched with functional cells and signaling molecules. The scaffold is planned to undergo degradation preparing space for new tissue deposits. The bottom-up approach is based on molecular self-assembly and attraction mechanism that lead to the formation of macromolecules. These can be used to encapsulate desired cells and incorporate signaling molecules that is aimed to deposit new tissues.

and proliferation of cells and allow adequate perfusion for cell function and differentiation. Hence, the desired cells are proposed to populate the scaffold and produce appropriate ECM micro-architecture in restoring the tissue. Examples of methods that implement the top-down approach in the production of nanoscale biomaterials include; nanolithography, chemical etching, template-assisted method, phase separation, and electrospinning [6–9]. The main advantages of top-down approach are the possibility to use a wide range of biomaterials (hydrogel to bioceramic) and the ability to produce porous scaffolds of high mechanical properties [10]. However, this approach possesses several challenges such as; difficult control of the cell density, nonuniform cell distribution, and inadequate diffusion and vascularization due to the limited access to the central part of the scaffold. Another main disadvantage is the lack of intrinsic bioactive properties of scaffolds, that is, osteoinductivity and cell recognition/attachment cues. These limitations result in an improper re-creation of intricate microenvironmental features of cells, and inadequate duplication of nano to microstructural properties of native tissues [11,12]. Different attempts have been made to overcome these limitations such as; incorporation of bioactive molecules (i.e., growth factor, protein, peptide), and improving cell seeding method (i.e., flow perfusion), etc. [13,14].

13.3 Bottom-up approach in tissue engineering

In this approach, the nano- to macroscale building blocks interact one by one at atomic and molecular level and self-assemble to organize into more complex supramolecular units that can further rearrange into functional structures of native tissue. In this way, this approach mimics the structural features of tissues in the opposite direction to top-down approach [15]. Different methods for assembly have been introduced including; random assembly [16], manual manipulation [17], multilayer photopatterning [18], and microfluidic-directed assembly [19]. The molecular self-assembly technique offers significant opportunities for development of novel and optimized biomaterials [20]. Examples of this approach include; self-assembly, laser pyrolysis, plasma spraying synthesis, micro-molding, photolithography, cell-sheeting, and 3D bioprinting [21–24]. 3D bioprinting of cells, tissue, and organ is a promising scaffold-free method that is based on rapid prototyping technology. This method has a great potential for production of complex 3D tissue constructs [25,26].

In fact, self-assembly is a spontaneous phenomenon employed by nature's bottom-up process in the development and construction of natural tissue. In this natural process, the organic molecules such as amino acids, nucleotides, lipid molecules, and sugars act as a building block by self-assembly into functional structures such as proteins, nucleic acids, metabolic intermediates, and ECM components. Furthermore, other inorganic counterparts such as simple elements (e.g., metal ions, salt) also act as complementary building blocks in nature.

Molecular self-assembly involves mostly insignificant noncovalent bonds which are weak individually. However, the collective outcome of these bonds plays an important role in all biological systems and their dynamic interaction. These bonds include; hydrogen bond, ionic bond (electrostatic interaction), van der Waals interaction, and hydrophobic interaction. Furthermore, since all these materials interact with water, the water-mediated hydrogen bond is also an integral part of the biological structure. For example, it is crucial in maintaining the 3D structure of triple helix collagen matrix in the biological tissue [27].

The bottom-up technology allows step-by-step build-up of biomaterials by organizing one molecule at a time similar to nature. Although extremely complex and difficult, this approach also allows incorporation of specific features and other functional biologic motifs. For example, peptides could be as a building block for production of several other peptide scaffolds to support cell proliferation and differentiation [28]. They can be designed to incorporate other building units such as mineral ions for alignment of inorganic materials, in a similar way to biomineralization process.

Considering bottom-up approach and molecular self-assembly, it should be mentioned that for complex 3D structures, there are almost no methods for proper control of the interconnected architecture at the atomic or molecular scale [27]. However, the bottom-up tissue engineering approach holds a great promise in the production of functional repeating units that could mimic the complex native tissue. There are many synthetic building blocks that have been produced through the latest advances in materials science and nanotechnology which utilize self-assembly as a new scope for a wide range of applications. However, bottom-up approach suffers from several challenges and shortcomings such as technical sensitivity, difficult control and manipulation, and unpredictable biodegradation pattern. Furthermore, this approach suffers from lack of scalability which may lead to the formation of random, uncontrolled structures due to poor control over assembly mechanisms [29]. These shortcomings challenge the production of constructs with homogenous shape and size. Another challenge is poor mechanical stability of available products (i.e., hydrogel) due to their high water content that limits their applications to nonload-bearing sites [10]. As such, this new era needs a deep understanding of molecular assembly and dynamic behaviors along with further development and updates to overcome current limitations.

13.4 Bone regeneration via bottom-up approach: Biomimetic mineralization

13.4.1 Rationale and significance

Currently, the autogenous bone grafting procedure is considered the gold standard bone graft materials despite its additional pain and morbidity for the patients [30]. This is because autograft involves the basic elements for tissue engineering (e.g., cells, framework, and signals) with both osteoconductive (permits bone ingrowth into the scaffold) and osteoinductive (permits recruitment and stimulation of undifferentiated cells to form bone-forming cell lineage) potentials. Other primary alternatives such as allograft materials lack the osteoactive (bioactive) capacity and carry the risk of infection or immune rejection [31]. The common biomaterials in use as artificial bone are the calcium and phosphate (Ca–P)-based bioceramics. However, these biomaterials cannot provide an ideal microenvironment for cell survival and bone regeneration. This is because cells and related extracellular microenvironment are organized at nanoscale while these biomaterials are designed at microscale so they cannot function as native ECM. Furthermore, the central area of biomaterial scaffold and bone defect usually suffers from inadequate vascularization and poor oxygenation due to the difficult accessibility of cells and biofluid. This would highlight the concern about development of central necrosis after scaffold implantation [32,33]. Moreover, it should be noted that the biologic bone is biphasic in composition, where the organic phase (i.e., collagen, glycoprotein, and proteoglycan) forms only 30% of the bone structure. The major component of bone ECM is inorganic in nature where hydroxyapatite (HA)—biological apatite—constitutes about 70% of the mineral phase of natural bone by dry weight [34] (Fig. 13.2).

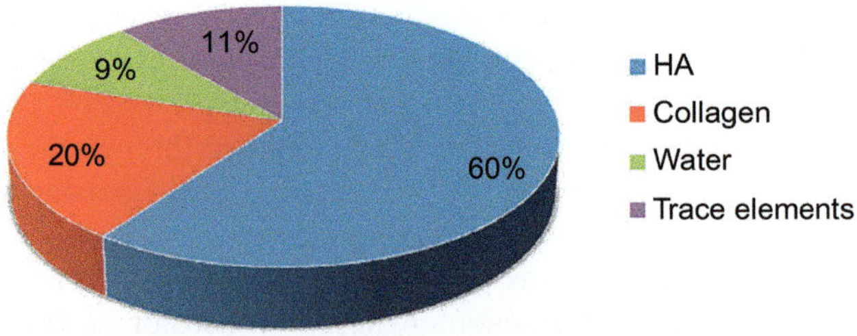

Fig. 13.2 Main bone constitutes in wt%. The biomimetic approach in the production of ideal tissue constitutes should consider different phases of natural bone composition.

Some studies have focused on improving the efficacy of autograft and allograft by incorporation of growth factors concentrates, that is, platelet-rich plasma (PRP) and platelet-rich fibrin (PRF) or bone marrow aspirates to enhance bone repair [35–37]. Recently, there is a growing interest in design and application of synthetic self-assembling peptide amphiphile (SAPA) scaffolds to facilitate 3D bone regeneration throughout the bone defect. In general, bioceramic scaffolds do not allow sufficient deep penetration of cells and vascular endothelial ingrowth into the scaffolds, while SAPA scaffolds can promote angiogenesis and bone regeneration by supporting colonization and deep ingrowth of these cells [38]. Based on the mechanism of SAPA, many types of ECM components are used as alternatives to promote cell function such as; laminin, gelatin, glycosaminoglycan, and collagen [39]. More recently newly developed and commercialized ECM alternatives are also available (e.g., Matrigel and PuraMatrix) which are complex structures used in 3D cell culture and transplantation.

However, bulks of the research works are toward enhancing the bioactivity of biomaterials by the development of new alternatives to overcome complications related to traditional strategies. The primary goal is the production of synthetic bone substitutes that simulate main key features of natural bone namely, osteoconductivity and osteoinductivity. Osteoinductivity can be achieved in two forms; active (extrinsic) osteoinductivity that is induced by addition of growth factors such as bone morphogenetic protein (BMP) molecules, and passive (intrinsic) osteoinductivity where scaffolds nano-microstructure and topography could support osteogenic potential [40]. Moreover, it should be noted that the bone growth and repair (osteogenesis) involves osseointegration with the formation of direct and intimate physicochemical contact between the bone substitutes and natural bone. Therefore, the ideal bone graft substitutes should provide basic features of natural bone and support osseointegration. In this context, the study of biomineralization process of natural bone and the involved biological processes is the first essential step for the future research. This provides a base to overcome the current limitations aiming to better implement the biomimetic strategy in clinical practice.

13.4.2 Natural biomineralization process

Mineralized tissues have developed well-organized hierarchical structures that can adapt to different stimuli to achieve the desired functions. Organic phases of mineralized tissues play a key role in biomineralization by providing a template for inorganic phase precipitation and crystals growth. From the biological aspect, the mineralized crystals form and grow in an organic matrix of proteins via a precise synthetic mechanism in dynamic equilibrium with the surrounding environment [41]. Understanding this complex interaction between organic and inorganic phases is of great importance to biomaterials science for the development of better biomaterials for tissue engineering.

The natural apatite crystal formation is a unique mechanism that involves intimate dynamic interaction between microenvironment and organic/inorganic phases. Biological apatite contains high carbonate substitutions and imperfections in its crystal structure and it also permits adsorption of other ions/molecules. These imperfections occur as a result of the body that utilizes bone as a reservoir for homeostasis and ions balance that subsequently cause variation in the molar ratio of Ca/P in biomineralized tissues. Calcium-deficient and carbonated HA (Ca_{10-x} $(PO_4CO_3)_6OH_2$) is the most common phosphate minerals in mammalian bones [42].

Collagen molecule functions as the main ECM protein of the biologic bone model. There are different types of collagen available based on their molecular sequences. Collagen type I is the most abundant ECM protein in the connective tissue that arranges in triple helix pattern and contributes to most of the structural integrity of the bone structure. The entire collagen molecule consists of about 3000 amino acids of three main types namely; glycine, proline, and hydroxyproline. A single collagen molecule (also known as tropocollagen) is a rigid protein 300nm in length, 1.5nm in diameter, and approximately 300kDa in molecular weight [43].

Tropocollagen (the basic fundamental unit of collagen fibrils) is composed of three helically arranged polypeptide strands (α-chains) that are each coiled into a left-handed helix. Type I collagen is composed of two α-1 (I) chains and one α-2 (II) chain (Fig. 13.3). These three chains are then wrapped around each other into a right-handed triple helix; thereby the final structure is in a triple helical rope-like pattern. The triple-helical domain has a characteristic primary structure, where glycine in every third amino acid generates repeating (Gly X–Y)n units; X is alanine or proline, and Y is hydroxyproline [44].

Five tropocollagen units arrange in a parallel manner to form a collagen microfibril that is about 300nm long and 4nm thick. Microfibrils, in turn, are bundled into collagen fibrils as a result of covalent bonding between many tropocollagen molecules. Subsequently, one collagen fiber (with a diameter of about 50nm), is created by the bundling of hundreds of fibrils that forms the structural framework of bone. This highly ordered arrangement results in the formation of periodic gaps (hole zone) which are evenly spaced spiraling bands of 40nm at the ends of each collagen fibrils units (between C-terminal and N-terminal), in addition to a 27nm overlap between adjacent units [45] (Fig. 13.3).

The mineralized collagen fiber is considered as the primary building block of bone structure at the nanoscale level. Calcium and phosphates are present in the form of biological apatite as HA plate-shaped crystals. Atomic structure of HA consists of four distinctive parts; (1) calcium, (2) oxygen, (3) phosphate tetrahedral, and (4) hydroxyl in channel sites. HA crystals are about 40–60nm long, 20nm wide, and 1.5–5nm thick. HA crystals are deposited parallel to the collagen fibrils such as that they reinforce fibrils with their c-axis aligned with the long axis of the fibril. The HA crystals are embedded in collagen matrix with the volume ratio of mineral to matrix on the order of 1:2, demonstrating stratified stacks with slightly dislocated centers [46–48].

The interface between collagen and biological apatite crystals is still poorly understood. However, it is revealed that the mineralized crystals are concentrated particularly in the hole zone (between opposing ends) of the collagen fibrils and pores which present along the lengths of adjacent microfibrils [49]. Therefore, these hole zones are proposed to be the initial sites of mineral nucleation, growth, and proliferation with the possibility of constraining the mineral growth [50].

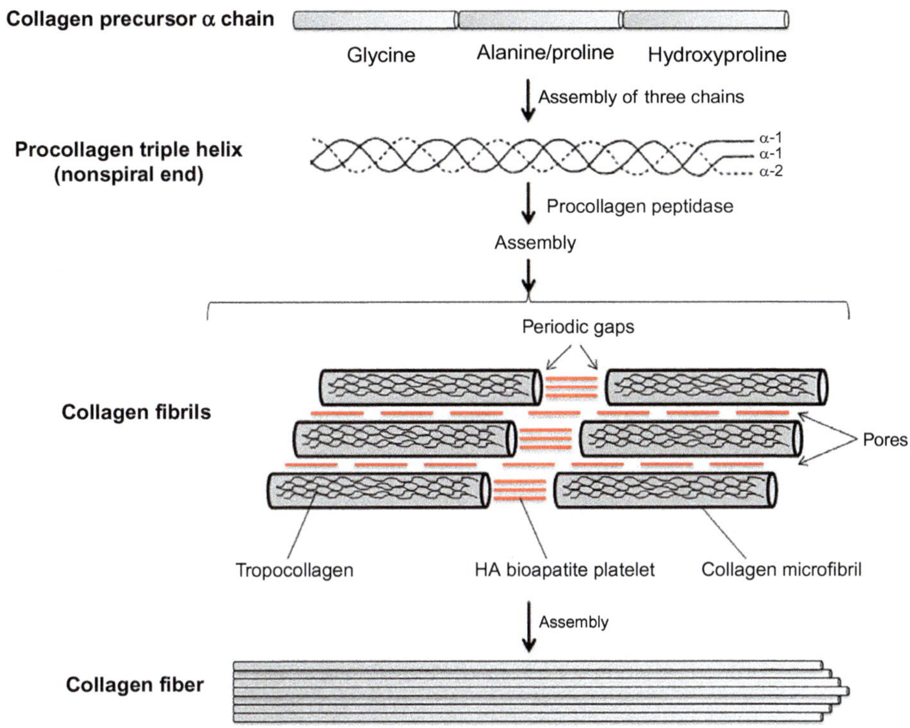

Fig. 13.3 The mineralized collagen fibers as primary building unit of natural bone. Tropocollagen is the basic fundamental unit of collagen fibers which is composed of triple helix polypeptide chains. Five tropocollagen units arrange in a parallel manner to form a collagen microfibril that is 300 nm long and 4 nm thick. Microfibrils, in turn, are bundled into collagen fibrils. Subsequently, bundling of hundreds of fibrils results in the formation of one collagen fiber (with a diameter of about 50 nm) which forms the structural framework of bone. The periodic gaps are evenly spaced spiraling bands of 40 nm at the ends of each collagen fibrils units (between c-terminal and N-terminal) and a 27 nm overlap between adjacent units. The apatite crystallites start to form in the area of structural voids or gaps in two forms; (i) between opposing ends of fibers (periodic gaps or holes) and (ii) along the length of adjacent microfibrils (channels/pores).

Other noncollagenous proteins (e.g., bone sialoprotein, osteonectin, osteopontin, and osteocalcin) represent <10% of total protein content. Aspartic acid and glutamic acid constitute the primary amino acid sequence of these proteins that possess a high affinity to calcium ions [51]. Although their specific functions are not revealed completely, but it is known that they influence ions homeostasis and cells signaling in addition to their important role in crystal nucleation and growth during biomineralization process [52].

In general, the process of bone formation may occur via intramembranous or endochondral ossification. The intramembranous ossification is involved in the embryonic development of flat bones of the cranium, parts of mandible, and clavicle, while the endochondral ossification occurs in development of rest of human skeleton [53].

Endochondral ossification involves formation and then mineralization of a cartilaginous template through the cylindrical arrangement of collagen fibers into osteons. Osteon or Haversian system is the fundamental functional unit of lamellar bone that is typically several millimeters long with about 200 μm diameter. The process of biomineralization requires a local microenvironment with adequate supersaturation of mineral precursors for nucleation and growth of apatite crystals [54] (Fig. 13.4). In biological tissue, the bone forming cells (osteoblasts and osteocytes) produce and lay down crystals of calcium phosphate on bone matrix template. It is believed that the nucleation of HA first start at intracellular vesicles of bone forming cells, but as the crystallite grows, it break through the vesicle and release from the outer membrane of cells with subsequent exposure to ECM. Therefore, the bone cells control the initial mineral nucleation (phase I) while the collagen in ECM mediates the mineral growth and propagation (phase II) [55]. However, the crystal growth and nucleation are not limited to the sites of matrix vesicles, but different matrix proteins can also contribute to this process [56]. For example, in addition to collagen, there are other functional motifs within the noncollagenous protein matrix that influence biomineralization process [57]. Several of these proteins have been studied for their influence on apatite crystal formation such as; albumin [58], fibronectin [59], fibrinogen [60], natural biopolymers (i.e., polysaccharides; starch, hyaluronic acid, chitin, and chitosan) [34], and other synthetic polymers. Despite, the mineralization mechanism of collagen fibril is still the main research focus of many groups as this structure provides a scaffold which resembles autogenous bone graft. However, the process of formation of plate-shaped HA crystals in collagen fibrils is still not fully understood. Furthermore, the presence of transient precursors as the intermediate phase is an ongoing debate in mineralization of HA in biological tissue [61]. Determining the precise mechanism of biomineralization necessitates further research at molecular and cellular levels.

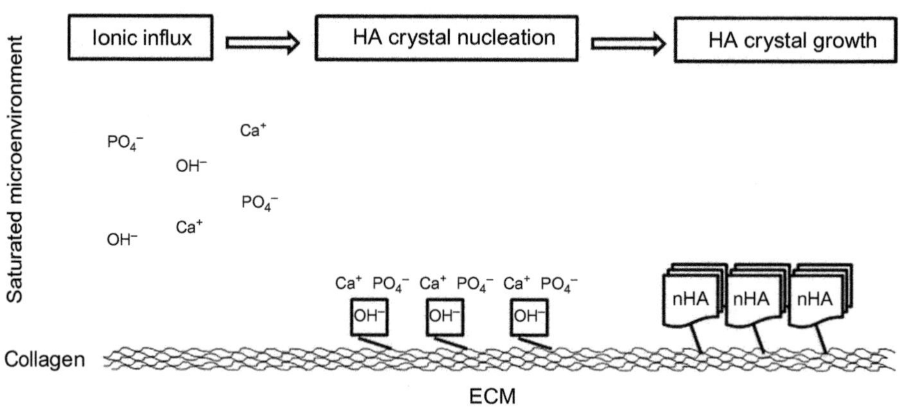

Fig. 13.4 Saturation of microenvironment with ions such as calcium and phosphates initiates the mineralization process. Nucleation and growth of HA crystals depend on the availability of ions and binding sites on ECM.

13.4.3 Biomimetic mineralization

The impact of ECM on biomineralization process, cellular responses, and tissue regeneration has been highlighted in the literature. It is known that mineralization of hard biological tissue is under the control of specific regulatory proteins [41]. Direct delivery of these proteins to the site of interest is not practical due to rapid clearance before any potential effect. Therefore, synthetic and natural protein polymers have been prepared to restore and maintain the therapeutic agents or bioactive motifs in the network of hydrogel biomaterials. Both synthetic and natural polymers can be utilized to fulfill the function of ECM during mineralization phase. These polymers can be used as a scaffold for mineralization of HA, stimulation of new bone formation, and as an aid for drug delivery to induce bioactivity [41]. Furthermore, the biocompatibility and biodegradability of these polymers are interesting features that need to be optimized for better clinical efficacy in bone regeneration. Natural polymers such as collagen are generally biocompatible and safe for clinical application, but the limited source and economic issue are the main challenges. On the other hand, the synthetic polymers are less biocompatible, but they offer a better option for programing and chemical functionality. It has been shown that synthetic polymers such as macromolecules can simulate to a large extent the fibrous feature of collagen fibril, and have the ability to induce mineralization. Also, the functional motifs (i.e., carboxylic acids and phosphorylated amino acids) present in the macromolecules structure can induce or inhibit crystallite nucleation and growth during HA biomineralization. In this context, several synthetic methods have been introduced for incorporation of functional groups into the natural and synthetic polymers [62].

Organic macromolecules can induce apatite crystal nucleation and modify its size and morphology. For example, it has been reported that the nature of utilized polymer (amino acid) can significantly control the apatite crystals morphology. Poly "L-glutamic acid" generates nanoscale crystallites while poly "L-lysine" generates large flat and thin crystals [63].

One of the important aspects of the bottom-up approach in nanotechnology and tissue engineering is molecular self-assembly. It is a process by which a disorganized system of poorly defined molecular pattern adopts an organized well-defined arrangement without external guidance as a consequence of specific interactions among molecules themselves. Furthermore, the process of self-assembly offers additional opportunities to formulate functional macromolecules with complex structure and tunable properties. This supramolecular platform could mimic the architecture of natural fibrous matrix and would allow designing of nanostructures with the possibility of incorporation of bioactive motifs which particularly target the biomineralization process [64].

Different groups [65–67] worked on developing peptide amphiphile molecular designs capable of self-assembly into defined nanostructures (i.e., nanofibers, nanotubes, vesicles) under certain pH and temperature. These structures express specific bioactive ligands on their surface that can play an important role in the biological process by modifying cell behavior [68]. The nanofiber SAPA molecule consists of five domains (regions) that are responsible for rearrangement of the molecule into the well-defined supramolecular structure. These domains are; (1) alkyl (hydrophobic) tail for hydrophobic interaction with peptide region to produce amphiphilic molecules, (2) hydrophobic peptide segment (i.e., cysteine residues) with strong tendency for β-sheet formation via di-sulfide bonds, (3) charged amino acids group (i.e., glycine residues) as flexible hydrophilic linker for solubility, (4) mineralization motive/inducer—a single phosphorylated serine residue—to direct precipitation of HA by strong interaction with calcium ions, and (5) bioactive peptide epitope such as RGD (Arg-Gly-Asp) that act as cell adhesion ligand for cell activity. The process of self-assembly or gelation is induced by addition of mono or multivalent metal ions, drying or combination of differently charged peptide amphiphile (PA) under neutral and physiological pH conditions. Self-assembly is controlled by hydrophobic interactions between alkyl tails (lipid chain) and intermolecular hydrogen bonds between peptide segments. Because of higher hydrophobicity, the amino acid domain forms the core of nanofibers during self-assembly while the amino acid β-sheet domain enhances self-assembly of molecules into fibrous structures [69]. This structural arrangement at the nanoscale results in exposure of functional peptide segments at the outer aqueous interface on the surface of the nanostructure which is able to grow in length. This functional peptide epitope can comprise cell signaling biological motifs. The resulted molecular architecture highly mimic the fibrous component of ECM such as collagen fibrils. Furthermore, several SAPA nanofiber molecules can be co-assembled for presenting various bioactive motifs for further control of cell behavior [70].

Stupp and coworkers have investigated the impact of PA nanofibers as a biomimetic matrix for tissue engineering and regenerative medicine. These studies include the possible use of SAPA for cartilage regeneration [71], neuronal differentiation [72], angiogenesis [73], and ischemic tissue repair [74]. Other possible applications include; skin care, carrier system for hydrophobic drugs, antimicrobial peptide, gene delivery, and scaffolds for 3D cell culture and biomineralization [71,75–78].

The idea of molecular self-assembly has been associated with biomineralization process for fabrication of various composite materials in bone tissue engineering (BTE). Natural bone is a complex composite with a macro to nano scales of hierarchical organization. Cortical and cancellous bones are at the highest level of this hierarchy as macrostructures, while the arrangement of collagen molecules (fibrils) and bone mineral crystals (HA) represents the lowest level [79]. In native tissue, the self-assembly of collagen triple helices results in the development of collagen fibrils. Subsequently, HA crystals arrange themselves within the collagen fibrils with their *c*-axes parallel to long axes of the fibrils [49]. Therefore, considering the biomimetic mineralization, fabrication of composite materials in BTE should consider multiple levels of hierarchical organization of natural bone. Simulation of natural bone using synthetic biomaterials at the nanoscale is a persistent challenge as it involves association of two dissimilar phases (organic/inorganic) in a specific ratio and spatial relationship.

One promising approach has been the preparation of synthetic organic phase at the nanoscale that can initiate and control inorganic crystal nucleation and growth. With regards to BTE, biomimetic mineralization of the nanofiber matrix has been investigated as an interesting method in a bottom-up approach. The ability of matrix to initiate nucleation of HA crystals is a biomimetic feature that resembles the situation in natural bone. It has been reported that PA nanofibers bearing phosphoserine residues can allow nucleation of thin HA crystals along their *c*-axis parallel to nanofibers, in a way similar to the crystallographic orientation along collagen fibrils biologically [65]. Furthermore, the mineralization can also be promoted in vitro using osteogenic media containing organophosphates and alkaline phosphatase (ALP) enzyme [80]. This nanostructured matrix may be similar in function to noncollagenous proteins such as bone sialoprotein (or phosphophoryn). Bone sialoprotein is phosphorylated serine residue with high affinity to calcium and phosphate ions which plays a critical role in bone mineralization and induction of osteoblastic differentiation of mesenchymal stem cell (MSC) [41].

Recently, researchers investigate the application of these SAPA biomimetic systems as a 3D matrix to enhance bone regeneration. PA nanostructures self-assemble utilizing calcium chloride and have a high affinity to calcium ions. In addition to induction of biomimetic nucleation of HA, they also help in augmentation of mineralization within the defect and contribute to biomineralization process [57]. Furthermore, phosphate ions can also be harvested from phosphorylated artificial PA matrix by enzymes such as ALP that has important role in early bone regeneration [81]. With regards to calcium phosphate minerals, in addition to their osteoconductivity, several groups have also shown their osteoinductive effects by guiding MSC differentiation toward osteoblastic cell lineage [82,83]. Furthermore, they may stimulate osteoclast activity, which in turn influence bone deposition by osteoblasts [84]. Further synergy in bone regeneration may also occur, as promoted osteoblastic activity enhances expression of ALP enzyme [85].

Furthermore, co-assembly of different SAPA molecules is also a promising approach to study the combined effect of bioactive motifs. This approach allows formulation of complex supramolecular systems offering multiplex biological signals along with the capacity to create a suitable niche for nucleation and propagation of mineralized tissues. For example, a study has explored the bone bioactivity of a supramolecular system bearing two bioactive motifs; SAPA of fibronectin epitope RGDS sequence (Arg-Gly-Asp-Ser) that promote integrin-mediated cell attachment and phosphoserine residues that allow matrix mineralization [78].

Currently, there is lack of a general and ideal synthetic approach in preparing inorganic–organic hybrid complex in the same hierarchical order as that of natural bone. A deeper understanding of biology is required to find out the roles and interplay of multiple proteins in controlled biomineralization process, as well as their function to organize highly structured complex in natural hard tissues.

13.5 Integrated approach: A new era in tissue engineering

Simulating the nature by bottom-up approach resulted in major improvements of biomaterials bioactivity and tissue responses. However, the main challenge is the low mechanical resistance that limits the application to the nonload-bearing anatomical sites. On the other hand, the top-down approach provides biomaterials with higher mechanical properties that can better resist the functional load, but they suffer from disadvantages such as insufficient tensile strength and low bioactivity.

In order to enhance the bone regeneration process to its maximum potential, it is sometimes necessary to combine different approaches to overcome the shortcomings of single approach. The advantages of one single approach can be optimized by incorporation of different concepts from another approach. In this context, the combination of both bottom-up and top-down approaches may provide new horizon to overcome most of the limitations of each approach. However, the main challenge is the potential approach which should be able to integrate these both different directions [10,86,87].

Both bottom-up and top-down approaches have their limitations for clinical application. The major problem in these approaches is management and control of mechanical properties of final products. With regards to BTE, both approaches can be combined to produce multiphasic biomaterials possessing similar compositional structure and bioactive signaling as the natural tissue. A combined approach may enhance biological apatite formation, provide better mechanical support, allow functional load, allow 3D bioencapsulation of cells, facilitate cells migration and interconnection throughout the graft biomaterials, facilitate distribution of bone cells, and induce earlier deposition of ECM. However, further investigations are required to explore possible practical strategies in combination approach with clinical implications.

13.6 Current focus and challenges and the future directions

The current focus in the development of new biomaterials for BTE is to guide cellular response toward enhanced and improved new tissue formation using a biomimetic approach. The bone ECM provides a complex microenvironment which consists of inorganic HA nanocrystals reinforcing the network of the organic matrix of mostly collagen. Native bone ECM serves as a mean for 3D bioencapsulation of cells that provides support, regulates dynamic cellular responses, and controls intercellular communications [88,89]. This is achieved by ECM presenting specific signaling biomolecules that regulate cell-ligand interactions. It is critical to design biomaterials with similar physiological features as natural bony tissue ECM. However, implementing this strategy remains the major challenge in BTE due to the inability to synthesize biomaterials that fully resemble structural and functional features of local microenvironment as that of natural bone ECM.

Different efforts have been made to synthesize bone ECM microenvironment with required functionality; however, the two main obstacles remain challenging namely; bioencapsulation of cells under physiological condition and optimization of the incorporated bioactive signaling molecules. In this context, both natural (i.e., collagen and polysaccharides) and synthetic polymeric hydrogels (i.e., polymethacrylate and polyethylene glycol) were among the first and most studied biomaterials for bone tissue applications [90–92]. An ideal hydrogel for BTE should mimic the properties of bone ECM. It should offer 3D encapsulation of cells within a nanofiber matrix that allows proper diffusion, instructive bioactive signaling, cells migration and communication, and biodegradation for native tissue ingrowth [93]. Recently, development of synthetic peptide-based hydrogel biomaterials (i.e., peptide amphiphile) capable of self-assembling into higher complex nanofiber structures has introduced new promises. These bioactive materials are capable of self-assembly into supramolecular nanostructures that resemble fibrous ECM [94]. Furthermore, modification of peptide sequence allows customization of these materials for different tissue engineering applications. This is due to the nature of self-assembly process producing cylindrical nanostructures with bioactive peptide signals exposed at outer surface interacting with cells and microenvironment. Furthermore, these self-assembled macromolecules can be enriched with bioactive signaling molecules for additional instructions. However, the incorporation of these additional molecules and growth factors need to be optimized, and the issues such as incorporation methods and release profile need to be addressed in future studies.

Use of nanostructure SAPA hydrogel can support mineralization of matrix via the incorporated active functional groups, but this process may be time consuming and cannot provide sufficient mechanical support to the grafted area. In comparison to bone repair after applications of bioceramics, the bone strength after repair by SAPA is low. Further long-term animal studies are required to compare the full mechanical properties of repaired bone. Furthermore, PA hydrogel is considered single-phase biomaterials that lack inorganic component. An interesting approach in BTE is the development of biphasic PA hydrogel by incorporating inorganic phase to simulate both composite phases of the natural bone. There are different reports on the use of PA with incorporated bioceramics (i.e., HA) to improve composite properties. It has been reported that considering its high biocompatibility, incorporation of HA reinforce PA nanomatrix and improves mechanical properties, integrity, and handling of composite implants [95,96]. Furthermore, HA could potentially improve the osteoinductivity by enhancing osteogenic differentiation of encapsulated MSC [97–99]. However, evidence of osteoinductive potential of HA is still a controversial issue in the literature that requires further investigations [100,101].

The new era for future studies may include an emphasis on combination approach by integrating two different tissue engineering strategies; top-down and bottom-up approach to address the limitations of the available approach. Recent advances in nanotechnologies offer additional new potentials to fabricate tissue-engineered biomimetic constructs with both physical and biochemical hierarchical order. However, integration of different tissue engineering approaches necessitates further advancement and investigations in the field. Currently, no successful standard approach is proposed for combination of top-down and bottom-up approaches. However, the main aim is the development of bioactive scaffold with nano-micro scale topographical features and the potentials to provide ideal supportive and instructive environment for functional cells and tissue regeneration. Therefore, more sophisticated technologies need to be investigated that allow assembling of ideal complex tissue constructs considering the clinical requirements. This should consider the biomimetic principle that needs to be implemented and optimized using current advancement in nanotechnology and molecular self-assembly tailored with specific ECM ligands for development of instructive matrix to program the cellular fate and differentiation pattern.

13.7 Conclusion

Earlier and enhanced bone tissue regeneration (improved quantity and quality) together with sufficient mechanical support for functional rehabilitation is the ultimate goal in BTE. Majority of the currently available biomaterials for BTE are single phase in composition. Thus, they differ from the natural bone in either physical features (compositional structure) or chemical properties (bioactive signaling). Therefore, there is a major need to engineer nanocomposite biomaterials that provide biomimetic instructive microenvironment capable of controlling the regeneration process at cellular and molecular levels. Furthermore, this approach should have the potential of translation into real clinical bone regeneration. Recent integrated approaches and the new advancements in nanotechnology and molecular self-assembly would be promising era in the future of tissue engineering.

References

[1] S. Sethuraman, L.S. Nair, S. El-Amin, M.-T. Nguyen, A. Singh, Y.E. Greish, H.R. Allcock, P.W. Brown, C.T. Laurencin, Development and characterization of biodegradable nanocomposite injectables for orthopaedic applications based on polyphosphazenes. J. Biomater. Sci. Polym. Ed. 22 (2011) 733–752. https://doi.org/10.1163/092050610X491670.

[2] D.S. Katti, K.W. Robinson, F.K. Ko, C.T. Laurencin, Bioresorbable nanofiber-based systems for wound healing and drug delivery: optimization of fabrication parameters. J. Biomed. Mater. Res. B. Appl. Biomater. 70 (2004) 286–296. https://doi.org/10.1002/jbm.b.30041.

[3] A. Subramanian, U.M. Krishnan, S. Sethuraman, Fabrication of uniaxially aligned 3D electrospun scaffolds for neural regeneration. Biomed. Mater. 6 (2011) 25004. https://doi.org/10.1088/1748-6041/6/2/025004.

[4] L.S. Nair, S. Bhattacharyya, C.T. Laurencin, Development of novel tissue engineering scaffolds via electrospinning. Expert. Opin. Biol. Ther. 4 (2004) 659–668. https://doi.org/10.1517/14712598.4.5.659.

[5] H.-Q. Mao, S.H. Lim, S. Zhang, G. Christopherson, K. Kam, S. Fischer, The nanofiber matrix as an artificial stem cell niche. Tissue Eng. Biomater. 2 (2010) 89–118. https://doi.org/10.1007/8415_2010_5.

[6] A. Pimpin, W. Srituravanich, Review on micro- and nanolithography techniques and their applications. Eng. J. 16 (2012) 37–56. https://doi.org/10.4186/ej.2012.16.1.37.

[7] I.-H. Jo, K.-H. Shin, Y.-M. Soon, Y.-H. Koh, J.-H. Lee, H.-E. Kim, Highly porous hydroxyapatite scaffolds with elongated pores using stretched polymeric sponges as novel template. Mater. Lett. 63 (2009) 1702–1704. https://doi.org/10.1016/j.matlet.2009.05.017.

[8] C.J. Luo, M. Nangrejo, M. Edirisinghe, A novel method of selecting solvents for polymer electrospinning. Polymer (Guildf) 51 (2010) 1654–1662. https://doi.org/10.1016/j.polymer.2010.01.031.

[9] T. Lu, Y. Li, T. Chen, Techniques for fabrication and construction of three-dimensional scaffolds for tissue engineering. Int. J. Nanomedicine (2013) 337. https://doi.org/10.2147/IJN.S38635.
[10] Y. Kang, E. Jabbari, Y. Yang, Integrating top-down and bottom-up scaffolding tissue engineering approach for bone regeneration, in: Micro and Nanotechnologies in Engineering Stem Cells and Tissues, John Wiley & Sons, Inc., Hoboken, NJ, 2013, pp. 142–158. https://doi.org/10.1002/9781118574775.ch6.
[11] J.W. Nichol, A. Khademhosseini, Modular tissue engineering: engineering biological tissues from the bottom up. Soft Matter 5 (2009) 1312. https://doi.org/10.1039/b814285h.
[12] R. Tiruvannamalai-Annamalai, D.R. Armant, H.W.T. Matthew, A glycosaminoglycan based, modular tissue scaffold system for rapid assembly of Perfusable, high cell density, engineered tissues. PLoS One 9 (2014) e84287. https://doi.org/10.1371/journal.pone.0084287.
[13] X. He, J. Ma, E. Jabbari, Effect of grafting RGD and BMP-2 protein-derived peptides to a hydrogel substrate on osteogenic differentiation of marrow stromal cells. Langmuir 24 (2008) 12508–12516. https://doi.org/10.1021/la802447v.
[14] J.F. Alvarez-Barreto, B. Landy, S. VanGordon, L. Place, P.L. DeAngelis, V.I. Sikavitsas, Enhanced osteoblastic differentiation of mesenchymal stem cells seeded in RGD-functionalized PLLA scaffolds and cultured in a flow perfusion bioreactor. J. Tissue Eng. Regen. Med. 5 (2011) 464–475. https://doi.org/10.1002/term.338.
[15] B. Zamanian, M. Masaeli, J.W. Nichol, M. Khabiry, M.J. Hancock, H. Bae, A. Khademhosseini, Interface-directed self-assembly of cell-laden microgels. Small 6 (2010) 937–944. https://doi.org/10.1002/smll.200902326.
[16] A.P. McGuigan, M.V. Sefton, Vascularized organoid engineered by modular assembly enables blood perfusion. Proc. Natl. Acad. Sci. USA 103 (2006) 11461–11466. https://doi.org/10.1073/pnas.0602740103.
[17] Y. Nahmias, R.E. Schwartz, C.M. Verfaillie, D.J. Odde, Laser-guided direct writing for three-dimensional tissue engineering. Biotechnol. Bioeng. 92 (2005) 129–136. https://doi.org/10.1002/bit.20585.
[18] V. Liu Tsang, A.A. Chen, L.M. Cho, K.D. Jadin, R.L. Sah, S. DeLong, J.L. West, S.N. Bhatia, Fabrication of 3D hepatic tissues by additive photopatterning of cellular hydrogels. FASEB J. 21 (2007) 790–801. https://doi.org/10.1096/fj.06-7117com.
[19] D.A. Bruzewicz, A.P. McGuigan, G.M. Whitesides, Fabrication of a modular tissue construct in a microfluidic chip. Lab Chip 8 (2008) 663–671. https://doi.org/10.1039/b719806j.
[20] S. Zhang, Fabrication of novel biomaterials through molecular self-assembly. Nat. Biotechnol. 21 (2003) 1171–1178. https://doi.org/10.1038/nbt874.
[21] A. Khademhosseini, J. Yeh, S. Jon, G. Eng, K.Y. Suh, J.A. Burdick, R. Langer, Molded polyethylene glycol microstructures for capturing cells within microfluidic channels. Lab Chip 4 (2004) 425–430. https://doi.org/10.1039/b404842c.
[22] A. Khademhosseini, S. Jon, K.Y. Suh, T.-N.T. Tran, G. Eng, J. Yeh, J. Seong, R. Langer, Direct patterning of protein- and cell-resistant polymeric monolayers and microstructures. Adv. Mater. 15 (2003) 1995–2000. https://doi.org/10.1002/adma.200305433.
[23] J. Yeh, Y. Ling, J.M. Karp, J. Gantz, A. Chandawarkar, G. Eng, J. Blumling, R. Langer, A. Khademhosseini, Micromolding of shape-controlled, harvestable cell-laden hydrogels. Biomaterials 27 (2006) 5391–5398. https://doi.org/10.1016/j.biomaterials.2006.06.005.

[24] T. Shimizu, M. Yamato, A. Kikuchi, T. Okano, Cell sheet engineering for myocardial tissue reconstruction. Biomaterials 24 (2003) 2309–2316. http://www.ncbi.nlm.nih.gov/pubmed/12699668.
[25] N.E. Fedorovich, J.R. De Wijn, A.J. Verbout, J. Alblas, W.J.A. Dhert, Three-dimensional fiber deposition of cell-laden, viable, patterned constructs for bone tissue printing. Tissue Eng. Part A 14 (2008) 127–133. https://doi.org/10.1089/ten.a.2007.0158.
[26] T.J. Hinton, Q. Jallerat, R.N. Palchesko, J.H. Park, M.S. Grodzicki, H.-J. Shue, M.H. Ramadan, A.R. Hudson, A.W. Feinberg, Three-dimensional printing of complex biological structures by freeform reversible embedding of suspended hydrogels. Sci. Adv. 1 (2015) e1500758. https://doi.org/10.1126/sciadv.1500758.
[27] S. Zhang, Building from the bottom up. Mater. Today 6 (2003) 20–27. https://doi.org/10.1016/S1369-7021(03)00530-3.
[28] S. Zhang, D.M. Marini, W. Hwang, S. Santoso, Design of nanostructured biological materials through self-assembly of peptides and proteins. Curr. Opin. Chem. Biol. 6 (2002) 865–871. http://www.ncbi.nlm.nih.gov/pubmed/12470743.
[29] N.N. Kachouie, Y. Du, H. Bae, M. Khabiry, A.F. Ahari, B. Zamanian, J. Fukuda, A. Khademhosseini, Directed assembly of cell-laden hydrogels for engineering functional tissues. Organ 6 (2010) 234–244. https://doi.org/10.4161/org.6.4.12650.
[30] J.E. Fleming, C.N. Cornell, G.F. Muschler, Bone cells and matrices in orthopedic tissue engineering. Orthop. Clin. North Am. 31 (2000) 357–374. http://www.ncbi.nlm.nih.gov/pubmed/10882463.
[31] J.I. Dawson, R.O.C. Oreffo, Bridging the regeneration gap: stem cells, biomaterials and clinical translation in bone tissue engineering. Arch. Biochem. Biophys. 473 (2008) 124–131. https://doi.org/10.1016/j.abb.2008.03.024.
[32] K.J. Burg, S. Porter, J.F. Kellam, Biomaterial developments for bone tissue engineering. Biomaterials 21 (2000) 2347–2359. http://www.ncbi.nlm.nih.gov/pubmed/11055282.
[33] H. Misawa, N. Kobayashi, A. Soto-Gutierrez, Y. Chen, A. Yoshida, J.D. Rivas-Carrillo, N. Navarro-Alvarez, K. Tanaka, A. Miki, J. Takei, T. Ueda, M. Tanaka, H. Endo, N. Tanaka, T. Ozaki, PuraMatrix facilitates bone regeneration in bone defects of calvaria in mice. Cell Transplant. 15 (2006) 903–910. http://www.ncbi.nlm.nih.gov/pubmed/17299995.
[34] A.J. Salgado, O.P. Coutinho, R.L. Reis, Bone tissue engineering: State of the art and future trends. Macromol. Biosci. 4 (2004) 743–765. https://doi.org/10.1002/mabi.200400026.
[35] E.T. Daif, Effect of autologous platelet-rich plasma on bone regeneration in mandibular fractures. Dent. Traumatol. 29 (2013) 399–403. https://doi.org/10.1111/edt.12021.
[36] A. Albanese, M.E. Licata, B. Polizzi, G. Campisi, Platelet-rich plasma (PRP) in dental and oral surgery: from the wound healing to bone regeneration. Immun. Ageing 10 (2013) 23. https://doi.org/10.1186/1742-4933-10-23.
[37] Y. Zhang, S. Tangl, C.D. Huber, Y. Lin, L. Qiu, X. Rausch-Fan, Effects of Choukroun's platelet-rich fibrin on bone regeneration in combination with deproteinized bovine bone mineral in maxillary sinus augmentation: a histological and histomorphometric study. J. Craniomaxillofac. Surg. 40 (2012) 321–328. https://doi.org/10.1016/j.jcms.2011.04.020.
[38] M.E. Davis, J.P.M. Motion, D.A. Narmoneva, T. Takahashi, D. Hakuno, R.D. Kamm, S. Zhang, R.T. Lee, Injectable self-assembling peptide nanofibers create intramyocardial microenvironments for endothelial cells. Circulation 111 (2005) 442–450. https://doi.org/10.1161/01.CIR.0000153847.47301.80.
[39] J.D. Sipe, Tissue engineering and reparative medicine. Ann. N. Y. Acad. Sci. 961 (2002) 1–9. http://www.ncbi.nlm.nih.gov/pubmed/12081856.

[40] G. Daculsi, B.H. Fellah, T. Miramond, M. Durand, Osteoconduction, Osteogenicity, Osteoinduction, what are the fundamental properties for a smart bone substitutes. IRBM 34 (2013) 346–348. https://doi.org/10.1016/j.irbm.2013.07.001.

[41] L.C. Palmer, C.J. Newcomb, S.R. Kaltz, E.D. Spoerke, S.I. Stupp, Biomimetic systems for hydroxyapatite mineralization inspired by bone and enamel. Chem. Rev. 108 (2008) 4754–4783. https://doi.org/10.1021/cr8004422.

[42] M.J. Olszta, X. Cheng, S.S. Jee, R. Kumar, Y.-Y. Kim, M.J. Kaufman, E.P. Douglas, L.B. Gower, Bone structure and formation: a new perspective. Mater. Sci. Eng. R Reports 58 (2007) 77–116. https://doi.org/10.1016/j.mser.2007.05.001.

[43] H. Lodish, A. Berk, S.L. Zipursky, et al., Collagen: The Fibrous Proteins of the Matrix. W.H. Freeman, New York, 2000, p. 22.3. http://www.ncbi.nlm.nih.gov/books/NBK21582/.

[44] M.D. Shoulders, R.T. Raines, Collagen structure and stability. Annu. Rev. Biochem. 78 (2009) 929–958. https://doi.org/10.1146/annurev.biochem.77.032207.120833.

[45] H. Terheyden, C. Knak, S. Jepsen, S. Palmie, D.R. Rueger, Mandibular reconstruction with a prefabricated vascularized bone graft using recombinant human osteogenic protein-1: an experimental study in miniature pigs. Part I: Prefabrication. Int. J. Oral Maxillofac. Surg. 30 (2001) 373–379. https://doi.org/10.1054/ijom.2001.0032.

[46] W.J. Landis, Tomographic imaging of collagen-mineral interaction: implications for osteogenesis imperfecta. Connect. Tissue Res. 31 (1995) 287–290. http://www.ncbi.nlm.nih.gov/pubmed/15612370.

[47] W.J. Landis, Mineral characterization in calcifying tissues: atomic, molecular and macromolecular perspectives. Connect. Tissue Res. 34 (1996) 239–246. http://www.ncbi.nlm.nih.gov/pubmed/9084632.

[48] W.J. Landis, K.J. Hodgens, J. Arena, M.J. Song, B.F. McEwen, Structural relations between collagen and mineral in bone as determined by high voltage electron microscopic tomography. Microsc. Res. Tech. 33 (1996) 192–202. https://doi.org/10.1002/(SICI)1097-0029(19960201)33:2<192::AID-JEMT9>3.0.CO;2-V.

[49] W. Traub, T. Arad, S. Weiner, Three-dimensional ordered distribution of crystals in Turkey tendon collagen fibers. Proc. Natl. Acad. Sci. USA 86 (1989) 9822–9826. https://doi.org/10.3109/03008209209014230.

[50] K.E. Kadler, D.F. Holmes, J.A. Trotter, J.A. Chapman, Collagen fibril formation. Biochem. J. 316 (Pt 1) (1996) 1–11. http://www.ncbi.nlm.nih.gov/pubmed/8645190.

[51] A. Oldberg, A. Franzén, D. Heinegård, The primary structure of a cell-binding bone sialoprotein. J. Biol. Chem. 263 (1988) 19430–19432. http://www.ncbi.nlm.nih.gov/pubmed/3198635.

[52] J.Q. Feng, L.M. Ward, S. Liu, Y. Lu, Y. Xie, B. Yuan, X. Yu, F. Rauch, S.I. Davis, S. Zhang, H. Rios, M.K. Drezner, L.D. Quarles, L.F. Bonewald, K.E. White, Loss of DMP1 causes rickets and osteomalacia and identifies a role for osteocytes in mineral metabolism. Nat. Genet. 38 (2006) 1310–1315. https://doi.org/10.1038/ng1905.

[53] E.J. Mackie, Osteoblasts: novel roles in orchestration of skeletal architecture. Int. J. Biochem. Cell Biol. 35 (2003) 1301–1305. http://www.ncbi.nlm.nih.gov/pubmed/12798343.

[54] S. Weiner, An overview of biomineralization processes and the problem of the vital effect. Rev. Mineral. Geochem. 54 (2003) 1–29. https://doi.org/10.2113/0540001.

[55] H.C. Anderson, Molecular biology of matrix vesicles. Clin. Orthop. Relat. Res. (1995) 266–280. http://www.ncbi.nlm.nih.gov/pubmed/7634645.

[56] A.L. Boskey, Matrix proteins and mineralization: an overview. Connect. Tissue Res. 35 (1996) 357–363. http://www.ncbi.nlm.nih.gov/pubmed/9084675.

[57] G. He, A. Ramachandran, T. Dahl, S. George, D. Schultz, D. Cookson, A. Veis, A. George, Phosphorylation of phosphophoryn is crucial for its function as a mediator of biomineralization. J. Biol. Chem. 280 (2005) 33109–33114. https://doi.org/10.1074/jbc.M500159200.

[58] P.A.A.P. Marques, A.P. Serro, B.J. Saramago, A.C. Fernandes, M.C.F. Magalhães, R.N. Correia, Mineralisation of two phosphate ceramics in HBSS: role of albumin. Biomaterials 24 (2003) 451–460. http://www.ncbi.nlm.nih.gov/pubmed/12423600.

[59] G. Daculsi, P. Pilet, M. Cottrel, G. Guicheux, Role of fibronectin during biological apatite crystal nucleation: ultrastructural characterization. J. Biomed. Mater. Res. 47 (1999) 228–233. http://www.ncbi.nlm.nih.gov/pubmed/10449634.

[60] S. Areva, T. Peltola, E. Säilynoja, K. Laajalehto, M. Lindén, J.B. Rosenholm, Effect of albumin and fibrinogen on calcium phosphate formation on sol–gel-derived Titania coatings *in Vitro*. Chem. Mater. 14 (2002) 1614–1621. https://doi.org/10.1021/cm0107272.

[61] M.D. Grynpas, S. Omelon, Transient precursor strategy or very small biological apatite crystals? Bone 41 (2007) 162–164. https://doi.org/10.1016/j.bone.2007.04.176.

[62] R. Filmon, F. Grizon, M.F. Baslé, D. Chappaard, Effects of negatively charged groups (carboxymethyl) on the calcification of poly(2-hydroxyethyl methacrylate). Biomaterials 23 (2002) 3053–3059. http://www.ncbi.nlm.nih.gov/pubmed/12069348.

[63] S.I. Stupp, P.V. Braun, Molecular manipulation of microstructures: biomaterials, ceramics, and semiconductors. Science 277 (1997) 1242–1248. http://www.ncbi.nlm.nih.gov/pubmed/9271562.

[64] J.D. Hartgerink, E. Beniash, S.I. Stupp, Self-assembly and mineralization of peptide-amphiphile nanofibers. Science 294 (2001) 1684–1688. https://doi.org/10.1126/science.1063187.

[65] J.D. Hartgerink, Self-assembly and mineralization of peptide-amphiphile nanofibers. Science 294 (2001) 1684–1688. https://doi.org/10.1126/science.1063187.

[66] S.E. Paramonov, H.-W. Jun, J.D. Hartgerink, Self-assembly of peptide-amphiphile nanofibers: the roles of hydrogen bonding and amphiphilic packing. J. Am. Chem. Soc. 128 (2006) 7291–7298. https://doi.org/10.1021/ja060573x.

[67] E. Kokkoli, A. Mardilovich, A. Wedekind, E.L. Rexeisen, A. Garg, J.A. Craig, Self-assembly and applications of biomimetic and bioactive peptide-amphiphiles. Soft Matter 2 (2006) 1015. https://doi.org/10.1039/b608929a.

[68] E. Beniash, J.D. Hartgerink, H. Storrie, J.C. Stendahl, S.I. Stupp, Self-assembling peptide amphiphile nanofiber matrices for cell entrapment. Acta Biomater. 1 (2005) 387–397. https://doi.org/10.1016/j.actbio.2005.04.002.

[69] H. Jiang, M.O. Guler, S.I. Stupp, The internal structure of self-assembled peptide amphiphiles nanofibers. Soft Matter 3 (2007) 454. https://doi.org/10.1039/b614426h.

[70] Y.S. Velichko, S.I. Stupp, M.O. de la Cruz, Molecular simulation study of peptide amphiphile self-assembly. J. Phys. Chem. B 112 (2008) 2326–2334. https://doi.org/10.1021/jp074420n.

[71] R.N. Shah, N.A. Shah, M.M. Del Rosario Lim, C. Hsieh, G. Nuber, S.I. Stupp, Supramolecular design of self-assembling nanofibers for cartilage regeneration. Proc. Natl. Acad. Sci. 107 (2010) 3293–3298. https://doi.org/10.1073/pnas.0906501107.

[72] G.A. Silva, Selective differentiation of neural progenitor cells by high-epitope density nanofibers. Science 303 (2004) 1352–1355. https://doi.org/10.1126/science.1093783.

[73] M.J. Webber, J. Tongers, C.J. Newcomb, K.-T. Marquardt, J. Bauersachs, D.W. Losordo, S.I. Stupp, Supramolecular nanostructures that mimic VEGF as a strategy for ischemic tissue repair. Proc. Natl. Acad. Sci. 108 (2011) 13438–13443. https://doi.org/10.1073/pnas.1016546108.

[74] M.J. Webber, J. Tongers, M.-A. Renault, J.G. Roncalli, D.W. Losordo, S.I. Stupp, Development of bioactive peptide amphiphiles for therapeutic cell delivery. Acta Biomater. 6 (2010) 3–11. https://doi.org/10.1016/j.actbio.2009.07.031.

[75] K.A. Brogden, Antimicrobial peptides: Pore formers or metabolic inhibitors in bacteria? Nat. Rev. Microbiol. 3 (2005) 238–250. https://doi.org/10.1038/nrmicro1098.

[76] J.J. Panda, A. Varshney, V.S. Chauhan, Self-assembled nanoparticles based on modified cationic dipeptides and DNA: novel systems for gene delivery. J. Nanobiotechnol. 11 (2013) 18. https://doi.org/10.1186/1477-3155-11-18.

[77] N.V. Berezhnoy, N. Korolev, L. Nordenskiöld, Principles of electrostatic interactions and self-assembly in lipid/peptide/DNA systems: applications to gene delivery. Adv. Colloid Interface Sci. 205 (2014) 221–229. https://doi.org/10.1016/j.cis.2013.08.008.

[78] A. Mata, Y. Geng, K.J. Henrikson, C. Aparicio, S.R. Stock, R.L. Satcher, S.I. Stupp, Bone regeneration mediated by biomimetic mineralization of a nanofiber matrix. Biomaterials 31 (2010) 6004–6012. https://doi.org/10.1016/j.biomaterials.2010.04.013.

[79] S. Weiner, H.D. Wagner, THE MATERIAL BONE: structure-mechanical function relations. Annu. Rev. Mater. Sci. 28 (1998) 271–298. https://doi.org/10.1146/annurev.matsci.28.1.271.

[80] E.D. Spoerke, S.G. Anthony, S.I. Stupp, Enzyme directed templating of artificial bone mineral. Adv. Mater. 21 (2009) 425–430. https://doi.org/10.1002/adma.200802242.

[81] A.K. Majors, C.A. Boehm, H. Nitto, R.J. Midura, G.F. Muschler, Characterization of human bone marrow stromal cells with respect to osteoblastic differentiation. J. Orthop. Res. 15 (1997) 546–557. https://doi.org/10.1002/jor.1100150410.

[82] P. Müller, U. Bulnheim, A. Diener, F. Lüthen, M. Teller, E.-D. Klinkenberg, H.-G. Neumann, B. Nebe, A. Liebold, G. Steinhoff, J. Rychly, Calcium phosphate surfaces promote osteogenic differentiation of mesenchymal stem cells. J. Cell. Mol. Med. 12 (2008) 281–291. https://doi.org/10.1111/j.1582-4934.2007.00103.x.

[83] T. Osathanon, M.L. Linnes, R.M. Rajachar, B.D. Ratner, M.J. Somerman, C.M. Giachelli, Microporous nanofibrous fibrin-based scaffolds for bone tissue engineering. Biomaterials 29 (2008) 4091–4099. https://doi.org/10.1016/j.biomaterials.2008.06.030.

[84] P. Narducci, V. Nicolin, Differentiation of activated monocytes into osteoclast-like cells on a hydroxyapatite substrate: an *in vitro* study. Ann. Anat. 191 (2009) 349–355. https://doi.org/10.1016/j.aanat.2009.02.009.

[85] P.V. Giannoudis, H. Dinopoulos, E. Tsiridis, Bone substitutes: an update. Injury 36 (Suppl. 3) (2005) S20–S27. https://doi.org/10.1016/j.injury.2005.07.029.

[86] H.W. Ouyang, S.L. Toh, J. Goh, T.E. Tay, K. Moe, Assembly of bone marrow stromal cell sheets with knitted poly (L-lactide) scaffold for engineering ligament analogs. J. Biomed. Mater. Res. B Appl. Biomater. 75 (2005) 264–271. https://doi.org/10.1002/jbm.b.30281.

[87] T.D. Sargeant, M.O. Guler, S.M. Oppenheimer, A. Mata, R.L. Satcher, D.C. Dunand, S.I. Stupp, Hybrid bone implants: self-assembly of peptide amphiphile nanofibers within porous titanium. Biomaterials 29 (2008) 161–171. https://doi.org/10.1016/j.biomaterials.2007.09.012.

[88] C. Streuli, Extracellular matrix remodelling and cellular differentiation. Curr. Opin. Cell Biol. 11 (1999) 634–640. http://www.ncbi.nlm.nih.gov/pubmed/10508658.

[89] F. Rosso, A. Giordano, M. Barbarisi, A. Barbarisi, From cell-ECM interactions to tissue engineering. J. Cell. Physiol. 199 (2004) 174–180. https://doi.org/10.1002/jcp.10471.

[90] S.K. Sarkar, B.T. Lee, Hard tissue regeneration using bone substitutes: an update on innovations in materials. Korean J. Intern. Med. 30 (2015) 279–293. https://doi.org/10.3904/kjim.2015.30.3.279.

[91] K. Gkioni, S.C.G. Leeuwenburgh, T.E.L. Douglas, A.G. Mikos, J.A. Jansen, Mineralization of hydrogels for bone regeneration. Tissue Eng. Part B Rev. 16 (2010) 577–585. https://doi.org/10.1089/ten.TEB.2010.0462.
[92] C. Vinatier, J. Guicheux, G. Daculsi, P. Layrolle, P. Weiss, Cartilage and bone tissue engineering using hydrogels. Biomed. Mater. Eng. 16 (2006) S107–S113. http://www.ncbi.nlm.nih.gov/pubmed/16823101.
[93] C.E. Semino, Self-assembling peptides: From bio-inspired materials to bone regeneration. J. Dent. Res. 87 (2008) 606–616. https://doi.org/10.1177/154405910808700710.
[94] J.D. Hartgerink, E. Beniash, S.I. Stupp, Peptide-amphiphile nanofibers: a versatile scaffold for the preparation of self-assembling materials. Proc. Natl. Acad. Sci. 99 (2002) 5133–5138. https://doi.org/10.1073/pnas.072699999.
[95] M. Supová, Problem of hydroxyapatite dispersion in polymer matrices: A review. J. Mater. Sci. Mater. Med. 20 (2009) 1201–1213. https://doi.org/10.1007/s10856-009-3696-2.
[96] A.C. Lawson, J.T. Czernuszka, Collagen—calcium phosphate composites. Proc. Inst. Mech. Eng. H 212 (1998) 413–425. https://doi.org/10.1243/0954411981534187.
[97] J.M. Anderson, J.L. Patterson, J.B. Vines, A. Javed, S.R. Gilbert, H.-W. Jun, Biphasic peptide amphiphile nanomatrix embedded with hydroxyapatite nanoparticles for stimulated osteoinductive response. ACS Nano 5 (2011) 9463–9479. https://doi.org/10.1021/nn203247m.
[98] J.B. Vines, D.-J. Lim, J.M. Anderson, H.-W. Jun, Hydroxyapatite nanoparticle reinforced peptide amphiphile nanomatrix enhances the osteogenic differentiation of mesenchymal stem cells by compositional ratios. Acta Biomater. 8 (2012) 4053–4063. https://doi.org/10.1016/j.actbio.2012.07.024.
[99] T.D. Sargeant, C. Aparicio, J.E. Goldberger, H. Cui, S.I. Stupp, Mineralization of peptide amphiphile nanofibers and its effect on the differentiation of human mesenchymal stem cells. Acta Biomater. 8 (2012) 2456–2465. https://doi.org/10.1016/j.actbio.2012.03.026.
[100] J. Xie, M.J. Baumann, L.R. McCabe, Osteoblasts respond to hydroxyapatite surfaces with immediate changes in gene expression. J. Biomed. Mater. Res. A 71 (2004) 108–117. https://doi.org/10.1002/jbm.a.30140.
[101] S.-W. Kang, J.-S. Lee, M.S. Park, J.-H. Park, B.-S. Kim, Enhancement of *in vivo* bone regeneration efficacy of human mesenchymal stem cells. J. Microbiol. Biotechnol. 18 (2008) 975–982. http://www.ncbi.nlm.nih.gov/pubmed/18633301.

Index

Note: Page numbers followed by *f* indicate figures and *t* indicate tables.

C

N

CPI Antony Rowe
Chippenham, UK
2019-01-18 12:43